BIOACTIVE COMPOUNDS FROM MULTIFARIOUS NATURAL FOODS FOR HUMAN HEALTH

Foods and Medicinal Plants

Innovations in Plant Science for Better Health: From Soil to Fork

BIOACTIVE COMPOUNDS FROM MULTIFARIOUS NATURAL FOODS FOR HUMAN HEALTH

Foods and Medicinal Plants

Edited by

Hafiz Ansar Rasul Suleria, PhD

Megh R. Goyal, PhD

Huma Bader Ul Ain, PhD

First edition published 2022

Apple Academic Press Inc.
1265 Goldenrod Circle, NE,
Palm Bay, FL 32905 USA

4164 Lakeshore Road, Burlington,
ON, L7L 1A4 Canada

CRC Press
6000 Broken Sound Parkway NW,
Suite 300, Boca Raton, FL 33487-2742 USA

2 Park Square, Milton Park,
Abingdon, Oxon, OX14 4RN UK

Apple Academic Press exclusively co-publishes with CRC Press, an imprint of Taylor & Francis Group, LLC

Library and Archives Canada Cataloguing in Publication

Title: Bioactive compounds from multifarious natural foods for human health : foods and medicinal plants / edited by Hafiz Ansar Rasul Suleria, PhD, Megh R. Goyal, PhD, Huma Bader Ul Ain, PhD.
Names: Suleria, Hafiz, editor. | Goyal, Megh R., editor. | Ul Ain, Huma Bader, editor.
Series: Innovations in plant science for better health.
Description: First edition. | Series statement: Innovations in plant science for better health : from soil to fork | Includes bibliographical references and index.
Identifiers: Canadiana (print) 2021038462X | Canadiana (ebook) 20210384727 | ISBN 9781774637159 (hardcover) | ISBN 9781774637975 (softcover) | ISBN 9781003189763 (ebook)
Subjects: LCSH: Functional foods. | LCSH: Medicinal plants. | LCSH: Plant bioactive compounds. | LCSH: Bioactive compounds.
Classification: LCC QP144.F85 B56 2022 | DDC 613.2—dc23

Library of Congress Cataloging-in-Publication Data

Names: Suleria, Hafiz, editor. | Goyal, Megh R., editor. | Ul Ain, Huma Bader, editor.
Title: Bioactive compounds from multifarious natural foods for human health: foods and medicinal plants / Hafiz Ansar Rasul Suleria, Megh R. Goyal, Huma Bader Ul Ain.
Other titles: Innovations in plant science for better health.
Description: First edition. | Palm Bay, FL, USA : Apple Academic Press, 2022. | Series: Innovations in plant science for better health : from soil to fork | Includes bibliographical references and index. | Summary: "Divided into two sections, the volume first examines health claims of food-based bioactive compounds, which are extranutritional constituents that typically occur in small quantities in foods. This section lays out the concepts of extraction of food-based bioactive molecules, along with both conventional and modernized extraction techniques, as well as the available sources, biochemistry, structural composition, and potential biological activities of bioactive compounds. The book goes to present new research on health claims of bioactive compounds from medicinal plants, their importance, and health perspectives. Both sections cover the various pharmacological and therapeutic aspects of bioactive compounds, along with their methods of extraction, their phytochemistry, their pharmacological and biological activities, their medicinal properties, and their applications for disease management and prevention. Among the specific foods and plants included are soybean, durum wheat, avocado, watermelon, blueberries, macro and micro algae, bitter cucumber (or Citrullus colocynthis), black myrobalan, clove, flaxseed, and even industrial waste from cereal bran. This book volume sheds new light on the potential of natural and plant-based foods for human health from different technological aspects, contributing to the ocean of knowledge on food science and technology. This compendium will be useful for students, researchers, and industry professsionals in the study of functional foods"-- Provided by publisher.
Identifiers: LCCN 2021058131 (print) | LCCN 2021058132 (ebook) | ISBN 9781774637159 (hardcover) | ISBN 9781774637975 (paperback) | ISBN 9781003189763 (ebook)
Subjects: LCSH: Medicinal plants. | Plant bioactive compounds. | Bioactive compounds. | Functional foods.
Classification: LCC QP144.F85 B532 2022 (print) | LCC QP144.F85 (ebook) | DDC 581.6/34--dc23/eng/20211221
LC record available at https://lccn.loc.gov/2021058131
LC ebook record available at https://lccn.loc.gov/2021058132

ISBN: 978-1-77463-715-9 (hbk)
ISBN: 978-1-77463-797-5 (pbk)
ISBN: 978-1-00318-976-3 (ebk)

OTHER BOOKS ON PLANT SCIENCE FOR BETTER HEALTH BY APPLE ACADEMIC PRESS, INC.

Book Series: *Innovations in Plant Science for Better Health: From Soil to Fork*
Editor-in-Chief: Hafiz Ansar Rasul Suleria, PhD

- **Assessment of Medicinal Plants for Human Health: Phytochemistry, Disease Management, and Novel Applications**
  Editors: Megh R. Goyal, PhD, and Durgesh Nandini Chauhan, MPharm
- **Bioactive Compounds of Medicinal Plants: Properties and Potential for Human Health**
  Editors: Megh R. Goyal, PhD, and Ademola O. Ayeleso
- **Bioactive Compounds from Plant Origin: Extraction, Applications, and Potential Health Claims**
  Editors: Hafiz Ansar Rasul Suleria, PhD, and Colin Barrow, PhD
- **Bioactive Compounds from Multifarious Natural Foods for Human Health**
  Editors: Hafiz Ansar Rasul Suleria, PhD, Megh R. Goyal, PhD, and Huma Bader Ul Ain
- **Cereals and Cereal-Based Foods**
  Editors: Megh R. Goyal, PhD, Kamaljit Kaur, PhD, Jaspreet Kaur, PhD
- **Health Benefits of Secondary Phytocompounds from Plant and Marine Sources**
  Editors: Hafiz Ansar Rasul Suleria, PhD, and Megh Goyal, PhD
- **Herbs, Spices, and Medicinal Plants for Human Gastrointestinal Disorders**
  Editors: Megh R. Goyal, PhD, Preeti Birwal, PhD, and Durgesh Nandini Chauhan, MPharm
- **Human Health Benefits of Plant Bioactive Compounds: Potentials and Prospects**
  Editors: Megh R. Goyal, PhD, and Hafiz Ansar Rasul Suleria, PhD
- **Phytochemicals and Medicinal Plants in Food Design: Strategies and Technologies for Improved Healthcare**
  Editors: Megh R. Goyal, PhD, Preeti Birwal, PhD, and Santosh K. Mishra, Ph

- **Phytochemicals from Medicinal Plants: Scope, Applications, and Potential Health Claims**
  Editors: Hafiz Ansar Rasul Suleria, PhD, Megh R. Goyal, PhD, and Masood Sadiq Butt, PhD
- **Plant- and Marine-Based Phytochemicals for Human Health: Attributes, Potential, and Use**
  Editors: Megh R. Goyal, PhD, and Durgesh Nandini Chauhan, MPharm
- **Plant-Based Functional Foods and Phytochemicals: From Traditional Knowledge to Present Innovation**
  Editors: Megh R. Goyal, PhD, Arijit Nath, PhD, and Hafiz Ansar Rasul Suleria, PhD
- **Plant Secondary Metabolites for Human Health: Extraction of Bioactive Compounds**
  Editors: Megh R. Goyal, PhD, P. P. Joy, PhD, and Hafiz Ansar Rasul Suleria, PhD
- **The Role of Phytoconstitutents in Healthcare: Biocompounds in Medicinal Plants**
  Editors: Megh R. Goyal, PhD, Hafiz Ansar Rasul Suleria, PhD, and Ramasamy Harikrishnan, PhD
- **The Therapeutic Properties of Medicinal Plants: Health-Rejuvenating Bioactive Compounds of Native Flora**
  Editors: Megh R. Goyal, PhD, PE, Hafiz Ansar Rasul Suleria, PhD, Ademola Olabode Ayeleso, PhD, T. Jesse Joel, and Sujogya Kumar Panda

ABOUT THE EDITORS

Hafiz Ansar Rasul Suleria, PhD, is the McKenzie Fellow at the School of Agriculture and Food in the Faculty of Veterinary and Agricultural Science, The University of Melbourne, Australia. He was formerly the Alfred Deakin Research Fellow at Deakin University, Victoria, Australia. He did his postdoctoral fellowship at the Department of Food, Nutrition, Dietetic and Health at Kansas State University, USA. He is also an Honorary Fellow of the Diamantina Institute, Faculty of Medicine, The University of Queensland (UQ), Australia. Dr. Suleria has been awarded an International Postgraduate Research Scholarship and the Australian Postgraduate Award for his PhD research at the UQ School of Medicine and the Translational Research Institute in collaboration with the Commonwealth and Scientific and Industrial Research Organization, Australia. Before joining the UQ, he worked as a lecturer in the Department of Food Sciences, Government College University Faisalabad, Pakistan. He also worked as a research associate in the PAK-US Joint Project funded by the Higher Education Commission, Pakistan, and Department of State, USA, with collaboration of the University of Massachusetts, USA, and the National Institute of Food Science and Technology, University of Agriculture Faisalabad, Pakistan.

Dr. Suleria has published more than 80 peer-reviewed scientific papers in professional journals and has co-edited several books. He is also affiliated with more than ten universities where he is working as a co-supervisor/special member for PhD and postgraduate students. His major research focus is on food nutrition, particularly in screening of bioactive molecules, including isolation, purification, and characterization using various cutting-edge techniques from different plants, marine, and animal sources; *in vitro*, in vivo bioactivities; and cell culture and animal modeling.

Readers may contact him at: hafiz.suleria@uqconnect.edu.au.

Megh R. Goyal, PhD, PE, is, currently a retired professor of agricultural and biomedical engineering from the General Engineering Department at the College of Engineering at the University of Puerto Rico–Mayaguez Campus (UPRM); and Senior Acquisitions Editor and Senior Technical Editor-in-Chief for Agricultural and Biomedical Engineering for Apple Academic Press Inc.

During his long career, he has worked as a Soil Conservation Inspector; Research Assistant at Haryana Agricultural University and Ohio State University; Research Agricultural Engineer/Professor at the Department of Agricultural Engineering of UPRM; and Professor of Agricultural and Biomedical Engineering in the General Engineering Department of UPRM. He spent a one-year sabbatical leave in 2002–2003 at the Biomedical Engineering Department of Florida International University, Miami, USA.

Dr. Goyal was the first agricultural engineer to receive the professional license in agricultural engineering from the College of Engineers and Surveyors of Puerto Rico. In 2005, he was proclaimed the "Father of Irrigation Engineering in Puerto Rico for the Twentieth Century" by the American Society of Agricultural and Biological Engineers, Puerto Rico Section, for his pioneering work on micro irrigation, evapotranspiration, agroclimatology, and soil and water engineering.

During his professional career of 52 years, he has received many awards, including Scientist of the Year, Membership Grand Prize for the American Society of Agricultural Engineers Campaign, Felix Castro Rodriguez Academic Excellence Award, Man of Drip Irrigation by the Mayor of Municipalities of Mayaguez/Caguas/Ponce and Senate/Secretary of Agriculture of ELA, Puerto Rico, and many others. He has been recognized as one of the experts "who rendered meritorious service for the development of [the] irrigation sector in India" by the Water Technology Centre of Tamil Nadu Agricultural University in Coimbatore, India, and ASABE who bestowed on him the 2108 Netafim Microirrigation Award. VDGOOD Professional Association of India awarded Lifetime Achievement Award at 12th Annual Meeting on Engineering, Science and Medicine that was held on 20-21 of November of 2020 in Visakhapatnam, India.

Dr. Goyal has authored more than 200 journal articles and edited more than 100 books.

Dr. Goyal received his BSc degree in Engineering from Punjab Agricultural University, Ludhiana, India, and his MSc and PhD degrees from the Ohio State University, Columbus, Ohio, USA. He also earned a Master of Divinity degree from the Puerto Rico Evangelical Seminary, Hato Rey, Puerto Rico, USA.

Readers may contact him at goyalmegh@gmail.com.

Huma Bader Ul Ain, PhD, is an Assistant Professor at Riphah International University, Faisalabad, Punjab-Pakistan. She holds a PhD in Food Science and Technology with a specialization is cereal technology. Dr. Ul Ain has published more than 30 peer-reviewed scientific papers in different reputed/ impacted journals and 10 international book chapters. She also has 16 international abstracts in international conferences. She has attended various trainings and conferences and has memberships in national and international organizations and platforms. She has done three-month internship at Fatima Memorial Hospital, Lahore, Pakistan, and a six-month internship at the School of Nutrition and United Hospital Faisalabad.

Dr. Ul Ain received her BS degree in Food Science and Technology from the University of Sargodha, Pakistan, and her MPhil and PhD degrees from the Government College University Faisalabad, Punjab-Pakistan. During her PhD program, she worked as a visiting lecturer for a year at the Government College University, Faisalabad, Punjab-Pakistan. Her PhD specialization and topic is "Isolation and characterization of cereal cell wall with special reference to amelioration of soluble dietary fiber."

Readers may contact her at: huma.bader@riphahfsd.edu.pk.

CONTENTS

CONTRIBUTORS

Rana Muhammad Aadil
Assistant Professor, National Institute of Food Science and Technology, University of Agriculture Faisalabad, University Main Rd., Faisalabad, Punjab – 38000, Pakistan, Mobile: +92-333-6437789, E-mail: dilrana89@gmail.com

Muhammad Afzaal
Assistant Professor, Government College University Faisalabad, New Campus Jhang Road Faisalabad, Pakistan, Mobile: +923003159543, E-mail: muhammadafzaal@gcuf.edu.pk

Shabbir Ahmad
Assistant Professor, Department of Food Science and Technology, MNS-University of Agriculture, Multan, Pakistan, E-mail: shabbir.ahmad@mnsuam.edu.pk

Huma Bader Ul Ain
Assistant Professor, Faculty of Rehabilitation and Allied Health Sciences, Riphah International University Faisalabad, Punjab, Pakistan, E-mail: huma.bader@riphahfsd.edu.pk

Aqsa Akram
Lecturer, Department of Diet and Nutritional Sciences, Faculty of Health and Allied Science, Imperial College of Business Studies, Lahore, Pakistan, Mobile: +92-346-4036927, E-mail: aqsaakram93@gmail.com

Haseeb Anwar
Assistant Professor, Neurochemical Biology and Genetics Laboratory (NGL), Department of Physiology, Faculty of Life Sciences, Government College University, Faisalabad, Pakistan, Mobile: +92-300-9666112, E-mail: drhaseebanwar@gcuf.edu.pk

Muhammad Umair Arshad
Post-Doc, Canada, Associate Professor, Government College University Faisalabad, New Campus Jhang Road Faisalabad, Pakistan, Mobile: +923137986776, E-mail: umairfood1@gmail.com

Ayesha Aslam
Senior Lecturer, University Institute of Diet and Nutritional Sciences, Faculty of Allied Health Sciences, The University of Lahore, 1-Km Defense Road, Near Bhuptian Chowk, Lahore, Punjab – 54000, Pakistan, Mobile: 0334-9493330, E-mail: ayeshaaslam2016@gmail.com

Vinitha Saseendra Babu
PhD Candidate, Department of Botany, University of Kerala, Kariavattom, Thiruvananthapuram, Kerala – 695581, India, Mobile: +91-9061575820, E-mail: vinithasbabu55@gmail.com

Shahid Bashir
Associate Professor, The University of Lahore, 1-KM Defense Road, Lahore, Pakistan, Mobile: +923017929073, E-mail: shahid.bashir@rsmi.uol.edu.pk

Daniel Commane
Faculty of Health and Life Sciences, Northumbria University, Newcastle, UK, E-mail: daniel.commane@northumbria.ac.uk

Syed Amir Gilani
Professor and Dean Faculty of Allied Health Sciences, The University of Lahore,
1-KM Defense Road, Lahore, Pakistan, Mobile: +923008460876, E-mail: profgilani@gmail.com

Megh R. Goyal
Retired Faculty in Agricultural and Biomedical Engineering from College of Engineering at University of Puerto Rico-Mayaguez Campus; and Senior Technical Editor-in-Chief in Agricultural and Biomedical Engineering for Apple Academic Press Inc., USA, E-mail: goyalmegh@gmail.com

Halah Hafiz
Hugh Sinclair Unit of Human Nutrition and Institute for Cardiovascular and Metabolic Research, University of Reading, Reading, UK, E-mail: H.a.m.Hafiz@pgr.reading.ac.uk

Ghulam Hussain
Assistant Professor, Neurochemical Biology and Genetics Laboratory (NGL), Department of Physiology, Faculty of Life Sciences, Government College University, Faisalabad, Pakistan, Mobile: +923006654101, E-mail: ghulamhussain@gcuf.edu.pk

Muzzamal Hussain
PhD Candidate, Government College University Faisalabad, New Campus Jhang Road Faisalabad, Pakistan, Mobile: +923247298101, E-mail: muzamalhussain121@gmail.com

Iahtisham-Ul-Haq
Head of Department/Assistant Professor, Department of Diet and Nutritional Sciences, Faculty of Health and Allied Sciences, Imperial College of Business Studies, Lahore, Pakistan, Mobile: +92-300-7275255, E-mail: iahtisham@hotmail.com

Aiman Ijaz
PhD Candidate, Senior Lecturer, The University of Lahore, 1-KM Defense Road, Lahore, Pakistan, Mobile: +923097829789, E-mail: aiman.ijaz@dnsc.uol.edu.pk

Ali Ikram
PhD Candidate, Institute of Home and Food Sciences, Government College University Faisalabad, Pakistan, Mobil: +92-3416768630, E-mail: aliikram630@gmail.com

Ali Imran
Assistant Professor, Institute of Home and Food Sciences, Government College University, Faisalabad, Pakistan, Mobile: +923336575583, E-mails: dr.aliimran@gcuf.edu.pk; aliimran.ft@gmail.com

Anum Ishaq
Assistant Professor, The Superior College Lahore (University Campus), 17-km Raiwind Rd., Kot Araian, Lahore, Punjab – 54000, Pakistan, Mobile: +92-332-6652662,
E-mail: anum1797@gmail.com

Kim Jackson
Hugh Sinclair Unit of Human Nutrition, Institute for Food, Nutrition and Health and Institute for Cardiovascular and Metabolic Research, University of Reading, Reading, UK,
E-mail: k.g.jackson@reading.ac.uk

Muhammad Kashif
Associate Professor, Faculty of Rehabilitation and Allied Health Sciences, Riphah International University Faisalabad, Punjab, Pakistan, E-mail: kashif.shaffi@gmail.com

Ahood Khalid
Senior Lecturer, University Institute of Diet and Nutritional Sciences, Faculty of Allied Health Sciences, The University of Lahore, 1-Km Defense Road, Near Bhuptian Chowk, Lahore, Punjab – 54000, Pakistan, Mobile: 0307-7058404, E-mail: ahoodkhalid@hotmail.com

Anees Ahmed Khalil
Assistant Professor, University Institute of Diet and Nutritional Sciences, Faculty of Allied Health Sciences, The University of Lahore, 1-Km Defense Road, Near Bhuptian Chowk, Lahore, Punjab – 54000, Pakistan, Mobile: +92-333-7000757,
E-mails: aneesahmedkhalil@gmail.com; anees.khalil@dnsc.uol.edu.pk

Shoaib Ahmad Malik
Institute of Home and Food Sciences, Government College University, Faisalabad, Pakistan

Javeria Maqbool
PhD Scholar, Neurochemical Biology and Genetics Laboratory (NGL), Department of Physiology, Faculty of Life Sciences, Government College University, Faisalabad, Pakistan,
Mobile: +92-320-2542265, E-mail: javeria_maqbool@yahoo.com

Haroon Munir
Lecturer, Institute of Home and Food Sciences, Government College University Faisalabad, Pakistan, Mobile: +92-3216636404, E-mail: haroon.munir@gcuf.edu.pk

Ambreen Naz
Food Science and Technology, Assistant Professor, Department of Food Science and Technology, MNS-University of Agriculture, Multan, Pakistan, E-mail: amber1912@yahoo.com

Puthuparambil Madhavan Radhamany
Professor, Department of Botany, University of Kerala, Kariavattom, Thiruvananthapuram – 695581, Kerala, India, Mobile: +91-9446215095, E-mail: radhamany_m@rediffmail.com

Ubaid Ur Rahman
Assistant Professor, School of Food and Agricultural Sciences, University of Management and Technology, Lahore, C-II Block C 2 Phase 1 Johar Town, Lahore, Punjab – 54000, Pakistan, Mobile: +92-300-8021788, E-mail: ubaid1783@gmail.com

Azhar Rasul
Assistant Professor, Department of Zoology, Faculty of Life Sciences, Government College University, Faisalabad, Pakistan, Mobile: +923006650122, E-mail: drazharrasul@gmail.com

Nighat Raza
Lecturer, Department of Food Science and Technology, MNS-University of Agriculture Multan, Pakistan, E-mail: nighat.raza@mnsuam.edu.pk

Farhan Saeed
PhD Post-Doc, Australia, Associate Professor, Government College University Faisalabad, New Campus Jhang Road Faisalabad, Pakistan, Mobile: +923338040311, E-mail: f.saeed@gcuf.edu.pk

Muhammad Zia Shahid
Candidate, Senior Lecturer, The University of Lahore, 1-KM Defense Road, Lahore, Pakistan, Mobile: +923006203767, E-mail: zia.shahid@dnsc.uol.edu.pk

Qurat Ul Ain Shahid
Senior Lecturer, University Institute of Diet and Nutritional Sciences, Faculty of Allied Health Sciences, The University of Lahore, 1-Km Defense Road, Near Bhuptian Chowk, Lahore, Punjab – 54000, Pakistan, Mobile: +92-345-0011660, E-mail: quratulain351@gmail.com

Maryam Shahzadi
MPhil Candidate, The University of Lahore, 1-KM Defense Road, Lahore, Pakistan,
Mobile: +923076886839, E-mail: dr.marium117@gmail.com

Shahrina Shakir
MSc (Hons.) Scholar, Department of Food Science and Technology, MNS-University of Agriculture, Multan, Pakistan, E-mail: shahrinashakir@yahoo.com

Hafiz Ansar Rasul Suleria
McKenzie Fellow Department of Food Science, University of Melbourne, Australia,
Mobile: +61-470-439-670, E-mail: hafiz.suleria@unimelb.edu.au

Tabussam Tufail
PhD Candidate, Senior Lecturer, The University of Lahore, 1-KM Defense Road, Lahore, Pakistan,
Mobile: +923057678764, E-mail: tabussam.tufail@dnsc.uol.edu.pk

Gemma Walton
Food Microbial Sciences Unit, Department of Food and Nutritional Sciences, University of Reading, Reading, UK, E-mail: g.e.walton@reading.ac.uk

ABBREVIATIONS

16α-OHE1	16α-hydroxylation
AA	arachidonic acid
ABCA1	ATP-binding cassette carrier 1
ABI	acquired brain injuries
AChE	acetylcholinesterase
ACNs	anthocyanins
AD	Alzheimer's disease
ADSCs	adipose-derived mesenchymal stem cells
AGE	advanced glycation end
AIDS	acquired immunodeficiency syndrome
ALP	alkaline phosphatase
ALS	amyotrophic lateral sclerosis
ALT	alanine aminotransferase
A-PAGE	acidic polyacrylamide gel electrophoresis
APC	allophycocyanin
ART	antiretroviral therapy
AST	aspartate aminotransferase
ATP	adenosine triphosphate
BAC	bioactive compounds
BAP	bone-specific alkaline phosphatase
BAT	brown adipose tissues
BBB	blood-brain barrier
BBR	berberine
BDNF	brain-derived neurotrophic factor
BHA	butylated hydroxy-anisole
BMC	bone mineral content
BMD	bone mineral density
BMI	body mass index
BMSCs	based mesenchymal stem cells
BMSCs	bone marrow-derived stem cells
BMSC-SCs	bone marrow-derived stem cells-Schwann cells
BTM	bone turnover markers
C.C.	*C. colocynthis*

CAD	coronary artery disease
CHD	coronary heart disease
CL	*Citrullus lanatus*
CLSO	*Citrullus lanatus silymarin* oil
CMV	cytomegalovirus
CNS	central nervous system
CNTF	ciliary neurotrophic factor
Col I	collagen I genes
COX	cyclo-oxygenase
CRC	colorectal cancer
CRP	C-reactive protein
CTL	cyto-poisonous T lymphocyte
CTX-I	C-telopeptide of type I collagen
Cu	copper
CVD	cardiovascular disease
DBS	deep brain stimulation
DDT	desired dough temperature
DHA	docosahexaenoic acid
DHEND	dihydroxyenterodiol
DHENL	dihydroxyenterolactone
DNA	deoxyribose nucleic acid
DPPH	2,2-diphenyl-1-picrylhydrazyl
ECM	extracellular matrix
END	enterodiol
ENF	enterofuran
ENL	enterolactone
eNOS	endothelial nitric oxide synthase
EPA	eicosatetraenoic acid
ER	estrogen receptor
ES	electrical stimulation
ESCs	embryonic stem cells
FAQ	fair average quality
FOS	fructooligosaccharides
FSH	follicle-stimulating hormone
FTIR	Fourier transform infra-red
GDNF	glial cell line-derived neurotrophic factor
GI	gastrointestinal
GI	gluten index

GYPS	glucose peptone yeast and sucrose
HCMV	human cytomegalovirus
HD	Huntington's disease
HDL	high density lipoprotein
HIV	human immunodeficiency virus
HOS-1	human osteosarcoma cell line
HPLC	high performance chromatography
HRT	hormone replacement therapy
HSV	herpes simplex virus
Hup A	Huperzine-A
HUVEC	human umbilical vein endothelial cells
Hz	hertz
IBS	irritable bowel syndrome
IFN-γ	interferon-gamma
IL	interleukin
IL-1β	interleukin 1 beta
IL-6	interleukin-6
iNOS	inducible nitricoxidostasis
KOH	potassium hydroxide
LAB	lactic acid bacteria
Lari	lariciresinol
LDH	lactic acid dehydrogenase
LDL	low density lipoprotein
LGG	*Lactobacillus rhamnosus* GG
LH	luteinizing hormone
LOX	lipoxygenase
MAT	matairesinol
MCF	Michigan cancer foundation
M-CSE	macrophage colony-stimulating element
MDA	malondialdehyde
MI	myocardial infarction
MIC	minimum inhibitory concentration
MS	multiple sclerosis
MT	million tons
Mw	molecular weight
NASCIS	national acute spinal cord injury study
NCI	National Cancer Institute
NDDs	neurodegenerative diseases

NGF	nerve growth factor
NHANES	National Health and Nutrition Examination Survey
NMDA	N-methyl-D-aspartate
NMJs	neuromuscular junction
NMR	nuclear magnetic resonance imaging
NO	nitric oxide
NPY	neuropeptide-Y
NTX-I	nucleotide reverse transcriptase inhibitors
OC	osteocalcin
OCIF	osteoclastogenesis inhibitory factor
OPG	osteoprotegerin
OVX	ovariectomized
PAF	platelet-activating factor
PAG	periaqueductal gray
PC	phycocyanin
PCV	packed cell volume
PD	Parkinson's disease
PDGF	platelet-derived growth factor
PE	phycoerythrin
PENS	percutaneous electrical nerve stimulation
PGE	prostaglandin E_2
PICP	procollagen I carboxyterminal propeptide
Pin	pinoresinol
PL	pyloric ligation
PMS	premenstrual syndrome
PNI	peripheral nerve injuries
PNS	peripheral nervous system
PTH	parathyroid hormone
PUFA	polyunsaturated fatty acids
RA	rheumatoid arthritis
RANKL	receptor activator of nuclear factor kappa B ligand
RCTs	randomized controlled trials
ROS	reactive oxygen species
RT	reverse transcriptase
rTMS	repetitive transcranial magnetic stimulation
SC	Schwann cells
SCFAs	short-chain fatty acids
SCI	sciatic nerve injury

SD	spreading depolarization
SDS	sodium dodecyl sulfate
SDS-PAGE	SDS polyacrylamide gel electrophoresis
SECO	secoisolairiciresinol-diglycoside
SHBG	sex hormone-binding globulin
SHRSP	stroke-prone spontaneously hypertensive rats
SIBO	small intestine bacterial overgrowth
SNPs	silver nanoparticles
SOD	super oxide dismutase
T2D	type 2 diabetes
TAC	total antioxidant contents
TBARS	thiobarbituric acid reactive substances
TBI	traumatic brain injuries
TEM	transmission electron microscope
TENS	transcutaneous electrical nerve stimulation
TKW	thousand kernel weight
TLC	thin layer chromatography
TNF	tumor necrotic factor
TNFRSF11B	TNF receptor superfamily 11b
TNF-α	tumor necrosis factor-alpha
TPC	total phenolic contents
TRAP	tartrate-resistant acid phosphatase
TW	test weight
UCP1	uncoupling protein 1
USA	The United States of America
UV	ultraviolet
VDCC	voltage-dependent calcium channels
VEGF	vascular endothelial growth factor
VF-TENS	variable frequency
WAT	white adipose tissues
WHO	World Health Organization
Zn	zinc

PREFACE

We introduce this book volume under the book series Innovations in Plant Science for Better Health: From Soil to Fork. This book mainly covers the current scenario of the research on the importance of phytochemicals from plant-based therapeutics. It is organized under two main parts: Part I: Health Claims of Food-Based Bioactive Compounds and Part II: Health Claims of Bioactive Compounds from Medicinal Plants.

Part I: Health Claims of Food-Based Bioactive Compounds describes the concept of extraction of bioactive molecules from foods, especially fruits and cereals, various extraction techniques, biochemistry, structural composition, and potential biological activities. Natural products and their bioactive compounds are increasingly utilized in preventive and therapeutic medications. Bioactive compounds from different foods have been utilized in pharmaceutical and nutraceutical industries for the production of supplements.

Part II: Health Claims of Bioactive Compounds from Medicinal Plants covers the isolation of potentially bioactive molecules from plant sources and their health aspects against various diseases. This section provided an introduction of different medicinal plants with respect to physical characteristics, uniqueness, uses, distribution, importance, traditional importance, nutritional importance, bioactivities, and future trends. Bioactive compounds from different medicinal plants have a positive effect on health and cure various disease conditions such as metabolic disorders, cancer, and chronic inflammatory reactions.

The goal of this book volume is to guide the world science community on how bioactive compounds from medicinal plants and different foods can alleviate us from various conditions and diseases. This book volume sheds light on the potential of both foods and plants for human health for different technological aspects, and it contributes to the ocean of knowledge on food science and nutrition. We hope that this compendium will be useful for students and researchers as well as for persons working with the food, nutraceuticals, and herbal industries.

The contributions by the cooperating authors to this book volume have been most valuable in the compilation. Their names are mentioned in each chapter and in the list of contributors. We appreciate you all for having patience with our editorial skills. This book would not have been written without the valuable cooperation of these investigators, many of whom are renowned scientists who have worked in the field of food science, biochemistry, and nutrition throughout their professional career.

We will like to thank Apple Academic Press, Inc., for making every effort to publish this book when all are concerned with health issues.

We request that readers offer constructive suggestions that may help to improve the next edition.

We express our admiration to our families and colleagues for understanding and collaboration during the preparation of this book volume.

I thank Dr. Megh R. Goyal for his leadership qualities for inviting me to join his team. He is a world-known scientist and engineer with expertise in agricultural and biological engineering. Truly he is the giver and a model for budding scientists. I am on the board to learn.

—Hafiz Suleria

PART I

Health Claims of Food-Based Bioactive Compounds

CHAPTER 1

NUTRITIONAL ASPECTS OF SOYBEAN (*GLYCINE MAX*)

HUMA BADER UL AIN, FARHAN SAEED, MUHAMMAD KASHIF, TABUSSAM TUFAIL, and HAFIZ ANSAR RASUL SULERIA

ABSTRACT

Soybean is the most abundant known source of protein, isoflavone, dietary fiber (DF) and fatty acids especially alpha-linolenic acid (ALA). These functional ingredients have a diverse range of biological activities, such as isoflavone, which is essential for their estrogenic and hypocholesterolemic activities, improves digestive tract function, prevents cancer, improves lipid metabolism and bone health. Whereas, ALA is an essential fatty acid for its hypotriglyceridemic effect and improves heart health. However, protein also shows hypocholesterolemic, antiatherogenic effects and reduces body fat. Dietary fiber also plays its part by not only controlling the blood pressure but also aid in the reduction of blood cholesterol levels. From a functional perspective, dietary fiber is described as supporting laxation, attenuating blood glucose responses, and assisting in cholesterol-lowering.

1.1 INTRODUCTION

Soybean, having a botanical name called *Glycine max* Merrill, is an herbaceous plant and an important oilseed crop, which belongs to the family Fabaceae. Annual production of soybean was reported to be 285.89 MT (million tons) and the United States of America (USA) ranks as the world's leading producer of soybean, with the production of about 93.08 MT [66].

The plant has been native to China and East Asia since 5,000 years [36]. USA, Brazil, and Argentine are the main exporters, and Chine is the main importer of soybean seeds [27, 46]. Commercially, it is grown in 35 countries and mostly in temperate regions of the world. It is being grown globally for its high monetary value, at the national and international levels for the production of nutritious variant products both for humans and animals. Its products have been divided into two categories, based on its usage, i.e., oil beans and food beans. Its seed composition shows that it consists of two cotyledons, a seed coat of hull followed by two smaller components, the hypocotyls and the plumule [65, 67].

Soybean seeds are of primary use in the oil industry, where they are being subjected to different processing conditions [5, 57, 58], to produce various kinds of by-products including cakes, expellers, oilseed meal, which are of primary use in animal nutrition [4]. The raw soybean seeds are passed through many stages at an industrial scale, after which the oil is extracted from the processed seeds either by mechanical method or by means of solvent extraction, which has emerged as the principal method in this regard. Resultantly, the products obtained are raw oil and defatted flakes. Apart from these, a number of soybean products are being offered in the markets, including soymilk, its sprouts, soya-sauce, tofu, miso, soy-flour, which are quite popular among the masses, owing to their exceptional nutritional and functional properties [3, 30, 61].

In millennia, soybean has captured greater interest and attention for its utilization in feed and food industry as an important functional food. It is owing to its nutritional composition, surplus availability, low cost, and high quality. It is the richest source of protein, isoflavone, and dietary fiber. Both fats and proteins constitute the critical components of soybean seed cotyledons. Carbohydrate and antinutritional compounds are also present in cotyledon but in small quantity [64].

This chapter focuses on nutritional composition and benefits of soybean seeds.

1.2 PROXIMATE COMPOSITION

Dietary fiber, isoflavones, and protein are abundantly present in the soybean, as shown in Table 1.1. Soybean contains approximately 40 to 45% protein, 18 to 22% oil [18], 14% moisture including vitamins and minerals (e.g., calcium, folic acid, and iron) [51, 56]. Due to its non-gluten proteins, soybean has gained much of the interest. All essential amino acids, which are indispensable for the maintenance of human health, are being provided by the Soybean protein. Unlike many other sources like legumes and animal, soybean has higher protein content and delivers nearly 40% of the calories. Protein from 24 eggs or 3 liters of milk or 1 kg of meat is equivalent to protein delivers from just 250 grams of soybean. But it is the quality of soy protein, which is most notable.

Soybean contains a high amount of PUFA's like omega-3 and omega-6 fatty acids as found in fish oil and has low saturated fat content make it a readily available source of essential fatty acids. Unlike legumes, soybean has 2.5 times more mineral content especially phosphorus, iron, calcium, and zinc (Zn) and has a low quantity of sodium. The outer hull of soybean is an excellent source of dietary fiber as well as good sources of vitamin B complex. Rakasi [46] reported that to create a fiber additive for cereals, snacks, and bread soybeans are processed as well as to remove the hull.

1.3 SOYBEAN: A MULTIFARIOUS FUNCTIONAL FOOD

The demonstration of ethnomedicinal potential allied with the intake of soybean has stimulated the consumer attention in exploring functional ingredients present in the seed. As dietary sources of high-quality protein and lipids like PUFAS's, Soybeans are widely consumed. Isoflavones, phytic acid, saponins anthocyanins (ACNs), dietary fibers and phytosterols are the nutritionally functional moieties found in the soybeans which have numerous health-promoting benefits [29].

For example, Fonseca and Ward [15] illustrated that both the hormonal and antioxidant activities of soy isoflavones help to reduce the prevalence of certain cardiac diseases, menopausal symptoms, osteoporosis, and different forms of cancers [63]. The soy is treating asthma, lung functions, type 2 diabetes (T2D), cancers (i.e., endometrial, prostate, thyroid, breast, and lung cancer), kidney-related diseases, and muscle soreness. Bolla [7] reported that the use of soy against hot flashes for breast cancer, breast pain, menopausal complications, and premenstrual syndrome (PMS) in women.

TABLE 1.1 Chemical Composition of Soybean

Components	Pele et al. [75] (%)	Ramadan et al. [77] (%)	Van Eys et al. [82]; ENV/JM/MONO [73] 15 (%)
Moisture	7.23	6.15	5.6–11.5
Ash	4.76	5.80	4.86
Crude fiber	6.58	7.12	5.12
Crude fat	26.20	20.25	18.38
Crude protein	23.98	35.55	37.08
Minerals	**Plaza et al. [76] (μg/g d.w)**	**Edelman and Colt [71] (mg)**	**Abdullah and Baldwin [70] (mg/100 g)**
Zn	23.93	3.7	4.9
Ca	810	195	ND
Na	2,610	12.3	0.5
Mn	6.70	ND	2.5
K	2,020	2,387	1,766
Cu	11.43	ND	2.1
Mg	1,330	407	251
Fe	48.87	6	8.8
Vitamins	**Plaza et al. [76] (μg/g d.w)**	**Edelman and Colt [71] (mg)**	**USDA [81]**
A	18.79	114 IU	1 μg
E	0.89	1.8	47 μg
B1	0.47	0.6	0.874 mg
B2	1.29	1.1	0.87 mg
B6	7.79	0.5	0.377 mg
C	99.51	0	6.0 mg

Soy-based foods, due to their taste, nutritional content, versatility, environmental advantages, and health benefits, are becoming more familiar and popular choices to consumers worldwide. These foods, i.e., soy flour, soy sauces, soy tofu, soy curd, fortified soy products for women and infants, fermented soybeans, and others, are present worldwide widely and are used against Hypertension (high blood pressure), Hypercholesterolemia (elevated levels of cholesterol) and thus aid in averting the ailments related to heart and blood vessels. Literature demonstrated that omega-3 fatty acids, i.e., alpha-linolenic acid have anticancer, anti-inflammatory, cardio-protective, neuroprotective, anti-osteoporotic, and anti-oxidative effects. Even though limited data on the toxicological scenario of alpha-linolenic acid, no serious adverse effects have been reported. It may be concluded that alpha-linolenic acid could be safely consumed as a food ingredient and may also act as a nutraceutical or pharmaceutical contestant. Different components of soybean including the soluble fiber, could potentially affect the release of insulin and its way of action in sustaining glucose balance. Soybean had been seen to decrease the post-prandial blood glucose response in humans.

1.3.1 PROTEIN

Henkel [23] illustrated that soybean has an excellent source of proteins as shown in Table 1.2 [41]. Like animal proteins, soybeans have an approximately similar composition of amino acid and protein hence can be a substitute of meat protein. The daily intakes of soy protein are reported to be 1 g in the USA, 7 g in Hong Kong, 8 g in China, 20 g in Korea and 30 g in Japan [24, 39, 40].

Frias et al. [16] stated that owing to rich nutritional status with an excellent ratio of amino acids present coupled with low cost of soybean, their usage as animal feed have augmented. Soybean proteins constitute about 35–40% of the seed, on a dry weight basis, out of which 65–80% of total seed proteins are the storage proteins generally known as conglycinin (7S) and glycinin (11S) [54]. Lonnerdal [37] illustrated about the multimeric iron storage protein known as ferritin present in the soybean and is absorbed and bioavailable more in comparison with animal products. Thus, a diet containing soybean is recommended for anemic people.

According to source method of FDA's 'Protein Digestibility Corrected Amino Acid' besides of high-quality protein it has also therapeutic and preventive roles for numerous medicines [26] including antiatherogenic and hypocholesterolemic effects [62]. The glycemic index of soybean is 18–25 which is lower than most of the legumes. Therefore, diet containing soybean have a combating effect against diabetes. In addition to this, soy protein is more effective against osteoporosis as soy protein has less calcium leaching effect as compared to the animal protein and relieving menopause symptoms.

Rakasi [46] reported that for treating women who are in menopause stage, soybean might be considered as a natural substitute for hormone replacement therapy (HRT). In past years, seeds of soybean or its proteins were used to biosynthesize metallic nanoparticles. Single protein of soybean (molecular weight (Mw) 51 kDa) was strong to synthesize eco-friendly and nontoxic silver nanoparticles (SNPs) for medical applications [56]. Izadi et al. [26] were synthesized gold nanoparticles from soybean seed extract in acidic conditions.

TABLE 1.2 Important Amino Acids in Soybean

Amino Acids	ENV/JM/MONO [73] (%)	El-Shemy et al. [72] (%)	Edelman and Colt [71] (%)
Arginine	2.45–3.1	3.30	ND
Cystine	0.45–0.67	0.54	ND
Histidine	1.0–1.22	1.21	2.6
Isoleucine	1.76–1.98	1.65	4.6
Leucine	2.2–4.0	2.79	7.7
Lysine	2.5–2.66	2.05	6.3
Methionine	0.5–0.67	0.57	1.3
Phenylalanine	1.6–2.08	1.63	4.9
Threonine	1.4–1.89	1.26	4.1
Tryptophan	0.51–2.44	1.22	1.4
Valine	1.5–2.44	1.81	4.7

1.3.2 ISOFLAVONES

Polyphenolic compounds exist in all parts of plants, ubiquitously [40]. In the light of chemical structure, chalcones, catechins, flavanones, isoflavones, anthocyanidins, flavones, and flavonols are subcategories of flavonoids [22]. Soybeans (*Glycine max* (L.) Merrill) have concentrated source of isoflavones in human foods [21].

Kishida et al. [35] reported soy-based foods are renowned because of high concentration of oil and proteins especially isoflavones as well as have good functional and nutritional quality. Despite controversial effects on human health, the high content of isoflavones is the main reason for soybean popularity [8]. Due to its similarity in structure to mammalian estradiol, which can fix both the α and β isoforms of estrogen receptor (ER) and due to its estrogenic activity, isoflavones are also called phytoestrogens [9].

Isoflavones are a subclass of a large group called flavonoids. Isoflavones have four isoforms: acetyglucosides (acetylglycitin, acetylgenistin, and acetyldaidzin), malonylglucosides (malonylglycitin, malonylgenistin, and malonyldaidzin), glucosides (glycitin and daidzingenistin) and unconjugated structure aglycone (genistein 24.1 mg/kg, daidzein 37.6/kg mg and glycitein) have basically three types present in the soybean [32].

Glycetin, genistern, and diadzin present in 10%, 50% and 40% ratio in total isoflavones [28]. Different geographical locations and cultivars affect the relative concentration of genistin, glycerin, and diadzin [34, 52, 53]. Soybean and soy-based foods have high concentrations of isoflavones similar to legumes and other beans [39]. On dry basis, soy food generally contains 1.2–3.3 mg isoflavones per gram; soybean the specific quantity depends on both environmental as well genetic factors such as variety of soybeans, harvesting year, storage conditions, and geographical locations [39, 52, 69].

The isoflavones content in Asia, Canada, Romania, Korea, and United States have isoflavones content ranges from 699.7–2581.6 g/g, 360–2,241 g/g, 712 to 1,228 mg/g, 178.81 mg/100 g, and 159.98 mg/100 g respectively [6, 34, 52, 53]. Equol is a potentially significant metabolite of isoflavones which has higher estrogenic activity. It is also unique because only 35% of individuals have the intestinal flora responsible for its production.

P-ethyl phenol which is the metabolite of genistein shows no activity, whereas glycitein Metabolites have not been identified yet. Before entering the bloodstream, all three soy isoflavones are glucoronidated in the intestinal. Ingesting soy protein could aid hot flashes initiated by menopause. Though, it cannot lessen other symptoms of menopause, like itching or vaginal dryness. Results from human studies are supportive of the safety and beneficial effects of isoflavones [7].

1.3.3 DIETARY FIBER

As for fibers are concerned in soybean, both types including soluble and insoluble ones are present in it, as shown in Table 1.3. A significant difference between soluble and insoluble fiber is due to the presence of cellulose and non-cellulose polysaccharides contents. Guillon and Champ [19] reported that leguminous cotyledon has minor content of crude fiber as compared to seed coat, whereas in the dry matter the concentration of dietary fiber reaches up to 90%. Seed coat has more fiber in cellulosic form and has a low quantity of pectin and hemicellulose whereas cell wall of cotyledons contains an extensive array of polysaccharides comprising cellulose, non-starchy, non-cellulose glucans, and pectins [44, 68].

Insoluble dietary fiber increases stool bulk, may be useful in relieving symptoms of digestive disorders and may prevent colon cancer whereas Soluble dietary fiber helps in lower blood sugar as well as serum cholesterol. One serving of soy-based foods per day might be protective against many types of cancer suggested by human epidemiological studies [46]. Soy fiber along with soy protein and phospholipids are effective against hypocholesteremia when added to the diet with other foods [25, 48]. Most published trial reveals that soy protein had contained little or no fiber. Therefore, fibers of soybean are not only the main factor in lowering the lipid of soy foods.

TABLE 1.3 Dietary Fiber of Soybean

Dietary Fiber	Pisarikora and Zraly (g/100 g)	Redondo-Cuenca et al. [78] (%)	El-Shemy et al. [72] (g/100 g)
IDF	32.6	20.86	12.44
SDF	2.9	3.5	8.68
TDF	35.5	24.36	21.12

1.3.4 FATTY ACIDS

Now a day, growing interest related to fatty acid composition fetched the attention towards human health, as shown in Table 1.4. Approximately 19% of oil is present in soybean, and triglycerides are the chief component of the oil. A significant amount of PUFA's, i.e., approx. 55% and 8% linoleic acid and α-linolenic acid, respectively, are present in soybean oil [43].

Both linoleic acid and α-linolenic acid have essential physiological and nutritional functions and play a significant part in the regulation of various metabolic pathways, respectively. Phytosterols, lecithin which comprises of phospholipids and tocopherols are the minor constituent of crude soybean oil. Memory, learning abilities and lipid metabolism are being improved by lecithin [62].

Greenland Eskimos first described the beneficial health effects of omega 3 [2] against CVD, asthma, multiple sclerosis (MS), and type 1 diabetes mellitus was observed in rich seafood diet. In the meantime, positive effects of omega 3 fatty acid for the treatment of rheumatoid arthritis (RA), cancer, inflammatory bowel disease and psoriasis have been extended [11, 59, 60]. Similarly, elevated levels of omega 6 fatty acids are allied with the prevention of chronic diseases [17]. Though, modern western diet holds a high level of omega 6 fatty acid content in contrast with omega 3 fatty acids which limited the beneficial health endorsing effects [13].

TABLE 1.4 Fatty Acids in Soybean

Fatty Acids	Sharma et al. [79] (g/100 g Fatty Acids)	Ivanov et al. [74] (%)	ENV/JM/MONO [73] 15 (%)
Palmitic acid (16:0)	12.0	3.28	1.44–2.31
Stearic acid (18:0)	3.6	2.34	0.54–0.91
Oleic acid (18:1)	5.4	20.47	3.15–8.82
Linoleic acid (18:2)	54.2	68.02	6.48–11.6
Linolenic acid (18:3)	4.9	5.18	0.72–2.16

1.3.5 PHYTOSTEROLS

In the recent era, phytochemicals have fetched the attention of researchers around the globe, e.g., sterols obtained from the plant seed oils account for a significant base stock for the nutrition sector. They have been reported to reduce blood cholesterol level considerably, especially in those having mildly elevated levels of cholesterol [20].

β-Sitosterol was being ingested, curing hypercholesterolemia in USA. Phytosterol's role as emulsifiers in the cosmetic industry has also been reported. They are also the precursors to hormonal sterols, 75% of which have been derived from soybean oil, globally. These phytosterols are recovered as a co-product from the deodorization phase during the processing of raw soybean and other vegetable oils. Oil extracted from the soybean comprise of about 300–400 milligrams of plant sterols per 100 grams. The key constituents of soy sterols are β-sitosterol (53 to 56%), campesterol (20 to 23%), and stigmasterol (17 to 21%) [42]. These sterols are proved to exhibit the hypercholesterolemic activity, but the mechanism is still to be fully understood [38].

1.3.6 ANTI-NUTRITIONAL COMPOUNDS

Mikic et al. [55] reported that antinutritional elements, for instance, tannins, phytic acid, phenols, saponins, and trypsin inhibitors, etc., present in the raw soybean can cause health-related problems both in animal and human as well as lower the nutritive value of grain legumes when consumed in large quantity [55]. Trypsin or chymotrypsin, which is blocked by the Trypsin inhibitors results in the decrease of dietary protein hydrolysis as well as absorption and digestibility of amino acid [49].

The availability of calcium, iron, Zn, copper, and manganese is reduced by binding with the phytic acid [47]. Conjugation of protein and amino acid with tannins and phenols has a negative result on the digestibility [33]. To improve the organoleptic acceptability and nutritional quality of legumes, these antinutritional compounds should be removed for effective utilization as human food.

To improve the nutritional quality, antinutritional components exist in the raw seed are partially removed during domestic processing. Cooking, soaking, germination, etc., results in the number of biochemical, nutritional, physicochemical, and sensory changes in legumes. Prodanov et al. [45] reported the nutritional value of soybean could be enhanced by increasing the availability of protein, vitamins, and amino acid digestibility by different processing methods. During soaking antinutritional components, for example, phytic acid, α-galactosides, certain minerals and protease enzyme inhibitor are removed by the partial or by complete solubilization in the discarding solution [45].

Before the cooking process, soy-seeds are soaked in water before boiling in water, until they become soft, but this step can cause loss of few essential nutrients, e.g., water-soluble vitamins, that can seep into the cooking medium during the boiling process [12]. Both heat sensitive anti-nutritive factors and volatile compounds are decreased naturally by the boiling process [1].

Trypsin inhibitors are abundantly present in foods, including soybean, but the heat treatment destroys most of the activity of trypsin inhibitors from the soybean products. A minor amount of heat-stable Bowman-Birk inhibitor could impose a hypocholesterolemic effect by triggering the secretion of cholecystokinin, which would then arouse synthesis of bile acid from cholesterol and resultantly help to remove cholesterol through the GI (gastrointestinal) tract. Though, research studies on animals have not demonstrated a hypo-cholesterolemic effect the addition of trypsin inhibitor to the diet [14].

Phytic acid, hexaphosphate, and myoinositol is found in all non-fermented soy protein products and is quite stable while heating. Phytic acid along with oxalates chelates Zn and iron respectively intensely in the intestinal tract, thus subsiding its absorption [37]. Copper (Cu) deficiency or a high ratio of Zn to copper leads to rise in the blood cholesterol [14]. The current hypothesis is that soy foods comprise both phytic acid and copper; consequently, they may diminish cholesterol levels by reducing the ratio of Zn to copper. Antinutritional compounds have been shown in Table 1.5.

TABLE 1.5 Antinutritional Compounds of Soybean

Antinutritional Compounds	Pele et al. [75]	El-Shemy et al. [72]	Sharma et al. [79] (mg/g)
Tannin	29.3 mg/100 g	25.23 mg/100 g	11.1–18.8
Phytic acid	212 mg/100 g	8.05%	5.1–24.5
Trypsin	27.5 U/g DW	7.09%	30–102.5

1.4 SUMMARY

The soybean has many potential health effects, such as, hypocholesterolemic, antidiabetic, anticancer, antiatherogenic, immune-modulatory, and many more. The proteins in soybean have been used to synthesize silver and gold nanoparticles for medical applications. Further research studies are required to know their role as nutraceutical and recent advances to incorporate soybean into products so that people can consume more of these functional foods to avoid specific health disorders.

KEYWORDS

- **estrogen receptor**
- **isoflavone**
- **nanoparticles**
- **premenstrual syndrome**
- **protein**
- **rheumatoid arthritis**

REFERENCES

1. Akande, K. E., & Fabiyi, E. F., (2010). Effect of processing methods on some antinutritional factors in legume seeds for poultry feeding. *International Journal of Poultry Science, 9*, 996–1001.
2. Akbar, U., Yang, M., Kurian, D., & Mohan, C., (2017). Omega-3 fatty acids in rheumatic diseases: A critical review. *Journal of Clinical Rheumatology, 23*(6), 330–339.
3. Badger, T. M., Ronis, M. J., Simmen, R. C., & Simmen, F. A., (2005). Soy protein isolate and protection against cancer. *Journal of the American College of Nutrition, 24*(2), 146S–149S.
4. Banaszkiewicz, T., (2011). In: El-Shemy, H., (ed.), *Nutritional Value of Soybean Meal, Soybean and Nutrition* (p. 325). IntechOpen; InTech.
5. Berk, Z., (1992). Technology of production of edible flours and protein products from soybeans. In: *Soymilk and Related Product* (p. 58). Chapter 5; Rome: FAO Agricultural Services Bulletin 97.
6. Bhagwat, S., Haytowitz, D. B., & Holden, J. M., (2008). *USDA Database for the Isoflavone Content of Selected Foods*. Release 2.0; U.S. department of agriculture, agricultural research service, nutrient data laboratory home page. http://www.ars.usda.gov/nutrientdata/isoflav (accessed on 17 September 2021).
7. Bolla, K. N., (2015). Soybean consumption and health benefits. *International Journal of Science and Technology Research, 4*(7), 50–53.
8. Cavaliere, C., Cucci, F., Foglia, P., Guarino, C., Samperi, R., & Laganà, A., (2007). Flavonoid profile in soybeans by high-performance liquid chromatography/tandem mass spectrometry. *Rapid Communications in Mass Spectrometry, 21*(14), 2177–2187.
9. Cederroth, C. R., & Nef, S., (2009). Soy, phytoestrogens and metabolism: A review. *Molecular and Cell Endocrinology, 304*(1, 2), 30–42.
10. Grieshop, C. M., Kadzere, C. T., & Clapper, G. M., (2003). Chemical and nutritional characteristics of United States soybeans and soybean meals. *Journal of Agricultural and Food Chemistry, 51*, 7684–7691.
11. Connor, W. E., (2000). Importance of n-fatty acids in health and disease. *The American Journal of Clinical Nutrition, 71*, 171S–175S.
12. El-Adawy, T. A., (2002). Nutritional composition and antinutritional factors of chickpeas (*Cicer arietinum* L.) undergoing different cooking methods and germination. *Plant Foods for Human Nutrition, 57*, 83–97.
13. Enser, M., Richardson, R. I., Wood, J. D., Gill, B. P., & Sheard, P. R., (2000). Feeding linseed to increase the n-3 PUFA of pork: Fatty acid composition of muscle, adipose tissue, liver and sausages. *Meat Sciences, 55*, 201–212.
14. Erdman, J. W., (2000). Soy protein and cardiovascular disease. *Circulation, 102*, 2555–2559.
15. Fonseca, D., & Ward, W. E., (2004). Daidzein together with high calcium preserve bone mass and biomechanical strength at multiple sites in ovariectomized mice. *Bone, 35*, 489–497.

16. Frias, J., Song, Y. S., Martínez-Villaluenga, C., De Mejia, E. G., & Vidal-Valverde, C., (2008). Immunoreactivity and amino acid content of fermented soybean products. *Journal of Agricultural and Food Chemistry, 56*, 99–105.
17. Givens, D. I., Kliem, K. E., & Gibbs, R. A., (2006). The role of meat as a source of n-3 polyunsaturated fatty acids in the human diet. *Meat Sciences, 74*, 209–218.
18. Goyal, R., Sharma, S., & Gill, B. S., (2012). Variability in the nutrients, antinutrients and other bioactive compounds in soybean (*Glycine max* (L.) Merrill) genotypes. *Journal of Food Legumes, 25*, 314–320.
19. Guillon, F., & Champ, M. M. J., (2002). Carbohydrate fractions of legumes: Uses in human nutrition and potential for health. *British Journal of Nutrition, 88*, S293–S306.
20. Gylling, H., & Simonen, P., (2015). Phytosterols, phytostanols, and lipoprotein metabolism. *Nutrients, 7*(9), 7965–7977.
21. Hasanah, Y., Nisa, T. C., Armidin, H., & Hanum, H., (2015). Isoflavone content of soybean (*Glycine max* (L) cultivars with different nitrogen sources and growing season under dryland conditions. *Journal of Agriculture and Environment for International Development, 109*(1), 5–17.
22. Heim, K. E., Tagliaferro, A. R., & Bobilya, D. J., (2002). Flavonoid antioxidants: Chemistry, metabolism and structure-activity relationships. *The Journal of Nutritional Biochemistry, 13*(10), 572–584.
23. Henkel, J., (2000). Soy: Health claims for soy protein, questions about other components. *FDA Consumer, 34*(3), 13–15, 18–20.
24. Ho, S. C., Woo, J. L., Leung, S. S., Sham, A. L., Lam, T. H., & Janus, E. D., (2000). Intake of soy products is associated with better plasma lipid profiles in the Hong Kong Chinese population. *Journal of Nutrition, 130*(10), 2590–2593.
25. Høie, L. H., & Morgenstern, E. C., (2005). A double-blind placebo-controlled clinical trial compares the cholesterol-lowering effects of two different soy protein preparations in hypercholesterolemic subjects. *European Journal of Nutrition, 44*, 65–71.
26. Izadi, E., Rasooli, A., Akbarzadeh, A., & Davaran, S., (2017). Preparation and characterization of gold nanoparticles in the presence of citrate and soybean seed extract in an acidic conditions. *Drug Research, 67*(05), 266–270.
27. Jamet, J. P., & Chaumet, J. M., (2016). Soybean in China: Adapting to the liberalization. *OCL, 23*(6), 604. doi: 10.(1051)/ocl/(2016)044.
28. Jin, H. E. F., & Chen, J. Q., (2013). Consumption of soybean, soy foods, soy isoflavones and breast cancer incidence: Differences between Chinese women and women in Western countries and possible mechanisms. *Food Science and Human Wellness, 2*(3, 4), 146–161.
29. Jo, J. K., Ingale, S. L., & Kim, J. S., (2012). Effects of exogenous enzyme supplementation to corn-soybean based or complex diets on growth performance, nutrient digestibility and blood metabolites in growing pigs. *Journal of Animal Sciences, 89*, 1795–1804.
30. Jooyandeh, H., (2011). Soy products as healthy and functional foods. *Middle-East Journal of Scientific Research, 7*(1), 71–80.
31. Josipović, A., Sudar, R., Sudarić, A., Jurković, V., Kočar, M. M., & Kulundžić, A. M., (2016). Total phenolic and total flavonoid content variability of soybean genotypes in eastern Croatia. *Croatian Journal of Food Science and Technology, 8*(2), 60–65.

32. Kao, T. H., Huang, R. F. S., & Chen, B. H., (2007). Antiproliferation of hepatoma cell and progression of cell cycle as affected by isoflavone extracts from soybean cake. *International Journal of Molecular Science, 8*, 1095–1110.
33. Khandelwal, S., Udipi, S. A., & Ghugre, P., (2010). Polyphenols and tannins in Indian pulses: Effect of soaking, germination and pressure cooking. *Food Research International, 43*, 526–530.
34. Kim, W. J., Back, S. H., Kim, V., Ryu, I., & Jang, S. K., (2005). Sequestration of TRAF2 into stress granules interrupts tumor necrosis factor signaling under stress conditions. *Molecular and Cellular Biology, 25*, 2450–2462.
35. Kishida, T. H., Ataki, H., Takebe, M., & Ebihara, K., (2000). Soybean meal fermented by *Aspergillus* awamori increases the cytochrome p-450 content of the liver microsomes of mice. *Journal of Agricultural and Food Chemistry, 48*, 1367–1372.
36. Krishnan, H. B., & Jez, J. M., (2018). The promise and limits for enhancing sulfur-containing amino acid content of soybean seed. *Plant Science*.
37. La Frano, M. R., De Moura, F. F., Boy, E., Lönnerdal, B., & Burri, B. J., (2014). Bioavailability of iron, zinc, and provitamin A carotenoids in biofortified staple crops. *Nutrition Reviews, 72*(5), 289–307.
38. Law, M., (2000). Plant sterol and stanol margarines and health. *British Medical Journal, 320*, 861–864.
39. Liu, X. X., Li, S. H., Chen, J. Z., Sun, K., Wang, X. J., Wang, X. G., & Hui, R. T., (2012). Effect of soy isoflavones on blood pressure: A meta-analysis of randomized controlled trials. *Nutrition, Metabolism and Cardiovascular Diseases, 22*(6), 463–470.
40. Nagata, C., (2000). Ecological study of the association between soy product intake and mortality from cancer and heart disease in Japan. *International Journal of Epidemiology, 29*(5), 832–836.
41. Olanipekun, B., & Adelakun, O., (2015). Nutritional and microbiological attributes of soybean (*Glycine max*) during fermentation with *Rhizopus oligosporus*. *Food Science and Quality Management, 39*, 20–31.
42. Ozawa, Y., Sato, H., & Nakatani, A., (2001). Chemical composition of soybean oil extracted from hypocotyl-enriched soybean raw material and its cholesterol-lowering effects in rats. *Journal of Oleo Science, 50*(4), 217–223.
43. Park, H., (2012). Modifying the fatty acid profile of soybean oil for nutritional and industrial applications. *Theses, Dissertations, and Student Research in Agronomy and Horticulture, 71*, 200–202.
44. Petterson, D. S., (1998). Composition and food uses of lupin. In: *Lupin as Crop Plant: Biology, Production and Utilization* (pp. 353–384). CAB International, Wallingford, Australia.
45. Prodanov, M.,& Vierra, I., (2004). Influence of soaking and cooking on thiamin, riboflavin and niacin contents in legumes. *Food Chemistry, 84*, 271–277.
46. Rakasi, K., (2011). *Nutritional and Health Benefits of Soybeans and Micro-Enterprise Opportunities* (p. 6). Fifth congress on soy marcour: 14 al 16 de September; Rosario, Argentina.
47. Ramakrishna, V., Rani, P. J., & Rao, P. R., (2006). Antinutritional factors during germination in Indian bean (*Dolichos lablab* L.) seeds. *World Journal of Dairy and Food Science, 1*, 6–11.

48. Ramdath, D. D., Padhi, E. M. T., Sarfaraz, S., Renwick, S., & Duncan, A. M., (2017). Beyond the cholesterol-lowering effect of soy protein: A review of the effects of dietary soy and its constituents on risk factors for cardiovascular disease. *Nutrients, 9*(4), 324.
49. Roy, F., Boye, J. I., & Simpson, B. K., (2010). Bioactive proteins and peptides in pulse crops: pea, chickpea and lentil. *Food Research International, 43*, 432–442.
50. Sasikala, D., Govindaraju, K., Tamilselvan, S., & Singaravelu, G., (2012). Soybean protein: A natural source for the production of green silver nanoparticles. *Biotechnology and Bioprocess Engineering, 17*(6), 1176–1181.
51. Sauvant, D., Perez, J. M., & Tran, G. E. D., (2004). *Tables of Composition and Nutritional Value of Feed Materials* (p. 119). INRA, France.
52. Seguin, P., Zheng, W., Smith, D. L., & Deng, W., (2004). Isoflavone content of soybean cultivars grown in eastern Canada. *Journal of the Science of Food and Agriculture, 84*, 1327–1332.
53. Sertovic, E., Mujic, I., Jokic, S., Alibabic, V., & Saric, Z., (2012). Effect of soybean cultivars and the content of isoflavone on soymilk. *Romanian Biotechnological Letters, 17*, 7151–7159.
54. Sharma, S., Kaur, M., Goyal, R., & Gill, B. S., (2014). Physical characteristics composition of some new soybean (*Glycine max* (L.) Merrill) genotypes. *Journal of Food Science and Technology, 51*(3), 551–557.
55. Sharma, S., Kaur, M., Goyal, R., & Gill, B. S., (2011). Nutritional composition of some new soybean (*Glycine max* (L.) Merrill) genotypes. *Journal of Food Science and Technology*. doi: 10.(1007)/s(1319)7-011-0517-7.
56. Shashank, A., Tidke, D., & Ramakrishna, S., (2015). Nutraceutical potential of soybean: Review. *Asian Journal of Clinic Nutrition, 7*, 22–32.
57. Shurtleff, W., & Aoyagi, A., (2013). *History of Whole Dry Soybeans, Used as Beans, or Ground Mashed or Flaked (240 Bce To (2013): Extensively Annotated Bibliography and Sourcebook* (p. 210). Soy info Center, Lafayette, CA 94549-0234 USA.
58. Sikorski, Z. E., (2007). *Chemia Zywności* (p. 39). Praca Zbiorowa, WNT, Warszawa.
59. Simopoulos, A. P., (2008). The importance of the omega 6/omega 3 fatty acid ratio in cardiovascular disease and other chronic diseases. *Experimental Biology and Medicine (Maywood), 233*(6), 674–688.
60. Simopoulos, A. P., (2002). The importance of the ratio of omega-6/omega-3 essential fatty acids. *Biomedicine and Pharmacotherapy, 56*, 365–379.
61. Spector, D., Anthony, M., Alexander, D., & Arab, L., (2003). Soy consumption and colorectal cancer. *Nutrition and Cancer, 47*, 1–12.
62. Sugano, M., (2006). *Soy in Health and Disease Prevention* (p. 328). CRC Press, Boca Raton, FL.
63. Tikkanen, M. J., & Adlercreutz, H., (2000). Dietary soy-derived isoflavone phytoestrogens. Could they have a role in coronary heart disease prevention?. *Biochemistry and Pharmacology, 60*(1),1–5.
64. Torbica, A., Hadnađev, M., Dokić, P., & Sakač, M., (2008). Mixolab profiles of gluten free products ingredients. *Food Processing, Quality and Safety, 35*(1), 19–26.

65. Torres, N., Garcia, A. J., & Tovar, A. R., (2014). The use of soybean protein as functional food to control lipid metabolism in some chronic diseases. In: *Seeds as Functional Foods and Nutraceuticals* (pp. 110–119). Nova Science Publishers, Inc.
66. USDA (United States Department of Agriculture), (2012). *Crop Production*. National Agricultural Statistics Service. online.
67. Van, E. J. E., Offner, A., & Bach, A., (2004). Chemical analysis. *Manual of Quality Analysis for Soybean Products in the Feed Industry*. American Soybean Association. http://ussec.org/wp-content/uploads/2012/09/Manual-of-Quality-Analyses-2nd-edition (accessed on 17 September 2021).
68. Van, L. H., & Tamminga, S., (1999). Fermentation characteristics of cell wall sugars from soya bean meal, and from separated endosperm and hulls of soya beans. *Animal Feed, Science and Technology, 79*, 179–193.
69. Zhu, D., Hettiarachchy, N. S., Horax, R., & Chen, P., (2005). Isoflavone contents in germinated soybean seeds. *Plant Foods for Human Nutrition, 60*(3), 147–151.
70. Abdullah, A., & Baldwin, R. E. (2006). Mineral and Vitamin Contents of Seeds and Sprouts of Newly Available Small-Seeded Soybeans and Market Samples of Mungbeans. (2006). *Journal of Food Science 49*(2), 656–657. DOI:10.1111/j.1365–2621.1984.tb12495.x.
71. Edelman, M., & Colt, M. (2016). Nutrient value of leaf vs seed. *Front Chem. 4*, 32.
72. El-Shemy, H., Abdel-Rahim, E., Shaban, O., Ragab, A., Carnovale, E., & Fujita, K. (2000). Comparison of nutritional and antinutritional factors in soybean and fabacean seeds with or without cortex. Soil Science and Plant. *Nutrition 46*(2), 515–224. DOI: 10.1080/00380768.2000.10408804
73. ENV/JM/MONO2001 15. Unclassified. (2001). Series on the safety of novel foods and feeds 2 Consensus document on compositional considerations for new varieties of soybean: *Key Food and Feed Nutrients and Anti-Nutrients*, 30 November 2001.
74. Ivanov, D. S., Levic, J. D., & Sredanovic, S. A. (2011). Fatty acid composition of various soybean products. *Food and Feed Research. 37*(2), 65–20.
75. Pele, G. I., Ogunsua, A. O., Adepeju, A. B., Esan, Y. O., & Oladiti, E. O. (2016). Effects of processing methods on the nutritional and anti-Nutritional properties of soybeans (Glycine max). *African Journal of Food Science Technology 7*(1), 9–12.
76. Plaza, L., de Ancos, B. & Cano, P.M. (2003). Nutritional and health-related compounds in sprouts and seeds of soybean (Glycine max), wheat (*Triticum aestivum*.L) and alfalfa (*Medicago sativa*) treated by a new drying method. *Eur Food Res Technol 216,* 138–144. https://doi.org/10.1007/s00217–202–2640–2.
77. Ramadan, E. A., (2012). Effect of processing and cooking methods on the chemical composition, sugars and phytic acid of soybeans. *Food Public Health. 2,* 11–25.
78. Redondo-Cuenca, A., Villanueva-Suarez, M. J., & Mateos-Aparicio, I. (2008). Soybean seeds and its by-product okara as sources of dietary fibre. Measurement by AOAC and Englyst methods. *Food Chemistry. 108*(3), 1099–2105. DOI:10.1016/j.foodchem.2007.11.061
79. Sharma, S., Kaur, A., Bansal, A., & Gill, B. S. (2013). Positional effects on soybean seed composition during storage. *J Food Sci Technol. 50*(2): 353–359.
80. Sharma, S., Kaur, M., Goyal, R., & Gill, B. S. (2014). Physical characteristics and nutritional composition of some new soybean (*Glycine max* (L.) Merrill) genotypes. *J Food Sci Technol. 51*(3), 551–552. doi: 10.1007/s13197–211–2517–2.

81. *USDA Agricultural Projections to 2025, by Paul Westcott and James Hansen, USDA, Economic Research Service, February 2016.*
82. Van Eys, J. E., Offner, A., & Bach, A. (2004). Chemical Analysis. Manual of quality analysis for soybean products in feed industry. American Soybean Association.

CHAPTER 2

NUTRITIONAL ASPECTS OF DURUM WHEAT

SHAHRINA SHAKIR, SHABBIR AHMAD, AMBREEN NAZ, and NIGHAT RAZA

ABSTRACT

Durum wheat is the most nutritious crop, belongs to the family *Poaceae,* and it is cultivated all over the world, mainly it is produced in the EU, Canada, USA, and Italy. Durum wheat is tetraploid species, its grain is hard, vitreous, amber yellow in color and it is used for the formation of various products of food. The durum wheat grain is full of nutrition. Durum wheat is also associated with the reduction and prevention of a number of diseases. So, this wheat regarding health is very beneficial for consumption. Durum wheat has various types of quality parameters regarding its milling and manufacturing of different food products. Grain size, grain hardness, hard vitreous kernels, test weight (TW) and thousand kernel weights (TKW) are those parameters which describes the quality of durum and are utilized in its characterization. In composition of durum wheat, it contains different proportions of starch, protein, gluten, gliadins, glutenins, etc. For characterization of durum wheat, the important parameter is carotenoid pigment that gives color to durum, protein, and gluten. Durum wheat is milled into semolina and formation of semolina is a complex process. The production and quality of semolina is linked with the granulation of semolina and yellow color. The most consumable product of durum wheat is pasta and quality of pasta is described in terms of yellow color, wholesomeness, *al dente* trait, and stickiness.

2.1 INTRODUCTION

Durum wheat (*Triticum durum*) is one of the world's most important, nutritive crops that belongs to the family *Poaceae.* In 2018–2019, the world production of durum wheat was almost 800 million metric tons [2]. The biggest producers of durum wheat are the European Union, Australia, the USA, Canada, while Italy is producing about 4 million tons and has become the biggest contributor, i.e., 41% of EU's production and 10% of the world's production [17]. The main durum cultivating areas are North America (including Mexico), the former Soviet Union, North Africa, Southern Europe, Middle East, and India.

In the world, durum is grown and cultivated over 13 million hectares, and almost 30 million tons are produced every year. Among the world wheat production, durum constitutes only 5–8%. Durum wheat is tetraploid species of wheat having 28 chromosomes. The grain of durum wheat is very vitreous, hard, and amber-colored having high protein content. Durum wheat is utilized for the production of a variety of food products, for example, pasta, semolina, desserts, couscous, and burghul wheat [15].

2.2 NUTRITIONAL ASPECTS OF DURUM WHEAT

Durum wheat grain is hard, vitreous, yellow colored and has nutty taste. Durum wheat grain has protein (14–29%), amylose content (15 to 45%), arabinoxylan (4.1–7.5%), fructans content (1.7–2.5%), cellulose (2.7%), lignin (11.9–2.6%), lipid content (2.4 to 3.8%). Durum wheat also contain thiamin (4.7 μg/g), riboflavin (0.70 μg/g), pyridoxine (1.9 μg/g), sodium (0.01–0.05 mg/g), potassium (3.8–5.5 mg/g), phosphorus (1.8–5.2 mg/g), magnesium (1.0–1.5 mg/g), calcium (0.32–0.47 mg/g), iron (29.6–36.2 μg/g), zinc (Zn) (14.0–26.0 μg/g) and dietary fiber content (10.7–15.5%) [22].

2.3 HEALTH CONCERNS

Celiac disease is a typical digestion-related condition in which the person has an abnormal response to gluten. In gluten protein, the gliadin content is responsible for this type of disease [16]. The dietary fiber in durum wheat improves appetite regulation, prolonged satiety, and gut health. Dietary fiber helps in the prevention of gastric cancer, prevents colon diseases, reduces the risk of hiatal hernia and hemorrhoids, treats irritable bowel syndrome (IBS), prevents hypertension, hypercholesterolemia, reduces the risk of type 2 diabetes (T2D), gallbladder disease, and breast cancer [3].

Dietary fiber helps in reducing the digestion of starch in small intestine which in turn lowers level of glucose that enters into bloodstream, this lessens the demand of insulin and decreases glycemic index of consumed food. Furthermore, in small intestine the undigested starch moves to large bowl, fermented into fatty acids. These produced fatty acids are associated with large bowl function and help to increase satiety [14].

2.4 DURUM WHEAT QUALITY PARAMETERS

In durum wheat, there are various types of quality parameters that are of considerable importance in the case of milling and products preparation. The names and descriptions of those parameters are given below:

- Grain size;
- Grain hardness;
- Hard vitreous kernels;
- Test weight;
- Thousand kernel weights.

2.4.1 *GRAIN SIZE*

The grains that fall through 2 mm screen in milling are termed as undersized grains. When there is a high proportion of undersized grains, they become undesirable for milling because it lessens the semolina yield and produces non-uniform particle size. These have an effect on pasta quality and produce white specks in pasta because of uneven hydration [26].

2.4.2 *GRAIN HARDNESS*

The key determinant of milling performance for the production of semolina is grain hardness. The grain texture of wheat grain can be determined by proteins and these are known as puroindolines, friabilin and grain-softness protein. These all are related and friabilin is a puroindoline. But genes of puroindoline are not present in durum wheat, the absence of these is cause for hard texture of durum wheat [22].

2.4.3 HARD VITREOUS KERNELS

The kernel of durum wheat should be vitreous and translucent to give good semolina production and to lessen the quality-related issues of pasta like white specks, poor cooking quality, and weak strands, these issues occur when grain vitreousness is low, i.e., less than 50%. Starchy or non-vitreous kernels upon milling tend to give usual flour than semolina because non-vitreous kernels are soft and usual flour will produce white specks in pasta. If grain protein content is high, then hard vitreous kernels are not considered because millers of durum wheat use finer granulations in the milling process of durum wheat [26].

2.4.4 TEST WEIGHT (TW)

In the milling of durum wheat, the most important factor is test weight (TW). The measurement of the weight of a given volume is termed TW. The widely used specification in wheat grading is TW, which is affected by the shape and size of the grain because it is internationally recognized as an index of wheat milling potential and soundness of wheat. Two factors are responsible for lowering the TW, one is grain shriveling due to terminal drought and the second one is grain weathering caused by rain at the time of harvest. The high-protein, shriveled grain reduces the milling performance of semolina. Because small kernels have a lower percentage of endosperm, that is milled into semolina. In the case of durum wheat, the value of TW should be greater than 76 kilogram/hectoliter. The TW of weak wheat is 55–60 kg/hl and that of strong wheat is 80–83 kg/hl [26].

Pomeranz quotes to Geddes who reported that if the TW of wheat is higher than the endosperm percentage will also be higher and semolina yield will also be higher. If the kernel of wheat is undamaged by the environment, fully matured, free from diseases and plump then the TW is greater [13].

2.4.5 *THOUSAND KERNEL WEIGHT (TKW)*

There is another method also that is the measurement of weight of thousand kernel of wheat. The thousand kernel weight (TKW) of durum wheat is 30 to 40 g. The TKW is also the quality parameter for determining the weight of grains [26].

2.5 COMPOSITION OF DURUM WHEAT

In the chemical composition of durum wheat grain contains protein, lipids, starch, carotenoids, xanthophylls, etc., that affect the quality of durum products and are discussed here.

2.5.1 *STARCH*

Starch is a very important reserve and it is present in the form of semi-crystalline granule. Amylopectin and amylose are the major components of the starch. The amounts of amylopectin and amylose are 72 to 75% and 25 to 28%, respectively, in durum wheat. The diameter of these small granules is up to 10 mm (mean diameter is 5 mm) and the diameter of large granules is 20 mm [11]. Starch is composed of only one building block that is glucose [13]. Amylose retrogrades with cooling, stirring of pasta and pasta's viscosity rises to reach final value [22]. The sticky texture in pasta is contributed to the synthesis of a continuous starch phase linked with disrupted starch granules which are swollen on the surface of pasta [28].

2.5.2 *AMYLOSE AND AMYLOPECTIN*

The amylose content (0.7 to 22.9%) is obtained by substituting the durum starch with waxy hexaploid wheat. In the blends of flour, the amylose content was decreased by increasing the percentage (%) of waxy starch as a result stickiness and firmness of cooked pasta increased whereas cooking time decreased. During cooking, soluble carbohydrate like amylose arising from the granules of starch is reason for stickiness of pasta. So, to enhance the stickiness on pasta surface the carbohydrate in the form of amylopectin will be released in manufacturing of pasta [22].

The pasta firmness also increases by increasing the amylose content. The granules are tightly packed in high amylose starches, on swelling have resistance to deformation and rupture. This may be the explanation for greater ability for the production of firmer pasta. Due to the rise in amylose content, the cooking loss in pasta increased, and water uptake decreased. Hospers et al. fed humans pasta of 40% amylose content in comparison with control past of amylose content 25.9% and assumed that it lowers the insulin levels and postprandial blood glucose [22].

2.5.3 PROTEIN

Durum wheat contains about 12–16% of protein. Protein has its five subclasses like polymeric glutenin (16%), monomeric gliadin (33%), globulins (3%), albumins (15%) and the residue 33%. Proteins are made up of amino acids and every amino acid has a hydrocarbon group, amino group, and carboxyl acid group. In the proteins, the amino acid group and carboxyl acid group are joined with each other by peptide bonds [13]. In USDA World Wheat Collection, the protein content of 12,600 varieties were analyzed and concluded that protein ranges from 7%–22% (dry weight) [27].

In process of protein biosynthesis in durum wheat, it can be reported that the process of synthesis of protein and accumulation may lead to less or more aggregated state which affects the technological properties of semolina and pasta. Various authors like Favret et al. and Johnson et al. said that the protein content must be represented as the quantity in 1 seed, not as percentage of weight of seed [9]. Some scientists assumed that protein subunit (high molecular weight (Mw) glutenin subunit $2 \times 7 + 8$) was in correlation with improved dough strength as despite subunit null. On *Glu-B1* chromosome, the 6+8 subunit were in association with stronger gluten type than 13+16 and 7+8 but weak gluten characteristics is due to subunit 20 (*Glu-1B*). Bechere et al. reported that it is safe to use low Mw glutenin patterns in association with durum wheat quality because similar high Mw glutenins can be linked with two different patterns of low Mw glutenins.

In the program of breeding, it is useful to discard lines with HMW-GS 20 and LMW-1 and electrophoretic analysis is combined with Sodium dodecyl sedimentation test to develop durum wheat cultivars with good firmness and viscoelasticity of cooked pasta [11]. The protein content in durum wheat is an important quality parameter that should be greater than 13%. So that after milling more than 12% protein should be present in semolina important for producing pasta with good textural properties. High-protein semolina produces physically stronger and elastic pasta (dry) than low protein semolina. High protein content in semolina has potential to produce pasta firmer, less sticky and resistance to overcooking. There are some negative aspects also for extra high protein content like lesser TW and semolina yield that gives duller pasta. Semolina containing a low amount of protein produces fragile spaghetti having very low firmness. Durum wheat with high-protein content enables the pasta for swelling during cooking, allows retention of firmness with overcooking and lessens the cooking loss [26].

2.5.4 PROLAMINS

Durum wheat prolamins are classified on the basis of solubility, the alcohol-soluble proteins are called gliadin and insoluble protein are called glutenin. Glutenins are responsible for gluten strength and elasticity while gliadins are responsible for extensibility and cohesiveness of gluten. Gliadins are joined together with disulfide bonds and are classified as α/β- and γ-gliadins, those gliadins which are non-connected are named as ω-gliadins. Gliadins may act as chain terminators and LMW-GS can work as chain extenders. These proteins comprise 80 to 85% of gluten proteins and are very important in determining viscoelastic properties of dough important in the quality of pasta [4].

2.5.5 GLIADINS

Gliadins are known as single chain polypeptides and gliadins are classified into four classes α, β-, γ-, and ω-gliadins on the basis of electrophoretic mobility on sodium dodecyl sulfate (SDS)-polyacrylamide gel electrophoresis [4]. Gliadin-45 type band is linked with high elastic recovery and gliadin-42 type band is with poor elastic recovery. Wasik and Bushuk said that the quantity of glutenin polypeptides correlates with spaghetti-making quality. The Mw of these polypeptides is medium. The cause of relationship between gliadin 45 and gliadin 42 and viscoelastic properties of durum wheat is low Mw glutenin polypeptides. Payneeta reported that low Mw-2 group was present in durum varieties in gliadin-45 types and low Mw-1 group was in gliadin-42. It was suggested that the gliadins are causal agents of gluten quality and low Mw glutenins are of strength of gluten [8].

In pasta processing during dough formation, when water is added the disulfide bonds will break and unfolding of gliadin molecule occurs. This unfolding results in the formation of new bonds with glutenin molecules and making the gluten network stronger [13]. There is a close relationship between α- and β-gliadins. These are known as α-type gliadins. α-Gliadins are related with intolerance of gluten but glutenins and γ-gliadins are not. The sum of gliadins' Sulphur-rich and Sulphur-poor subunits ranged as 58.65%–64.43% in durum wheat [27].

2.5.6 GLUTENINS

Glutenins molecules are also classified into low Mw glutenin subunits and high Mw glutenin subunits based on SDS polyacrylamide gel electrophoresis [4]. The molecules having 30–55 kilo Dalton Mw range come in the category of low Mw and those molecules having 100–140 kilo Dalton come in the category of high Mw subunits. These subunits make heterogeneous mixtures of polymers with bonding of disulfide and peptide linkages. The Mw of glutenin proteins is up to tens of millions and these are considered as the largest protein in nature [27].

The ability to make disulfide bonds required for making up glutenin polymer structure is affected by differences in a number of cysteine residues, in size of glutenin subunits and polarity. The insoluble glutenin content is 26.76% of total proteins in durum wheat. Glutenin is also the major wheat storage protein [27]. In the process of pasta manufacturing, the viscosity (extensibility) of dough is increased due to the gliadin and glutenin is responsible for elasticity (strength) of the dough [13].

2.5.7 GLUTEN

The viscoelastic properties of the dough are determined by grain proteins, and the protein that is involved in network formation of dough is known as gluten. Gluten protein is measured when we wash the dough under running water. In this process, the particulate and soluble matter are washed out to give a proteinaceous mass that exhibits cohesive nature on stretching. On a dry weight basis the gluten constitutes 75% protein and the remaining are lipids and starch. Furthermore, the prolamins that are not soluble in alcohol-water mixtures due to their presence in polymers stabilized by interchain disulfide bonds are known as glutenins and gliadins, these both form gluten. The major proteins that are present in the endosperm of grain are prolamins [21].

The HMM subunits comprise of 12% total protein and it corresponds to 1 to 1.7% flour weight. The subunits are very important because these are responsible for determining dough viscoelasticity properties and are linked with high or low dough strength. The HMM subunits are present in polymers of glutenin, mainly in high molecular polymers and their amounts correlates with the strength of the dough. The extended chains that are formed give elastic energy to dough. This talks about the increase in dough resistance in the mixing process [21]. Gluten protein is responsible for the elasticity and *al dente* chewability of pasta [13].

The most important parameter of cooking quality of durum wheat is gluten viscoelasticity, and this is linked with protein termed as gamma gliadins coded. This protein has strong aggregative properties. This protein can differ type 42 and type 45 genotypes about 15% and 30%, respectively [9]. The viscoelastic properties of gluten are directly affected by the aggregation state of proteins. Dexter and Matsuo analyzed an improvement in cooking quality of pasta when maturation process goes on. It can be reported that if metabolism of nitrogen is not involved in the origin of various physicochemical properties, then the cooking quality potential will be poorer. John and Carnegie reported that disulfide bonding between proteins occurs only in the drying of grain [9].

The high Mw forms in protein with distribution of Mw same as in the case of gluten obtains from mature seed reported by Miflin et al. Pernollet said that the physical forces involving in the growth of granules of starch and in drying of grain that may arise for disruption of membrane of protein and for arising up new linkages in between proteins and in between other constituents and proteins [9]. The wet gluten content in durum wheat ranged as 20.00% (ZP 10/I)–32.20% (ZP 7858) in durum wheat [28]. The SDS-sedimentation test predicts the durum wheat gluten strength and explains that sprout damage has no influence on the strength of gluten.

SDS-sedimentation volume was in poor correlation with α-amylase activity [6].

Edwards et al. noted various range in strength of gluten among groupings of HMW-GS and said that specific allelic patterns indicate the ability for good performance but no guarantee for that performance. The exact cause of strength of gluten and firm pasta's optimum level is not clear. The gluten strength of durum is determined by gluten index (GI) test [22]. In the manufacturing process of pasta, gluten strength is very important because it influences the texture of pasta and has a role in cooking quality of pasta [26]. Dexter and Matsuo noted that Farinograph properties are strongly influenced by gluten strength; this has a relationship to enhance the cooking quality of pasta [20].

SDS sedimentation volume and GI tests are used for checking the quality of gluten towards the suitability for pasta cooking. SDS polyacrylamide gel electrophoresis (SDS-PAGE) and acidic polyacrylamide gel electrophoresis (A-PAGE) are such techniques which are used for the determination of glutenin and gliadin proteins which are also linked with the quality of gluten [21].

2.5.8 CAROTENOID PIGMENT

The natural carotenoid pigment content is responsible for giving the bright yellow color in durum products. The oxidative degradation of carotenoid by lipoxygenase (LOX) enzyme is also important in imparting yellow color to products. In pasta manufacturing process, the pigment loss mainly occurs. The pigment loss is related with semolina LOX enzyme activity. If durum wheat has higher pigment content, then the reaction of LOX will be hindered by beta-carotene and bleaching of semolina will be decreased. So, lower LOX activity and higher pigment content are selection parameters for giving yellow color in durum wheat product like pasta. If there exists high peroxidase activity then the pasta will show brownish color masking the yellow color. The lipid pigments present in endosperm comprise of xanthophylls. The main xanthophyll in durum wheat is lutein. The short mixing times (less than 1 minute) and high temperature drying (greater than 70°C) decreases the influences of LOX [22].

2.5.9 *LIPIDS*

The lipids in durum wheat are sterols, hydrocarbons, phospholipids, glycolipids, fatty acids and glycerides. Monoglycerides with saturated fatty acids when added in dough the stickiness of pasta is lowered and tolerance in overcooking was maintained. Free polar lipids can join with gliadin by hydrophilic bonds and polar lipids linked with glutenins. These bonds increase the interaction of proteins that gives structural support to the gluten network. The removal of non-polar lipids and total lipids had bad influence on quality of pasta like increase in cooking loss and pasta stickiness described by Matsuo [22].

2.6 MILLING OF DURUM WHEAT

Cleaning, tempering, milling, and purifying are the steps which are performed when durum wheat is milled into semolina. First of all, grains are cleaned from any other particle-like insects, stones, and other grains because these things affect the quality of semolina and can also decrease the efficiency of rolls in the grinding mill. After cleaning, tempering process is performed. Before grinding the durum, wheat grain is tempered to a moisture content of about 16% because durum wheat grain is hard. In the tempering process, the bran and endosperm separation can occur. The durum wheat after the tempering is ground on roller mill to separate endosperm from the bran portion. Afterwards, the milling of endosperm is carried out into semolina and semolina is purified from the bran particles and flour. In the milling process of durum wheat grains, the grain gives almost 25 to 20% bran, 65 to 70% semolina and 10% flour. The semolina fraction comprises of a variety of sizes; from around 100–650 μm, 65 to 70% of this is in size almost 250 μm [13].

The composition of durum wheat grain is:

- Germ (embryo) 2 to 4%;
- Endosperm 75 to 80%;
- Aleurone layer 5 to 8%;
- Pericarp 7 to 8% [1].

Gluten is present between the spaces of granules of starch. Before the breakdown of starch-protein interaction, the starch granules break in the milling process of hard wheat. In this case, the starch is termed as damaged starch. If the durum wheat grain is milled into usual flour, then more damaged starch would be produced. For pasta production, greater starch damage is not good so that is why durum wheat kernel is milled into semolina rather than usual flour [13]. When extraction of milling increases, more of endosperm near the aleurone layer and bran is removed results in increasing the protein content of semolina [20].

In the usual milling process, the aleurone layer and the pericarp together produces the bran streams. Semolina is produced by the endosperm (between 150 and 500 microns) but endosperm also produces flour (below 150 microns). At the cleaning stage, prior to milling, the major part of germ is usually removed from the grain [1]. The purpose of durum milling is to obtain much semolina with less production of flour for pasta manufacturing. Semolina has a higher selling price, so it is very cost-effective [13]. In wheat milling, a specific characteristic obtains from the presence of crease in grain. This is not possible to reach the bran present inside the crease because the aleurone layer and pericarp cannot be separated.

So, millers of durum wheat have to open the kernel then the stepwise removal of endosperm from the bran takes place instead of starting from outside and going inwards. This process is carried out with the help of break rolls that produces shearing and crushing action. The bran cannot be separated only from sieving in mill streams of durum wheat and cannot be scratched up like in normal flour and bread wheat technology. The milling process of durum wheat should use corrugated rolls and the semolina separation is carried out at every step. Then the purification process of semolina should be done on the basis of their density using purifiers and of their size using sifters [1].

In bread wheat milling, purifiers are not used. The important role of purifiers is striking between milling of 2 kinds of wheats. In durum wheat products, color is a crucial parameter in the case of quality. Color of pasta or of mill product is measured through brown index and yellow index colorimetric determinations. Low brown index and high yellow index shows higher quality. In streams of semolina, the yellow index of semolina remains constant while decreases for central to peripheral areas of the grains. In both the flour and semolina, the brown index is lower in the central streams and vice versa. The protein content is also lowest in the central streams and vice versa. In case of flours, the peripheral flours have lower gluten content in spite of higher protein content. The firmness of gluten increases from central to peripheral in both flours and semolina [1].

It is particularly higher in the case of central (3.55 mm) flours and of intermediates (3.21 mm). In the case of gluten recovery, it is lower in peripheral ones than in central in both flours and semolina. Gasiorowski and Obuchowski explained that in the milling of durum wheat, the starch granules are influenced by the occurrence of stresses because these are sensitive to mechanical action and are damaged. The granules of starch which are damaged absorb greater water [25].

2.6.1 GRANULATION OF SEMOLINA

The particle size of semolina has effect on water absorption in pasta dough and on drying process of pasta so it is very important in pasta manufacturing process. It is necessary to have small size of particle because small-sized particles absorb more water than larger ones. The white specks in pasta are due to the bigger particles. The usual size of semolina to produce short pasta like macaroni is finer than 350 μm and for long pasta like spaghetti, the particle size should be finer than 630 μm. Semolina flour is utilized for the production of noodles and coarsely milled semolina is for the production of Couscous. If semolina with very small particle size is utilized for the production of pasta, the starch will leach out into cooking water and mushy pasta will be produced. The reason is that in these small particles, the granules are more damaged [13].

2.6.2 SEMOLINA YIELD

The durum wheat milling performance is better predicted by TW than kernel weight. Purity parameter can be analyzed with the number of brown specks per unit area and ash content in semolina. The mill with higher extraction rates gives semolina with high ash content because semolina is contaminated with bran and this lowers the yellowness and brightness of semolina [26].

2.6.3 DOUGH

The dough formation process involves the gluten network development and, in this process, protein containing semolina particles exude proteinaceous fibrils, these make cohesive dough. During the process of mixing several changes occur and this is termed as dough development. At the molecular level, various changes occur like interaction of glutenin, gliadin, and gluten formation (Viscoelastic matrix). The irreversible protein-protein cross-linked are made when the hydrated gluten is subjected to heating. The temperature below 55°C during mixing, the starch acts as an inert filler and cannot absorb more water [22].

During heating (cooking) process, the starch becomes rubbery, loses its rigidness and can absorb water. As the granules swell, the viscosity increases and soluble material is released from the granule. If temperature reaches above 55°C, gluten becomes stiff, rough, and makes a gel. This type of gluten exhibits broken gel fragments in pasta and pasta strands will be weaker. So, maximum temperature is maintained below 55°C during dough making process. The dough of pasta absorbs almost 31 to 35% water. Edwards et al. by the use of reconstitution, reported that low Mw-2 protein is associated with the strengthening of the dough more than low Mw-1 protein [22]. Cubadda determined that the dough development is affected by particle size distribution with regulation of hydration during the stage of mixing. Improper hydration is dangerous to dough development because with increase in hydration the dough strength will decreased [17].

2.6.4 FARINOGRAPH

Farinograph is a process that is used to measure the dough development time, water absorption, dough softness and dough stability. The starch damage and gassing power is increased due to the reduction in particle size of semolina granules results in decreased development time and increased Farinograph water absorption [22]. Brabender Farinograph stability, dough development time and water absorption affected by milling technique. The main and first parameter associated with Farinograph is damaged starch and the second one is content of protein. In research, it was reported that particularly at low protein content Farinograph stability and development time decreased. The dough development time increased with the slow reduction of hard wheat into flour. The dough's properties are weakened due to excessive water absorptions. When the protein content is high, the gluten protein absorbs the water released by damaged starch, maintaining the consistency of dough and masking the weakening effect of released water [7].

2.6.5 MIXOGRAPH

The development time of mixograph strongly linked to firmness of pasta [22]. The bigger particles of semolina would take a longer time for hydration which would delay the formation of gluten and resulted in delaying the time required for maximum consistency of dough. By increasing the damaged starch, the mixogram peak time is reduced. Before achieving the maximum consistency of dough, the flour particles should be completely hydrated. In a study, it was resulted that peak height was highest with semolina (6.1 cm) and durum flour (7.6 cm) [5].

2.6.6 ALVEOGRAPH

Alveograph is frequently utilized for the determination of dough properties of durum wheat. The Alveograph shows strength of gluten [22]. The ICC process recommends that based on 15% flour moisture content the 50% continuous water addition is needed. The dough becomes stiffer when the damaged starch (absorption of water) of flour increases, the height of the Alveograph curve increases, the area under the curve increases and the length decreases. In a study, it was resulted that for lowest protein samples the curves remained higher and shorter but for those of higher protein samples the height and length were also restored [7].

2.6.7 PRODUCTS OF DURUM WHEAT

Durum wheat is used in the formation of various products like bulgur, freekeh, bread, bread rolls, couscous, and durum pastries, etc. But the most important product that is made from durum wheat in the whole world is pasta [23].

2.7 PASTA

Pasta is healthful, nutritious, low-glycemic-index food and it has high levels of lutein that is vitamin A precursor and antioxidant [19]. Pasta is made in various shapes and types like penne pasta, noodles, spaghetti, spiral, twister, seashell, elbow, butterfly, flower, etc. [10].

2.7.1 PASTA QUALITY

The important and necessary quality parameters of pasta are:

- Wholesomeness;
- Yellow color;
- Low stickiness;
- "*Al dente*" trait [24].

The color in pasta does not depend on the yellow pigment in semolina, but it is affected by the activity of enzyme LOX in semolina during the processing of pasta. Mechanical texture is described by a variety of terms like bulkiness, chewiness, stickiness, elasticity, and firmness [26]. The ash in semolina gives dark color to pasta therefore the grain of durum wheat should not contain high levels of ash. The pasta should be yellow due to the naturally occurring carotenoid pigments because in pasta manufacturing artificial dyes are not used. Two components, i.e., mineral content and ash content are very important for quality analysis of durum wheat because these components tell whether the wheat is suitable for pasta processing or not [18].

The *al dente* texture is depended on soft but external coherent zone and firm core that provides some level of cutting resistance [28]. It is very important for the manufacturing of high-quality pasta that the size of semolina granules should not be too small or too large. There is another important need for high quality pasta is that the activity of LOX should be very low because it will oxidize the carotenoid pigment, i.e., lutein, and colorless pasta will be produced. The semolina requires less water for dough formation because this less water in starting will ease the drying process and the water should be removed in drying process [13]. The water content, in cooked pasta decreases from 90–39 g/100 g [28]. If gluten and protein components are in lower amounts, the pasta will be fragile and brittle subjected to cracking during transport and drying and of low quality [18].

2.7.2 EFFECT OF DRYING TEMPERATURE ON PASTA QUALITY

At high temperature drying, differences in firmness are apparent, but at low temperature drying, differences in quality parameters are linked with stickiness. The presence of enzyme α-amylase in grain would lead to higher sugar levels and leads to arise in pasta redness during high temperature drying. Semolina α-amylase activity is not related to cooked pasta resilience, firmness, and stickiness [22]. Dexter and Matsuo represented that the extraction rate of semolina has no particular influence on the cooking quality of pasta. On the other hand, particle size distribution of semolina has significant role in making of pasta due to its influence on hydration rate [20].

2.7.3 COLOR

The yellowness in pasta is increased by the inactivation of endogenous LOX enzyme that prevents the bleaching of carotenoid pigment. Whereas pasta dried at high temperature at low product moisture content then it gives red-brown color: synthesis of melanoidins is associated with Maillard reaction. In a study, it was analyzed that all samples of pasta dried at 80°C and 100°C showed an increase in yellowness during the high temperature drying process [28]. The bright yellow color of pasta is due to the natural carotenoid pigment content and oxidative degradation of natural pigment by LOX. The less activity of LOX and high pigment content are desirable quality parameters for bright yellow color in pasta [22].

2.7.4 TEXTURE ANALYSIS OF PASTA

For the measurement of pasta quality two parameters are important, one is sensory tests like stickiness and firmness and the second is water uptake ability during cooking. During the process of cooking, it is important that the network of gluten is at a medium strength to give required cooking quality. The reason behind this fact is that the too strong gluten network will produce fragile pasta and too weak gluten network will produce mushy pasta on drying. The high protein content is an important parameter to enhance the cooking quality because high protein content gives resistance to overcooking. High quality pasta needs more big subunits of glutenin than smaller ones. Low gliadin contents and high glutenin contents are linked with high pasta hardness [13].

2.7.5 REDUCING SUGAR CONTENT

In the process of pasta/spaghetti manufacturing, the increase in reducing sugar content occurs during the process of extrusion due to shearing. In the spaghetti manufacturing process, the activity of α-amylase is limited due to less moisture content of pasta dough and instant moisture loss in spaghetti at the initial drying stages [6].

2.7.6 BREAKING STRENGTH

The breaking strength of pasta is affected by drying temperature because if drying temperature is low, the pasta will become weak after one week as compared to pasta dried at high temperature [6].

2.7.7 COOKING SCORE

Cooking score is determined in the sense of firmness and resilience and it is affected by storage time and drying procedure. The temperature for drying of pasta should be 70°C for obtaining good cooking quality. The high temperature drying has a good effect on pasta firmness and resilience [6].

2.7.8 STICKINESS

Pasta's stickiness will be less if it is dried at 70°C than at 39°C. Storage time has no influence on stickiness [6]. The quantitative information about the cooked pasta is obtained by adhesion test. It was resulted that the pasta dried at 100°C proved to be firmer in comparison with the pasta dried at low temperature. The pasta samples reduced the stickiness except dried at 80°C. High temperature drying leads to low stickiness and high firmness of cooked pasta [28].

2.8 SUMMARY

Durum wheat provides us essential nutrients like protein, iron, and Zn in higher amounts. So, durum wheat in nutritional perspective is very important. For characterization of durum wheat, different perspectives have been studied. Like undersized grains in milling gives non-uniform particle size and it reduces the quality of pasta. The virtuousness and translucency of durum wheat kernel gives good semolina production and lessens quality issues of pasta. Likewise, the higher TW of durum produces higher semolina. The stickiness in pasta is due to the higher amount of starch and amylose content. Similarly, the high firmness of pasta is due to the high amylose and protein content. The pasta's extensibility and elasticity are related with high amount of gluten in durum. The high amount of carotenoid pigment imparts bright yellow color to durum products. This study gives clear understanding for characterization of durum wheat on quality parameters. Nevertheless, further research will be done in future with new perspectives.

KEYWORDS

- **acidic polyacrylamide gel electrophoresis**
- **carotenoid**
- **durum wheat**
- **gluten index**
- **nutritional aspects**
- **pasta**

REFERENCES

1. Abecassis, J., Autran, J., & Kobrehel, K., (1987). Composition and quality of durum wheat mill streams. In: Morton, I. D., (ed.), *Cereal in a European Context* (pp. 301–312). Ellis Horwood Ltd, London.

2. UW Associates, (2018). *Wheat Classes.*
3. Babu, C. R., Ketanapalli, H., Beebi, S. K., & Kolluru, V. C., (2018). Wheat bran-composition and nutritional quality: A review. *Advances in Biotechnology and Microbiology, 9*, 1–7.
4. Balakireva, A., & Zamyatnin, A., (2016). Properties of gluten intolerance: Gluten structure, evolution, pathogenicity and detoxification capabilities. *Nutrients, 8*, 644.
5. Deng, L., (2017). *Whole-Wheat Flour Milling and the Effect of Durum Genotypes and Traits on Whole-Wheat Pasta Quality.* North Dakota State University.
6. Dexter, J., Matsuo, R., & Kruger, J., (1990). The spaghetti-making quality of commercial durum wheat samples with variable a-amylase activity. *Cereal Chemistry, 67*, 91–116.
7. Dexter, J., Preston, K., Martin, D., & Gander, E., (1994). The effects of protein content and starch damage on the physical dough properties and bread-making quality of Canadian durum wheat. *Journal of Cereal Science, 20*, 139–151.
8. Du Cros, D., (1987). Glutenin proteins and gluten strength in durum wheat. *Journal of Cereal Science, 5*, 3–12.
9. Galterio, G., Biancolatte, E., & Autran, J., (1987). Proteins deposition in developing durum wheat. Implications in technological quality. *Genetica Agraria, 41*, 461–480.
10. Giacco, R., Vitale, M., & Riccardi, G., (2016). In: Caballero, B., Finglas, P., & Toldra, F., (eds.), *Pasta: Role in Diet* (pp. 242–245). The encyclopedia of food and health.
11. Goesaert, H., Brijs, K., Veraverbeke, W., Courtin, C., Gebruers, K., & Delcour, J., (2005). Wheat flour constituents: How they impact bread quality, and how to impact their functionality. *Trends in Food Science & Technology, 16*, 12–30.
12. Gregova, E., Medvecka, E., Jomova, K., & Slikova, S., (2012). Characterization of durum wheat (*Triticum durum* desf.) quality from gliadin and glutenin protein composition. *The Journal of Microbiology, Biotechnology and Food Sciences, 1*, 610.
13. Haraldsson, J. (2010). Development of a Method for Measuring Pasta Quality Parameters. Retrieved on May 24, 2013 from lnu.diva-portal.org/smash/get/diva2:322522/FULLTEXT01.
14. Hazard, B., Zhang, X., Colasuonno, P., Uauy, C., Beckles, D. M., & Dubcovsky, J., (2012). Induced mutations in the starch branching enzyme II (SBEII) genes increase amylose and resistant starch content in durum wheat. *Crop Science, 52*, 1754–1766.
15. Kadkol, G. P., & Sissons, M. (2016). Durum wheat: overview. In Wrigley, C, Corke, H, Seetharaman, K & Faubion, J (eds) Encyclopedia of Food Grains, 2nd Edn, Academic Press, pp. 117–124. Oxford, OX5 1GB, UK.10.1016/B978-0-12-394437-5.00024-3.
16. Nilusha, R., Jayasinghe, J., Perera, O., & Perera, P., (2019). Development of pasta products with nonconventional ingredients and their effect on selected quality characteristics: A brief overview. *International Journal of Food Science.* https://doi.org/10.1155/2019/6750726.
17. Nosworthy, M. G., Neufeld, J., Frohlich, P., Young, G., Malcolmson, L., & House, J. D., (2017). Determination of the protein quality of cooked Canadian pulses. *Food Science & Nutrition, 5*, 896–903.
18. Rachon, L., Palys, E., & Szumilo, G., (2012). Comparison of the chemical composition of spring durum wheat grain (*Triticum durum*) and common wheat grain (*Triticum aestivum* ssp. vulgare). *Journal of Elementology, 17*(1),105–114.

19. Ranieri, R., (2000). *Durum Wheat Quality Management* (pp. 555–557). CIHEAM-options Mediterranean's.
20. Samaan, J., El-Khayat, G. H., Manthey, F. A., Fuller, M. P., & Brennan, C. S., (2006). Durum wheat quality: II. The relationship of kernel physicochemical composition to semolina quality and end product utilization. *International Journal of Food Science & Technology, 41*, 47–55.
21. Shewry, P. R., Halford, N. G., Belton, P. S., & Tatham, A. S., (2002). The structure and properties of gluten: An elastic protein from wheat grain. *Philosophical Transactions of the Royal Society of London, Series B: Biological Sciences, 357*, 133–142.
22. Sissons, M., (2008). Role of durum wheat composition on the quality of pasta and bread. *Food, 2*, 75–90.
23. Sissons, M. J., Abecassis, J., Marchylo, B., & Carcea, M., (2012). Durum wheat: Chemistry and technology. *AACC International*.
24. Steglich, T., (2013). *Pasta as an Example for Structure and Dynamics of Carbohydrate Rich Food Materials a Literature Review*. SIK Institutet for livsmedel och bioteknik.
25. Szumilo, G., Rachon, L., & Stankowski, S., (2010). The evaluation of grain and flour quality of spring durum wheat (*Triticum durum* Desf.). *Polish Journal of Agronomy.*
26. Wrigley, C. W., Corke, H., Seetharaman, K., & Faubion, J., (2015). *Encyclopedia of Food Grains*. Academic Press.
27. Zilic, S., Barac, M., Pesic, M., Dodig, D., & Ignjatovic-Micic, D., (2011). Characterization of proteins from grain of different bread and durum wheat genotypes. *International Journal of Molecular Sciences, 12*, 5878–5894.
28. Zweifel, C., Handschin, S., Escher, F., & Conde-Petit, B., (2003). Influence of high-temperature drying on structural and textural properties of durum wheat pasta. *Cereal Chemistry, 80*, 159–167.

CHAPTER 3

FUNCTIONAL AND NUTRACEUTICAL PROPERTIES OF CEREAL BRAN INDUSTRIAL WASTE: AN OVERVIEW

MUZZAMAL HUSSAIN, TABUSSAM TUFAIL, FARHAN SAEED, HUMA BADER UL AIN, MARYAM SHAHZADI, and HAFIZ ANSAR RASUL SULERIA

ABSTRACT

The handling of industrial waste is both a requirement and an obstacle. The recovery and recycling of food waste and by-products have become an important topic of analysis to clarify the functional and nutraceutical characteristics of industrial cereal waste. The practice of Agro-industrial waste as natural resources can help to reduce emissions impact on the environment and also lower production costs. Agricultural waste is the origin of crude protein, fat, carbohydrates, and other functional compounds, that form raw ingredients for industrial use. Including functional ingredients and immense health properties, cereal brans offer high nutritional value. Agricultural wastes are rich in bioactive compounds (BACs) and despite their health benefits, they have not been abundantly exploited in the food system. These by-products may be used directly for the processing of different foods or for the isolation or purification of different components after some modifications. Agricultural waste is utilized for the production of biofuels, antioxidants, antibiotics, vitamins, enzymes, animal feed, and other chemicals by isolation or purification. The isolated substances (Phenolic compounds or non-starch polysaccharides) in food processing industries as functional ingredients and as nutraceutical are used, pharmaceuticals, and cosmetics industries. This review provides a summary of cereal bran characteristics as by-products for investigate the bran optimize milling fractions and enhance bran utilization.

3.1 INTRODUCTION

Waste is directly linked to human development, both technologically and socially [1]. If such wastes are emitted into the atmosphere without proper disposal procedure this could lead to contamination of the ecosystem and adverse effects on human health. The improvement and alteration of waste is becoming enormously valuable in the food processing industries. The goal is to use the raw material more fully and reduce emissions and waste management problems [2]. Food is an essential human need and one of the massive problems faced by a significant proportion of countries to produce sufficient feed for the increasing population of developing areas [1]. Nevertheless, the issue is to improve the global food production system on a sustainable basis in order to increase food security and food safety without compromising the environment, making the idea of sustainable functional food appropriate to consumers [3]. It is derived from different sources of forestry streams and agricultural waste and is significant in the production of renewable energy, biofuels, and biochemicals [4]. A leading challenge facing most developing world countries is increasing agricultural production without degrading the environment.

Major areas to consume maximum quantity of cereals and vegetables, and produce such large industrial waste. For example, Asians cereal waste was 55% of the cereal waste in the world. Curiously, most of these wastes are obvious such as straws, husks, bran, and hard nutshells (inedible portion of cereals) must be eliminated as waste [5]. Evaluating the food matrix discarded as waste, the probability of its reuse in the food chain, and does not impose restrictions. Of this purpose, the usage of the word "by-product" refers to certain wastes which are Substrates for the improvement of food products as well as for removal of functional compounds. Through the current milling process used the two main by-products formed in the manufacture of cereals, bran, and germ [6]. Healthy amounts of essential health-benefiting compounds such as in bran are present vitamins, minerals, dietary fiber, phenolics, and antioxidants. Nevertheless, due to certain technological issues, Bran is discarded in milling operations as a by-product and is not used in significant quantities with human food [7].

Recent research shows that natural antioxidant as well as functional ingredients in foods, utilization of natural substances has been increasing past few years. Efficient use of raw residues, food waste and alternate sources of food for human consumption would contribute to an improved nutritional value of the products derived from existing resources. Food processing waste is the end products of different food processing sectors which were not recovered and used for other applications. During processing, crude raw materials (cereals, fruits, and vegetables) into final products, the food industries produced a huge amount of waste materials. Such as, 50% of the weight of the raw materials as waste material in the form of fibers, peels or stones the fruit and vegetable industries produced [1]. Therefore, they are discarded as waste, being the milling industries produced a large quantity of raw materials non-product flows, the economic prices of which are lower than the processing and recovery costs for recycle. However, when there are sufficient technological methods to recover them and if the benefit of the resulting products is to surpass the reprocessing costs, these wastes may be considered useful by-products. Sustainability offers the agriculture sector both opportunity and challenge. It is an incentive due to its possibility of using by-products from food processing to extract bioactive compounds (BACs) and nutrients has provided enormous potential for waste minimization and indirect income [9].

The purpose of this study is to discuss research results on the application of diverse cereal industry waste material used as an origin of targeted compounds. Thus far, most accessible research on the evaluation of cereal milled bran investigates particular by-products and their utilization in different foods. In this chapter, the aim is to develop waste / by-products in cereal brans, and how to use by-products to boost their technical, health-promoting properties and the origin of targeted compounds in their production. There is limited information on the use for health promotion food processing as a basis of useful nutrients or dietary by-products, their attributes as well as ability for utilization in the production of new useful foods. This review also discussed that analysis on bran (by-products food-processing), health-promoting possessions, possible uses and composition.

The review contains a discussion of the cereal brans' sensory possessions for prospective association, functional, microbiological, chemical, and physical as well as the utilization of this underused part of the milling to improve by the assessment. Considering the cereal brans useful and nutritional possessions were not applied in food industries due to the lack of literature on their quality attributes for informal description and evaluation during the formulation of food products regarding modifications introduced to reduce their harmful impact on technical and sensory properties of food. In relation to the increasing demand for nutraceutical ingredients as well as useful cereal brans, this study includes a general idea of cereal brans properties to give extensively accessible details that may improve the usage as well as the ability of cereal brans to generate income of the cereal milling manufacturing to underused by-product.

3.2 CEREAL INDUSTRIAL BYPRODUCTS

In most countries around the world, the fast-growing cereal milling industries generate various by-products such as cereals and pulse processing (Hull, husk, stalks, and chaff), Fruits, and vegetable processing (Peels, stones, fiber, pomace, pith, skin) Cereals (Straw, stalks, husk, cob, and bran), etc. Agricultural milling industries generated the huge amount of waste annually. Some of the agro-industrial waste is untreated and underused, and is therefore discarded of in maximum estimates by burning, dumping or unplanned landfill [10]. Many industrial by-products are produced during food processing they are prosperous source of bioactive moieties, functional as well as dietary. Production of natural bioactive products has increased and research into alternative sources is on the way, with increasing interest in nutritious functional foods [11, 12]. Cereal industrial by-products have a broad spectrum of bioactive compounds. The retrieval of compounds with significant added value has the competency to be used as additives to the food and functional ingredients. The method of waste management is not only costly but also has negative health impacts. There is a prerequisite for developing an environmentally friendly Technique to use the ability to convert by-products into edible products, e.g., recovering valuable biochemical. The agricultural waste products have excessive ecological importance that is an environmentally sustainable mechanism of usage [13].

Continuous economic growth may also be accomplished by, for example, the valorization or production of by-products with low value. The utilization of cereal milling waste to attain by-product bio-extracts may be based on the development of novel products with bioactive properties and the decrease of the environmental impact caused by accumulation [14]. The food production industry is growing very rapidly across the globe in the past years, and continues to grow. Also, with growing analytical capacity, we have become more experienced with functions of BAC and the biochemical composition as well as in various food products their impact on human health [11]. In developing countries, the fast-growing food processing industry is designed to produce more quantity of waste material products in coming years. The prominent waste materials are peel, pomace, pod, seed, hull, husk, stone, rind, core, stem, etc. These by-products, including dietary fiber, of more than a few BAC are known to be an important origin. Mostly as a source of functional fiber and nutrients with the utilization of food industry by-products for the many foods products manufacturing have produced huge scope for not direct profitability as well as waste reduction [11].

The production of cereals for animal feed is processed to eliminate the germ and bran, which are known grain endosperm by-products to meet consumer sensory requirements. This process stripped grains of significant nutrients that are beneficial to human well-being, that is, dietary fibers, minerals, vitamins, and phenolics, which also involve research as feedstocks for their potential [15]. Techniques for cereal processing by introducing more efficient processing of grain to generate a range of added value materials including biochemicals and bioenergy [16]. Supply chain or the food production milling of cereal is a very important fraction of these, however should not be abandoned the significance of cereals to the non-food sector. In cereal industries, milling is a primary technique, divided into wet and dry milling, as well as each one has its own characteristics. Distinguish the external fibers by dry milling as well as endosperm grain by-products recognized by the germs. With pearling might as well relate the dry milling, that eliminate the seed coat (testa and pericarp) steadily by an abrasive practice, the germ to attain refined kernel (barley, rice, and oat), sub aleurone and aleurone layers as well as by products with high concentration of BAC.

Contrarily, for gluten production and starch, mainly used wet milling having as sheer solids (prosperous in nutrients beneficial to the pharmaceutical manufacturing), coproducts germs (projected for the oil-crushing industry), and bran (combined outer layer of aleurone and pericarp) as coproducts. With huge health benefits, Cereal brans are high nutritive quality functional ingredients [17]. Other alcoholic beverages while enzymes soak up fermentable sugars as well as the production of beer by the method of malting and from grain the starch (in barley particularly), the used-up kernel leaving at the back [18]. Cereal by-products come from both field harvesting and grain production. Most of the cereal wastes and by-products are produced primarily during harvesting, post-harvest processing, and post-production points. The primary by-products produced are the straws during cereal grain field harvesting, as key raw materials for biofuel production/biogas are extremely recognized. Further, Corncob is discarded in the field during maize harvest as a result of the corn Stover (mixing leaves, stalks, husk, and cob).

Dry cereal milling may include flour production, wet milling to produce starch and glucose, and grain brewing [19]. Nowadays, satisfying technical and economic requirements is not enough for products; it is also valuable that their characteristics are profitable. This requirement helps to motivate academic and industrial efforts to determine creative technologies, from the huge figure of technical documents as be able to seen researching the application of cereal-industrial waste as raw materials for composites, nanomaterials, biofuels, and energy production [20]. Current studies have shown so as to different waste forms, including peel of pomegranate, lemon, orange, etc., and walnut husks, can be used as natural antimicrobial agents [10]. Some researchers like [21] describe the functional properties of industrial waste wheat straw.

3.3 CEREAL BRANS

Cereal brans have functional components with a high nutritional quality and immense nutritional characteristics. Despite their nutritional benefits, cereal brans are not abundantly exploited in food processing [22]. Part of the starchy endosperm, pericarp, aleurone, testa, and germ contain fractions of cereal brans. Widely consumed main cereals are the sorghum, millet, barley, maize, rye, rice, oat, and wheat [23]. Dry matter of the kernel about 3 to 30% makes up by the bran, based on the cereal crop type [24]. Differ in composition of cereal brans depending on the crop type, kernel size, shape, bran layer thickness, grain storage circumstance as well as period, prior to milling and process of friction grain conditioning technique [25]. Vitamins (particularly vitamin B and E), bioactive components, dietary fiber as well as minerals are high in cereal brans [26].

Relatively high in protein, ash, fat, phosphorus, and niacin is a main component of bran. Cereal bran provides product attributes, like thickening, emulsification gelling and stabilization [17]. Cereal brans health properties as well as nutritional and the global dietary recommendation make it possible to enhance food intake, such as more nutritious ingredients including entire grains or cereal brans [27]. In cereal brans as well contain the commercial benefit by using them in food production as dietary fiber origins and without additional cost of dietary fiber extraction [28].

3.4 CEREAL BRANS PROPERTIES

3.4.1 PHYSICAL PROPERTIES

Cereal brans physical properties in associations of their usability persuade the aspect of brans effectiveness. Optimal quality bran to consider maintaining during storage an imperative measurement is moisture content. The rate of development of the fungus that cause degradation of the cereal bran depends in part on the level of moisture. There will be no growth of fungi at low moisture content, but fungal growth starts at around 14% moisture content or slightly above [29]. Still, Due to the higher fat content, in the cereal bran oxidative rancidity moisture plays a critical function, rice bran particularly [29]. That qualitative limitation influences the utilization of cereal brans. To stabilize the cereal bran particular nutritional inhibitors as well as for inactivating enzymes suitable methods could be used. To stabilize cereal brans pH adjustment, wet extrusion, dry extrusion, refrigeration, microwave heating as well as treatment chemical with acidic acetic, acidic calcium hydroxide hydrochloric technologies were also used [30].

Commercially advanced, effective, and processing time lesser of the microwave technology, has minimal impact on the brans color, nutritional consistency, functional properties [30]. Instead of appropriate stability, bran is an essential fatty acid, protein cheap resource, additional nutrients as well as ferulic acid derivatives and tocopherols [31]. The color of cereal brans is a major physical factor in cereal grains, for the reason that color affects the acceptability of food products [30] assessed the color of cereal brans and also indicating the color of brans vary among types, demonstrating the occurrence of phytochemicals. The purple color of rice bran had the highest content of anthocyanin in cereal grains, which include maize, wheat, and barley, among other blue and purple bran [32]. A substantial portion of bound phenolics comprised rice bran light brown linking to whole phenolics [33].

Sustainable usual sources of anthocyanins (ACNs) which be able to utilized as a usual colorant as well as might be one of the mainly abundant, cereal brans natural colors, assured genotypes as well as purple rice bran, and used in nutraceuticals and functional foods for health promotion ingredients [33]. In the food products, colored cereal brans utilization might assist to reduce unidentified usage adverse effects in food produced by synthetic colorants, while helping to promote their nutritional characteristics. Cereal bran particle size is a measure of the degree over which their grains are milled. In food processing with a crucial outcome, the distribution of particles is a valuable as a basis of dietary fiber cereal bran possessions [34].

Its technical versatility affects by bran particle size because of improvements in the bran physicochemical, rheological, swelling, water absorption and fat binding properties. Cereal bran particle size caused by low water holding capacity and high oil binding capability [35]. Furthermore, wheat flour functionality influenced by the bran particle size [36]. Foremost to little gas retention lesser bread quantity results by the bran particle size, rough texture (dark crumb color, even crust). Because of the adverse effect of bran on wheat dough, a standard size assessment of the throughout the formulation of product particle of cereal bran is necessary. For bread production in wheat flour can be the most suitable little particle size (400 to 500 μm) [37]. In cookies preparation, medium bran particle sizes (220 to 430 μm) offered good spread concentrations, color as well as utilizing little particles (< 210 μm) organized cookies contrast satisfactory sensory scores [38]. Wheat cakes the uppermost suitability level given by the wheat and oat particle size of cereal bran (< 210 μm) [39]. Chemical compounds separation affects by the particle size of bran, as well as bran bioactive components, in addition to the consequence on functional attributes of bran particle size.

During storage food products shelf-life effects by particle size of cereal bran. Rate the wheat bran deterioration rises with particle size smaller; Exclude the finely milled bran (< 0.40 mm) by a halfway worsening phenomenon that the bran unfractionated demonstrating [40] While, on the composition of bran this depend in terms on fat and moisture content, as enzyme and/or microorganism activity may depend on the bran's available water and oxygen-taking ability. After grain milling, grading of cereal brans in different particle sizes is crucial before marketing or use, and to analyze their technical impacts in processed foods. This is because the consistency of the bran varies due to various milling processes and conditions that can influence of cereal brans milled chemical composition as well as physical possessions. As well useful to compare the quality of the product of incorporation with bran food products from other case studies regarding the explanation of bran particle size accuracy during product development.

3.4.2 CHEMICAL COMPOSITIONS

On the cereal grain category, the cereal brans chemical compositions are based, treatment, and the particle size. Rice bran has a high percentage of ash and fat among other cereal brans, whereas then other brans additional protein content in oat bran. Cereal brans nutritional profile protein, by its composition of amino acid illustrated, assess the rice bran protein includes healthier amino acid amount is higher than the wheat brans. As well as protein of oat brans, content of low tryptophan excepting even as elevated in leucine protein of maize bran as well as to other cereal brans contrast proline. Cereal bran proteins possess a nutritious and healthy amino acid composition [41, 42] in rice bran content of carbohydrate evaluate than cereals bran other forms higher although the highest value identified in maize bran. Cereal brans are the primary origin of cellulose, hemicellulose, and lignin as dietary fiber sources [42].

Insoluble dietary fiber of bran dietary fiber is the chief element that may affects digestibility and bioavailability as well as nutrition [17]. Insoluble dietary fiber, which is the maximum in bran of corn pursue by wheat bran as well as rice bran. Whereas that of other cereal brans, oat bran whole dietary fiber is relatively little, it is high in soluble dietary fiber [43]. Cereal dietary fibers' solubility and insolubility involve differences in their technological and biochemical properties [42]. In brans of cereal β-glucans, for instance, in oat bran with the uppermost quantity are wealthy the barley and oat (6 to 18%) [44]. Overview of the cereal bran, the composition of minerals in wheat bran has a higher content of bran of rice and oat to pursued the calcium, as content of calcium inferior in bran of maize. Among cereal brans, rice bran has higher levels of manganese, potassium, magnesium, iron and phosphorus. In bran of maize utmost is sodium while contrast to other bran of cereal selenium, copper, and zinc (Zn) uppermost quantities include in bran of wheat.

Compared to other bran cereal kind the phytin phosphorus comprises the uppermost quantity the bran of rice. However, phytic acid is perceived an anti-nutritional component since it decreases the bioavailability of certain minerals, Zn, iron as well as calcium including since the 1990s, biologically phytic acid illustrated for its health-promoting properties, in the deterrence of renal calculus, cancer as well as diabetes in precise [45]. Both multifaceted vitamins E as well as vitamins B elevated amounts contain the bran of rice than other types of cereal brans. In the bran of rice greater amount of vitamin E and vitamins B than most bran rice bran cereals serve as a possible origin for human body physiological processes and metabolic pathways. In rice bran to tocopherols contrast the tocotrienols elevated amount Vitamin E. Tocotrienols have higher antioxidant activities compared to tocopherols [46]. Likewise, tocotrienols, and 7 tocol isomers, specifically. α-, β-, π-, but not β-tocotrienols, especially in rice bran have been observed to be higher respectively along with wheat and barley compared to other bran of cereal [47].

Vitamin A is a wealthy source in maize bran relative to other brans of cereal. Cereal brans are rich sources of key fatty acids such as oleic, linoleic acids and palmitic with greater levels of palmitic acid in rice bran followed by smaller amount of oleic acid, linolenic, and linoleic acids. Squalene among other cereal bran oils from rice bran the extracted oil wealthy in ÿ-oryzanols, tocotrienols, tocopherols, phytosterols, and polyphenols [49]. Than to other vegetable oils, this oil is better, for the reason that it comprises particularly oryzanol, (omega 3 as well as omega 6) fatty acids and unsaponifiable greater amounts [17]. Due to its potential antioxidant activity and composition, in contrast to other cereal brans oils contain significant shelf life the rice bran oil, as well as low viscosity of it allows the absorption of oil easier during cooking and greatly decreases calories [50].

Cereal brans have higher antioxidant ability than other milled fractions of cereals [52]. p-hydroxy benzoic acids, p-coumaric acids, ferulic acid, vanillic acid, feruloyl oligosaccharides, and syringic contain bran of wheat [53]. For instance, avenanthramides antioxidants contain the bran of oat [54]. Flavonoids, anthocyanin, and phenolic acids contain the bran of Sorghum [55]. Folate, steryl ferulates, phenolic acids, alkylresorcinols as well as lignans contain the bran of rice [56]. p-coumaric acid, diferulic acid as well as ferulic acid include in the bran of maize, ferulated arabinoxylans and hydroxycinnamic acids with these conjugates [57]. The chief bioactive part is whole grains in the fraction of cereal grain such as bran [23]. Fatty products may be utilized to preserve, expand shelf life, and enhance their oxidative stability the antioxidants amplified activity suggested in cereal brans. Contained by the bran bioactive components composition also enables an important raw material for manufacturing advanced nutritional food products and novel elements [58]. Brans will act as a barrier to the providing of the bioactive components. This is approval, that is essential for optimal health when used in food production.

3.4.3 MICROBIOLOGICAL PROPERTIES

Cereal bran includes most microorganisms as a basis of spoilage of microorganisms their metabolites might be utilized (bacteria and fungi) [59]. In milling equipment, the increase of impurities as well as fraction in the milling procedure as well as for pathogenic organisms (particularly fungi, pathogens) produced in the course of cooling, fermentation, and heat treatment of mill products vital source is machinery too, [17, 58] indicated that cereal bran fermentation might help control indigenous microbes as well as enhance bran microbial health. In certain regions, there are also common procedures where to eradicate contaminants divide layers of bran outer 2–4%. In bran layers, the chemicals level to reduce it may use more to lessen the chemicals before processing.

3.5 FUNCTIONALITY OF CEREAL BRANS

Involve the food processing cereal brans with other innovative practical properties. As a dietary fiber source, the primary properties of cereal brans are associated to its solubility, viscosity, and capacity water-binding, organic molecule-binding capability, mineral, gel-forming ability as well as oil-binding ability that is why end product quality affect [60]. To absorb water and fat brans have adequate capacity. The hydration properties of cereals are a basis of dietary fiber, increase with temperature and are associated with higher solubility of dietary fibers [42]. Differently, cereal bran oil absorption is primarily related to the bran particle size to the possessions of outside, load compactness usual as well as the hydrophilic characteristic of the materials [61]. Ranging from 0.25 to 0.59 g/mL, 148.4 to 383.7%, 138.3 to 302.9% and 0.48 to 0.82, in bulk mass, absorption of water, absorption of fat as well as activity of water respectively properties of cereal brans [30]. Uppermost bulk compactness contains the bran of barley while bran of wheat contains the greatest potential to bind water and fat, and water intensity, according to the above researchers.

Cereal brans often differ in their levels of viscoelastic properties and solubility, also the Food industry could obtain from its physiochemical properties, in the production of novel food materials contain the building of structures, gelling as well as binding of water [62]. In cereals bran the solubility as well as insolubility of dietary fibers include differences, such as other dietary fiber types, biochemical effects as well as technical functionality relate to these [42]. To get better viscosity estimated on their capacity, the soluble dietary fibers, regulate plasma as well as glycemic reactions decrease while low compactness contains insoluble dietary fibers and approaches to enhance intestinal microflora development, enhance fecal bulk, and reduce bowel transit. In addition, most insoluble fibers are fermented in the large intestine, that is, helps to develop intestinal microflora for instance species of probiotic [62]. In the food industry because of their elevated oil absorption capacity propose great potential the cereal brans there is technical confirmation that may permit emulsions to even out as well as elevated fat foods whereas to modify viscosity might be used elevated water-holding capability as well as texture of formulated foods such as additives to inhibit syneresis [17].

On the flour physical as well as chemical possessions direct impact contain the uniqueness of cereal, leading to variations in the dough rheological properties if used as enriched flour. The bran of cereal major influence water-binding capability on dough like combined with differences in water absorption produced by gluten in the production of dough [63]. On the properties of the dough these outcomes such as during the addition stage absorption of water raise, in the consistency of the addition process refuse as well as enhancement time increase, during the proofing procedure alter the dough improvement, the dough extension (extensibility reduce) possessions alter as well as improve dough structures (stiffer dough or decreased stickiness) [64, 65]. Cereal brans the rheological possessions assessing, compared to barley brans and oat in rice brans as well as wheat increased elevated dough production time (DDT) reported when added to wheat flour between 0 and 40% while dough quality or respectively in oat and barley blends considerably reduced as of 8.5 to 4 as well as 7.0 to 3.5 min. Gradually declined extension resistance for mixtures in formulations by means of oat and barley bran amounts rising, whereas used up to 40% of wheat bran (3.11 to 10.00), rice bran (3.11 to 18.09) as well as barley bran (3.11 to 8.88) [66]. Literature demonstrated that the wheat bran incorporation in baking products significantly increased the water absorption followed by a significant decrease in dough stability, extensibility, and peak viscosity. May alter of baked products textural possessions as well as rheological the adding up of cereal bran reported by [64].

In baking products, the utilization of brans or their utilization in wheat flour present technological restrictions involved in dough development on the gluten system because of their negative effect as well as the gas preservation, which also subsequently influence crumb structure, texture, and loaf volume [58]. Cereal brans contain non-starch polysaccharides affect that develop a proper and strong gluten network, leading to rheological attributes and aggravation as well as the bread dough treatment, resulting in hard crumb, bitter flavor, lower loaf volume and dark color [17]. Mainly composition and physicochemical characteristics of arabinoxylans influence their functionality in the development cycle [62]. In bread production negative impact include in the Water-unextractable arabinoxylans, while on bread as well as dough have positive impacts of Water-extractable arabinoxylans, with texture of crumb as well as volume of loaf. Furthermore, the gluten dilution in bran fiber, particle sizes can improve the final product and dough characteristics because of surface area increased, as well as for instance ferulic acid the interaction among active components and gluten enhance too. So, impacting gluten development [64].

On the functionality of dough impact the particle size of bran. In different studies, there are generally contradictory reports about the result of bran fractions on dough development. For dough mixing reduced resistance of particle size of fine wheat bran reported by Zhang and Moore [67] as well as evaluated by the farinographic study reduced the necessity for dough mixing associated to bran coarse. Noort et al. [63] recommend the least detrimental impact of coarse bran the utilization of fractions observed. The results of the above researchers evaluate the impact of particular size could be partly related to variability in bran description, in physical uniqueness and bran structure natural difference, or in the bread-baking procedure difference. The technological influence of cereal bran in food production has been enhanced by heat treatment and fermentation, in particular cereal-based products [58]. Adding surfactants or critical gluten [68] for enhanced practice their dietary fibers solubility gets better. Surface of bran roughness also impacted the rheological properties of dough but has not adversely impacted their quality on bread preparation [69].

Degrading the cell wall utilize unlike enzymes wheat bran as well as rye were pretreated using bioprocessing methods including such fermentation with specific yeast, enzymatic treatments as well as lactic acid culture [27]. The utilization of enzymes and microbes improved bran technology in baking industries, the incorporation of bran in wheat flour increase loaf volume, enhance shelf life as well as crumb texture get better [58, 70]. For the dough carbon dioxide retention was enhanced by using fermented bran and reported that the utilization of enzymes enhanced the dough stability effects. Sanz, Collar, and Haros [71] suggested so as to amylotic and phytate degrading enzymes associated with cereal bran treatment increased the nutritional and technical consistency of bread. During bread preparation of wheat bran, the micronization technique has improved the technological possessions [72]. (The external bran division influenced by coarse fiber parts) Pearling enhanced the nutritional consistency of wheat bran exclusive of modifying the functional properties of 10% enrichment [17].

Previous to adding the dough cereal brans hydrated the quantity of the loaf improved as well as containing 12% bran in wheat bread improved quality of the bread [73]. The corn bran particle size substantially reduce might be by microfluidization as well as in bulk density resulting in a substantial decrease of the bran matrix eliminate the microstructure as well as in actual surface area enhance recommended by Wang, Hou, and Dubat [74]. Oil holding capability, swelling capability, water-holding capability, they observed too as well as cation exchange capability respectively 140%, 140%, 90%, and 90%, enhanced. For the cereal bran in processing methods, these enhancements during food processing will get better their use.

3.6 SENSORY PROPERTIES

Based on level of addition, bran particle size, and bran treatment, texture, taste, flavor, and color such these sensory properties of foods persuade by the cereal brans. The cereal bran comprises cellulose and lignin which adversely affect the taste and flavor that inhibit in a variety of consumer foods the cereal brans utmost use [58]. Mouth-feel as well as taste change, because of elevated fiber components, it results in gritty texture at higher substitution levels. Studies have shown that high-fiber bakery products are not well recognized because of their effect on the technical characteristics of final product quality [64]. Previous study shows so as to because of reduced loaf volume the baked product has relatively little acceptability scores, enhanced crumb hardness, crispness loss, changes in appearance (color and surface characteristics), and flavor. Because of their higher dietary fiber and antioxidant content, on the consumer market to contain a place carry on the cereal brans products [62]. Several other modifications have been identified in methods for processing cereal bran on the sensory properties of foods to resolve their unfavorable property. In bread to contain the most excellent sensory possessions pre-ferment bran sourdough bran has been recognized [75].

It was observed so as to wheat bran as well as rye fermentation is an important pretreatment process for enhancing the sensory attributes of the bran in bread. To soften the bread texture has been shown in the yeast with bran fermentation (for instance, α-amylase, xylanase, and lipase) the enzymes combined action [56]. Introduce fermentation with sourdough, in addition to enhancing texture properties improved the nutritional content product quality of the final (phenolic acids especially) [58] as well as protein by proteolysis from cereal bran flour also activate of bioactive peptides, when utilized lactic acid bacteria (LAB) [70]. For the production of bioactive components might be offer cereal brans as the raw material with the LAB bran fermentation it is evidence. Bread quality increased the wheat bran extrusion exclusive of adverse effects in wheat-based bread organoleptic products (up to 20% wheat bran) [39]. In bread preparation of wheat bran sensory properties of bran of wheat improved micronization [72]. With enhanced appearance as well as color the bitter flavor reduced by bran bleaching [17]. Additionally, improvements in the processing of cereals bran to maintain consumer acceptance in terms of sensory properties, in food production their functional attributes as well as technological, until an acceptable level to the product cereal bran must to be applied [8].

3.7 SUMMARY

By-products food waste reprocessing as well as recovery has become a vital issue of examination and to illuminate nutraceutical uniqueness as well as useful industrial cereal waste. As natural resource agro-industrial waste may supportive in surroundings to diminish the effect of emanation with the quality of lower production cost. Cereal bran provides a higher nutritional value. Agriculture wastes are extremely rich in BAC the origin of agriculture waste have carbohydrates, fats, crude protein as well as other compounds. In the food processing by-products might be utilize directly, after some alteration these may helpful in different component purification as well as these may also helpful in different component isolation. Agriculture wastes are utilized in the production of biofuels, agriculture wastes might be utilized in different ways as vitamins, antibiotics, antioxidants, as well as enzymes. As animal feed it can be also used. In industries of food processing phenolic compounds or non-Starch polysaccharides, these isolated substances in pharmaceuticals and cosmetics industries are used as useful ingredients as well as nutraceutical.

KEYWORDS

- **agricultural waste**
- **biochemicals**
- **cereal bran**
- **functional properties**
- **lactic acid bacteria**
- **nutritional properties**

REFERENCES

1. Ezejiofor, T. I. N., Enebaku, U. E., & Ogueke, C., (2014). Waste to wealth-value recovery from agro-food processing wastes using biotechnology: A review. *British Biotechnology Journal, 4*(4), 418–481.
2. Ravindran, R., & Jaiswal, A. K., (2016). Exploitation of food industry waste for high-value products. *Trends in Biotechnology, 34*(1), 58–69.

3. Fan, S., & Brzeska, J., (2016). Sustainable food security and nutrition: Demystifying conventional beliefs. *Global Food Security, 11*, 11–16.
4. Davidi, L., Morais, S., Artzi, L., Knop, D., Hadar, Y., Arfi, Y., & Bayer, E. A., (2016). Toward combined delignification and saccharification of wheat straw by a laccase-containing designer cellulosome. *Proc. Natl. Acad. Sci. U S A, 113*(39), 10854–10859.
5. Akanbi, T. O., Dare, K. O., & Aryee, A. N., (2019). High-value products from cereal, nuts, fruits, and vegetables wastes. *By-products from Agriculture and Fisheries: Adding Value for Food, Feed, Pharma, and Fuels,* 347–368.
6. Kiran, E. U., Trzcinski, A. P., Ng, W. J., & Liu, Y., (2014). Bioconversion of food waste to energy: A review. *Fuel, 134*, 389–399.
7. Pasha, I., Ahmad, F., Siddique, Z., & Iqbal, F., (2020). Probing the effect of physical modifications on cereal bran chemistry and antioxidant potential. *Journal of Food Measurement and Characterization,* 1–10.
8. Brockmole, C. L., & Zabik, M. E., (1976). Wheat bran and middlings in white layer cakes. *Journal of Food Science, 41*(2), 357–360.
9. Iriondo-DeHond, M., Miguel, E., & Del, C. M. D., (2018). Food by-products as sustainable ingredients for innovative and healthy dairy foods. *Nutrients, 10*(10), (1358).
10. Sadh, P. K., Duhan, S., & Duhan, J. S., (2018). Agro-industrial wastes and their utilization using solid-state fermentation: A review. *Bioresources and Bioprocessing, 5*(1), 1.
11. Sharma, S. K., Bansal, S., Mangal, M., Dixit, A. K., Gupta, R. K., & Mangal, A. K., (2016). Utilization of food processing by-products as dietary, functional, and novel fiber: A review. *Critical Reviews in Food Science and Nutrition, 56*(10), 1647–1661.
12. Ayala-Zavala, J. F., González-Aguilar, G., & Siddiqui, M. W., (2018). *Plant Food By-Products: Industrial Relevance for Food Additives and Nutraceuticals.* CRC Press.
13. Vargas, F., Gonzalez, Z., Sanchez, R., Jimenez, L., & Rodriguez, A., (2015). Cellulosic pulps of cereal straws as raw material for the manufacture of ecological packaging. *BioRes., 7*(3), 4161–4170.
14. Kowalska, H., Czajkowska, K., Cichowska, J., & Lenart, A., (2017). What's new in biopotential of fruit and vegetable by-products applied in the food processing industry. *Trends in Food Science Technology, 67*, 150–159.
15. Awika, J. M., Piironen, V., & Bean, S., (2011). *Advances in Cereal Science: Implications to Food Processing and Health Promotion.* American Chemical Society.
16. Tomás-Pejó, E., Fermoso, J., Herrador, E., Hernando, H., Jiménez-Sánchez, S., Ballesteros, M., & Serrano, D. P., (2017). Valorization of steam-exploded wheat straw through a biorefinery approach: Bioethanol and bio-oil co-production. *Fuel, 199*, 403–412.
17. Chinma, C. E., Ramakrishnan, Y., Ilowefah, M., Hanis-Syazwani, M., & Muhammad, K., (2015). Review: Properties of cereal brans: A review. *Cereal Chemistry Journal, 92*(1), 1–7.
18. Papageorgiou, M., & Skendi, A., (2018). Introduction to cereal processing and by-products. In: *Sustainable Recovery and Reutilization of Cereal Processing By-Products* (pp. 1–25). Woodhead Publishing.

19. Bastos, R., Coelho, E., & Coimbra, M. A., (2018). Arabinoxylans from cereal by-products: Insights into structural features, recovery, and applications. In: *Sustainable Recovery and Reutilization of Cereal Processing By-Products* (pp. 227–251). Woodhead Publishing.
20. Mendes, C. A. D. C., Adnet, F. A. D. O., Leite, M. C. A. M., Furtado, C. G., & Sousa, A. M. F. D., (2015). Chemical, physical, mechanical, thermal and morphological characterization of corn husk residue. *Cellulose Chemistry and Technology, 49*(9, 10), 727–735.
21. Pasha, I., Saeed, F., Waqas, K., Anjum, F. M., & Arshad, M. U., (2013). Nutraceutical and functional scenario of wheat straw. *Critical Reviews in Food Science and Nutrition, 53*(3), 287–295.
22. Lebesi, D. M., & Tzia, C., (2012). Use of endoxylanase treated cereal brans for development of dietary fiber enriched cakes. *Innovative Food Science Emerging Technologies, 13*, 207–214.
23. Gani, A., Wani, S. M., Masoodi, F. A., & Hameed, G., (2012). Whole-grain cereal bioactive compounds and their health benefits: A review. *J. Food Process Technol., 3*(3), 146–156.
24. Fišteš, A., Došenovic, T., Rakic, D., Pajin, B., Šereš, Z., Simovic, Š., & Loncarevic, I., (2014). Statistical analysis of the basic chemical composition of whole-grain flour of different cereal grains. *Acta Universitatis Sapientiae-Alimentaria, 7*, 45–53.
25. Górnaś, P., Radenkovs, V., Pugajeva, I., Soliven, A., Needs, P. W., & Kroon, P. A., (2016). Varied composition of tocochromanols in different types of bran: Rye, wheat, oat, spelt, buckwheat, corn, and rice. *International Journal of Food Properties, 19*(8), 1757–1764.
26. Patel, S., (2012). Cereal bran: The next superfood with significant antioxidant and anticancer potential. *Mediterranean J. Nutr. Metab., 5*(2), 91–104.
27. Delcour, J. A., Rouau, X., Courtin, C. M., Poutanen, K., & Ranieri, R., (2012). Technologies for enhanced exploitation of the health-promoting potential of cereals. *Trends in Food Science Technology, 25*(2), 78–86.
28. Alan, P. A., Ofelia, R. S., Patricia, T., & Rosario, M. R. S., (2012). Cereal bran and wholegrain as a source of dietary fibre: Technological and health aspects. *International Journal of Food Sciences and Nutrition, 63*(7), 882–892.
29. Hoseney, R. C., (1994). *Principles of Cereal Science and Technology* (p. 170). American Association of Cereal Chemists. Inc.; St. Paul.
30. Sharma, S., Kaur, S., Dar, B. N., & Singh, B., (2011). Storage stability and quality assessment of processed cereal brans. *Journal of Food Science and Technology, 51*(3), 583–588.
31. Galanakis, C. M., (2015). *Food Waste Recovery: Processing Technologies and Industrial Techniques*. Academic Press.
32. Abdel-Aal, E. S. M., Young, J. C., & Rabalski, I., (2006). Anthocyanin composition in black, blue, pink, purple, and red cereal grains. *Journal of Agricultural and Food Chemistry, 54*(13), 4696–4704.
33. Min, B., Anna, M. M., & Ming-Hsuan, C., (2006). Phytochemicals and antioxidant capacities in rice brans of different color. *J. Food Sci., 76*, 117–125.
34. Nelson, A. L., (2001). *High-Fiber Properties and Analyses* (pp. 29–44). High-fiber ingredients. St. Paul, Minn. American Association of Cereal Chemists.

35. Viuda-Martos, M., López-Marcos, M. C., Fernández-López, J., Sendra, E., López-Vargas, J. H., & Pérez-Álvarez, J. A., (2010). Role of fiber in cardiovascular diseases: Review. *Comprehv. Rev. Food Sci. Food Safety, 9*(2), 240–258.
36. Doblado-Maldonado, A. F., Pike, O. A., Sweley, J. C., & Rose, D. J., (2012). Key issues and challenges in whole wheat flour milling and storage. *J. Cereal Sci., 56*, 119–126.
37. Wang, N., Hou, G. G., & Dubat, A., (2017). Effects of flour particle size on the quality attributes of reconstituted whole-wheat flour and Chinese southern-type steamed bread. *LWT-Food Science and Technology, 82*, 147–153.
38. Ozturk, S., Ozboy, O., Cavidoglu, I., & Koksel, H., (2002). Effects of brewers' spent grain on the quality and dietary fibre content of cookies. *J. Inst. Brew., 108*, 23–27.
39. Gómez, M., Ruiz-París, E., & Oliete, B., (2010). Influence of flour mill streams on cake quality. *International Journal of Food Science and Technology, 45*(9), 1794–1800.
40. Galliard, T., & Gallagher, D. M., (1988). The effects of wheat bran particle size and storage period on bran flavor and baking quality of bran flour blends. *Journal Cereal Science, 8*, 147–154.
41. Di Lena, G., Vivanti, V., & Quaglia, G. B., (1997). Amino acid composition of wheat milling by-products after bioconversion by edible fungi mycelia. *Food/Nahrung, 41*(5), 285–288.
42. Elleuch, M., Bedigian, D., Roiseux, O., Besbes, S., Blecker, C., & Attia, H., (2011). Dietary fibre and fibre-rich by-products of food processing: Characterization, technological functionality and commercial applications: A review. *Food Chem., 124*(2), 411–421.
43. Carter, J. F., (1993). Potential of flaxseed and flaxseed oil in baked goods and other products in human nutrition. *Cereal Foods World, 38*, 753–759.
44. Guleria, P., Kumari, S., & Dangi, N., (2015). β-glucan: Health benefits and role in food industry-a review. *J. ERS. Tech. Eng., 8*, 255–263.
45. Canan, C., Cruz, F. T. L., Delaroza, F., Casagrande, R., Sarmento, C. P. M., Shimokomaki, M., & Ida, E. I., (2011). Studies on the extraction and purification of phytic acid from rice bran. *J. Food Comp. Analy., 24(7)*, 1057–1063.
46. Packer, L., (1995). Nutrition and biochemistry of the lipophilic antioxidants, vitamin E and carotenoids. In: Ong, A. S. H., Niki, E., & Packer, L., (eds.). *Nutrition, Lipids, Health, and Disease* (pp. 8–35). AOCS Press, Champaign.
47. Ko, S. N., Kim, C. J., Kim, H., Kim, C. T., Chung, S. H., Tae, B. S., & Kim, I. H., (2003). Tocol levels in milling fractions of some cereal grains and soybean. *J. American oil Chem. Soc., 80*(6), 585–589.
48. Juliano, B. O., (2003). *Rice Chemistry and Quality* (pp. 571 573). Philippine Rice Research Institute.
49. Ghosh, M., (2007). Review on recent trends in rice bran oil processing. *Journal of the American Oil Chemists' Society, 84*(4), 315–324.
50. Chakrabarty, M. M., (1989). Rice bran: A new source for edible and industrial oil. In: David, R. E., (ed.), *Proceedings of World Conference on Edible Fats and Oils Processing* (pp. 331–340). AOCS Press, Champaign.
51. Van, H. P., (2016). Phenolic compounds of cereals and their antioxidant capacity. *Critical Reviews in Food Science and Nutrition, 56*(1), 25–35.

52. Zhang, M. W., Zhang, R. F., Zhang, F. X., & Liu, R. H., (2010). Phenolic profiles and antioxidant activity of black rice bran of different commercially available varieties. *J. Agric. Food Chem., 58*, 7580–7587.
53. Wang, J., Sun, B., Cao, Y., & Wang, C., (2010). Wheat bran feruloyl oligosaccharides enhance the antioxidant activity of rat plasma. *Food Chemistry, 123*, 472–476.
54. Martinez-Tome, M., Murcia, M. A., Freha, N., Ruggieri, S., Jimenez, A. M., Roses, F., & Parras, P., (2004). Evaluation of antioxidant capacity of cereal brans. *Journal of Agricultural and Food Chemistry, 52,* 4690–4699.
55. Buitimea-Cantúa, N. E., Torres-Chávez, P. I., Ramírez-Wong, B., Serna-Saldívar, S. O., Rouzaud-Sández, O., Rosas-Burgos, E. C., & Salazar-García, M. G., (2017). Phenolic compounds and antioxidant activity of extruded nixtamalized corn flour and tortillas enriched with sorghum bran. *Cereal Chemistry, 94*(2), 277–283.
56. Katina, K., Salmenkallio-Marttila, M., Partanen, R., Forssell, P., & Autio, K., (2006). Effects of sourdough and enzymes on staling of high-fibre wheat bread. *LWT-Food Sci. Tech., 39*(5), 479–491.
57. Bento-Silva, A., Patto, M. C. V., & Do Rosário, B. M., (2018). Relevance, structure and analysis of ferulic acid in maize cell walls. *Food Chemistry, 246*, 360–378.
58. Katina, K., Juvonen, R., Laitila, A., Flander, L., Nordlund, E., Kariluoto, S., & Poutanen, K., (2012). Fermented wheat bran as a functional ingredient in baking. *Cereal Chem., 89*(2), 126–134.
59. Rosenkvist, H., & Hansen, Å., (1995). Contamination profiles and characterization of *Bacillus species* in wheat bread and raw materials for bread production. *Int. J. Food Microb., 26*(3), 353–363.
60. Tungland, B. C., & Meyer, D., (2002). Nondigestible oligo-and polysaccharides (Dietary Fiber): Their physiology and role in human health and food. *Comprehensive Reviews in Food Science and Food Safety, 1*(3), 90–109.
61. Caprez, A., Arrigoni, E., Amado, R., & Zeukom, H., (1986). Influence of different types of thermal treatment on the chemical composition and physical properties of wheat bran. *J. Cereal Sci., 4*, 233–239.
62. Foschia, M., Peressini, D., Sensidoni, A., & Brennan, C. S., (2013). The effects of dietary fibre addition on the quality of common cereal products. *Journal of Cereal Science*.
63. Noort, M. W. J., Van, H. D., Hemery, Y., Schols, H. A., & Hamer, R. J., (2010). The effect of particle size of wheat bran fractions on bread quality-evidence for fibre-protein interactions. *J. Cereal Sci., 52*, 59–64.
64. Ktenioudaki, A., & Gallagher, E., (2012). Recent advances in the development of high-fibre baked products. *Trends in Food Science Technology, 28*(1), 4–14.
65. Sudha, M. L., Ramasarma, P. R., & Rao, G. V., (2011). Wheat bran stabilization and its use in the preparation of high-fiber pasta. *Food Sci. Tech. Int., 17*(1), 47–53.
66. Schmiele, M., Jaekel, L. Z., Patricio, S. M. C., Steel, C. J., & Chang, Y. K., (2012). Rheological properties of wheat flour and quality characteristics of pan bread as modified by partial additions of wheat bran or whole-grain wheat flour. *Int. J. Food Sci. Tech., 47*(10), 2141–2150.
67. Zhang, D., & Moore, W. R., (1997). Effect of wheat bran particle size on dough rheological properties. *Journal of the Science of Food and Agriculture, 74*(4), 490–496.

68. Shogren, M. D., & Pomeranz, Y., (1980). Finney, K.F. Counteracting the deleterious effects of fiber in bread making. *Cereal Chem., 58*, 142–144.
69. Fendri, L. B., Chaari, F., Maaloul, M., Kallel, F., Abdelkafi, L., Chaabouni, S. E., & Ghribi-Aydi, D., (2016). Wheat bread enrichment by pea and broad bean pods fibers: Effect on dough rheology and bread quality. *LWT, 73*, 584–591.
70. Coda, R., Kärki, I., Nordlund, E., Heiniö, R. L., Poutanen, K., & Katina, K., (2013). Influence of particle size on bioprocess induced changes on technological functionality of wheat bran. *Food Microbiology*.
71. Sanz, P. J. M., Collar, C., & Haros, M., (2008). Effect of wheat bran and enzyme addition on dough functional performance and phytic acid levels in bread. *J. Cereal Sci., 48*(3), 715–721.
72. Rizzello, C. G., Coda, R., Mazzacane, F., Minervini, D., & Gobbetti, M., (2012). Micronized by- products from debranned durum wheat and sourdough fermentation enhanced the nutritional, textural and sensory features of bread. *Food Res. Int., 46*(1), 304–313.
73. Nelles, E., Randall, P., & Taylor, J., (1998). Improvement of brown bread quality by prehydration treatment and cultivar selection of bran. *Cereal Chem., 75*,536–540.
74. Wang, N., Hou, G. G., & Dubat, A., (2013). Effects of flour particle size on the quality attributes of reconstituted whole-wheat flour and Chinese southern-type steamed bread. *LWT-Food Science and Technology, 82*, 147–153.
75. Katina, K., & Poutanen, K., (2013). Nutritional aspects of cereal fermentation with lactic acid bacteria and yeast. In: *Handbook on Sourdough Biotechnology* (pp. 229–244). Springer, Boston, MA.
76. Singh, A., & Sharma, S., (2017). Bioactive components and functional properties of biologically activated cereal grains: A bibliographic review. *Critical Reviews in Food Science and Nutrition, 57*(14), 3051–3071.

CHAPTER 4

HEALTH BENEFITS OF AVOCADO

HUMA BADER UL AIN, MUHAMMAD KASHIF,
TABUSSAM TUFAIL, and HAFIZ ANSAR RASUL SULERIA

ABSTRACT

Avocado, *Persea americana*, also known as an alligator pear and butter fruit, is a perennial plant that has been cultivated and grown for centuries in tropical and subtropical climates. This fruit is a multifarious functional fruit with a broad spectrum of bioactive components and health benefits. In the present study, efforts are directed to highlight the nutritional and therapeutic potential of avocado. Literature revealed that the avocado is a versatile fruit with a good amount of monounsaturated fatty acids including linolenic, linoleic, oleic, stearic, myristic, palmitic, and capric acid and 20 important vitamins and minerals. The limelight of the present chapter is the overview of nutritional composition and health claims of avocado.

4.1 INTRODUCTION

Avocado, *Persea americana*, is a perennial, hardy, and medium to large evergreen plant that has been cultivated and grown for centuries in tropical and subtropical climates. It belongs to the family Lauraceous. For the last few decades, it has become a major international food item. It is also known as an alligator pear or king of fruits and butter fruit or poor man's butter. It was first time cultivated in Mexico and then Central America [1]. Avocado has various varieties (the most popular is Hass avocado), which vary in shape and color, are round or oval, and their skin is thick and rough green. Their weight varies from 8 ounces (220 grams) to 3 pounds (1.4 kg). This fruit is botanically known as single seed containing a large berry.

Its seeds and skin are discarded and the yellow-green flesh inside the fruit is eaten. This fruit is a multifarious functional fruit with a broad spectrum of bioactive components and health benefits [1, 2].

The avocado is a versatile fruit with a good amount of monounsaturated fatty acids including linolenic, linoleic, oleic, stearic, myristic, palmitic, and capric acid and 20 important vitamins and minerals. There are various health benefits regarding avocado including hypoglycemic, antihypertensive, antioxidant, anti-obesity, hypolipidemic, antilithiasis, anticonvulsant, antimicrobial, antiprotozoal, antimycobacterial, hepatoprotective, antiosteoarthritis, and chemo-protective effects. Moreover, they are good for skin and hair and protective against cancer, depression, and digestion. It protects the skin from environmental damage and reduces wrinkles and moisturize the skin owing to the presence of a large amount of fat in the form of fatty acids. The limelight of the present chapter is the overview of nutritional composition and health claims of avocado [3].

4.2 BIOACTIVE COMPOUNDS (BACS) IN AVOCADO

As far as the bioactive components of avocado are concerned, it has an array of various bioactive components. They contain more potassium than bananas. Avocado is loaded with heart-healthy monounsaturated fatty acids, powerful antioxidants, fiber, carotenoids, fatty acids, minerals, phenolics, and polyphenolic compounds, phytosterols, and phytostanols, proteins, seven-carbon sugars, and vitamins. Avocados are a great source of some fat- and water-soluble vitamins including ascorbic acid, phytonadione, tocopherol, niacin, pyridoxin, riboflavin, folate, pantothenic acid, magnesium, and potassium. Avocado also provides lutein, beta-carotene, and omega-3 fatty acids. Avocado fruit has low sugar content as compared to lipid content. Avocados are of great value due to the bioactive compounds (BACs) such as polyphenols (anthocyanins (ACNs), phenolic acids, tannins, and carotenoids [4, 5].

Avocados are well known for the presence of many nutritive (sugars, vitamins, and minerals) and non-nutritive compounds and BAC such as polyphenolic compounds like flavonoids, ACNs, flavonols, ellagitannins, and derivatives of hydroxycinnamic acid. These compounds are of greater interest due to their antioxidant properties and synergistic activities in cure of various disorders and positive effect in the promotion of human health.

The bioavailability of these compounds in blood is low (i.e., ACNs 1–3%). Likely, avocados have gained greater interest due to the high concentration of ACNs (galactosides, arabinosides of cyanidin and peonidin), flavonols, tannins (ellagitannins and proanthocyanidins), flavan-3-ols and phenolic acid derivatives. Some ACNs are present in negligible amount in cranberries such as cyanidin 3-*O*-glucoside, delphinidin 3-*O*-arabinoside, delphinidin 3-*O*-galactoside, peonidin 3-*O*-glucoside and delphinidin 3-*O*-glucoside [6].

4.3 HEALTH CLAIMS

In the recent era, avocados have gained much attention due to an array of phytochemicals, mainly polyphenols which have various health benefits including hypoglycemic, antihypertensive, antioxidant, anti-obesity, hypolipidemic, antilithiasis, anticonvulsant, antimicrobial, antiprotozoal, antimycobacterial, hepatoprotective, antiosteoarthritis, and chemo-protective effects. Moreover, they are good for skin and hair and protective against cancer, depression, and digestion. The potential of phytochemicals to fight against different disorders is attributed to their antioxidant capacity. It protects the skin from environmental damage and reduces wrinkles and moisturize the skin owing to the presence of a large amount of fat in the form of fatty acids.

Avocado oil is also used in cosmetics ranging from skin moisturizer, cleansing cream, sunscreen, lipstick, bath oil and hair conditioner. Their fat content may help to absorb nutrients from plant foods. In a study, the effect of fresh avocado on cardiovascular disease (CVD) was checked and avocado was found effective. Moreover, it was found that avocado is effective for skin, eye, joint, and cellular health. Avocado fruit has low sugar content as compared to lipid content. It is evident that the uptake of BAC in gastrointestinal (GI) tract can be affected and their chemical nature can be changed through specific GI conditions, enzymatic activity, and local microbiota action [22–25].

4.3.1 ANTI-OXIDANT EFFECTS

A serious imbalance between the production of reactive oxygen species (ROS) and their control results in the damage of body, nucleic acid, cell membrane and other parts of human body. Many serious issues develop in the human body such as cancer, cardiovascular, and other problems. For combating such issues, a trend of using antioxidant plants is increasing day by day like fruits, vegetables, and many important herbs as well. Various researchers have analyzed and extracted numerous BAC related to antioxidant capacity in various sources through different methods. In a study, Tulipani et al. [6] claimed that different BAC such as phenolic compounds and vitamin C which are known as strong oxygen radical scavengers are gaining importance owing to the antioxidant capacity of fruits [12]. These results revealed that avocados are at the top position for their total phenolics and total antioxidant capacity. The levels of total phenolics and antioxidant capacity in avocados are 4 times, 10 times, and 40 times their levels in other fruits, vegetables, and cereals, respectively [22].

Many plant-based substances (flavonoids, phenolic acid and phenolic diterpenes) have significant antioxidant effect. The antioxidant potential of phenolic compounds is due to their redox effect. The phenolic compounds have a key role in absorbing/neutralizing the free radical. Different parts (i.e., roots, seeds, pulp, leaves, peels) of avocado and their products have shown antioxidant property under appropriate conditions against various diseases through the neutralization of oxidative stress and cellular oxidation. Various studies showed the antioxidant activity of avocado and it was found that avocado inhibits the activity of two enzymes α-amylase and α-glucosidase and combats against diabetes. In a study, it was revealed that high carotenoid content in avocado is responsible for the antioxidant activity of avocado against hyperlipidemic. In this way, avocado protects liver, kidney, and vascular endothelium. It was reported in another study that the prevention and fight against CVD is mainly attributed to the antioxidant activity of avocado and its parts.

4.3.2 ANTI-INFLAMMATORY EFFECTS

A prolonged pro-inflammatory condition is the prevailing factor in the growth, succession, and obstacle of the ordinary sustained illnesses. This inflammatory condition can be neutralized by dietary antioxidants which are considered as proficient instrument. Earlier studies revealed that a lot of studies have done that demonstrated a strong connection between various few numbers of morphologies related to inflammation and a diet which correspond an excellent source of indispensable vitamins, nutritional fibers, and valuable biological active ingredients, especially effluent in vegetables and fruits. By virtue of immense nutritious components present in fruits, strawberries are considered as most beneficial for health subsistence such as fatty acids, fibers, vitamins, and minerals, in addition an extensive variety of polyphenolic compounds included lignans, flavonoids, tannins, and phenolic acids. Few active compounds (pyruvate, 2-linoleoylglycerol, and 5-hydroxymethylfurfural) have distinct anti-inflammatory effects. The results also indicated that Pyruvate has both anti-inflammatory as well as antioxidant activity [11, 17].

Commonly, anti-inflammatory activity is related to antioxidant activity but not true in all the cases like it is opposite in avocado. As far as the anti-inflammatory activity of avocado is concerned, it is attributed to any important bioactive compound known as eicosanoids (which is formulated from arachidonic acid (AA)). This eicosanoid comprises of lipoxygenase (LOX) derived leukotrienes and cyclo-oxygenase-derived prostaglandins. The mechanism behind the anti-inflammatory effect of eicosanoids is the release of AA by an enzyme phospholipase A from membrane phospholipids from the sn-2 position [13, 14].

Recent studies revealed that the bioactivities of phenolic compounds present in avocado are extended to various other routes including cellular survival and cellular growth, even though they are recognized primarily for their antioxidant and anti-inflammatory diseases. In cells treated with lipopolysaccharide, research demonstrated that pre-treatment with avocado decreases the apoptotic rate, level of reactive intracellular oxygen species, enhanced mitochondria functionality and carcinogenic defense. Incitement of the Nrf2 pathway and hindrance of NF-kB signaling pathway with AMPK-dependent mechanism, are responsible for these defensive biological activities of strawberry. These research outcomes affirm the medicinal advancements of avocado in inhibition of oxidative stress situations in lipopolysaccharide-treated cells.

4.3.3 ANTI-CANCER EFFECT

Designer food has listed the avocado fruit under the category of anti-cancer agent. The six basic characteristics (metastasis, resisted cell death, dodged growth suppressors, proliferation signaling, induced angiogenesis, activated invasion and enabled replicative mortality) undergoes during the development of human tumor. This multistep development is chocked by the functional components present in avocado. Literature demonstrated that there are various studies regarding the effect of avocado and its bioactive components on cancer and the mechanism behind this anti-cancer activity/effect of avocado [14, 17].

Avocado has good chemopreventive effect both in-vitro and in-vivo conditions. In a study conducted on human gastric cancer cells, avocado dose inhibited the cell proliferation and induced apoptosis in the aforementioned cells. In another study, conducted on Kunming mice that were inoculated with murine fore-gastric carcinoma cell lines. The results indicated that avocado inhibited the growth of inoculated tumors. Dong and co-authors reported that avocado has anticancer and antitumor properties. The results obtained from the study conducted on colon and gastric cells with use of avocado. Different studies conducted in China showed a direct linkage between the consumption of avocado and cancer. The results indicated the consumption of avocado greatly reduced the chances of cancer. In another study, stomach cancer risk chances are directly related to the consumption of functional foods like avocado.

4.3.4 CARDIOVASCULAR EFFECTS

Cardiovascular disorder is a chronic aberration and leading etiology of morbidity and death worldwide. According to the World Health Organization (WHO) statistics, an estimated 17.5 million people died from CVD s in 2007, representing 31% of all global deaths. CVD includes coronary heart disease (CHD) and diseases related to cerebral vessels. There are many different types of CVD. Four of the main types are CHD, strokes, and transient ischemic attack, peripheral arterial disease, and aortic disease. It can be treated in three different ways, i.e., diet, medicine, and lifestyle modification. Among these approaches, nutrition (diet) is the most preferred way to cure CVDs because it has no harmful effect [11, 12, 15, 16].

With respect to dietary approach, fruits, and vegetables are considered as important treatments in the prevention of CVDs. Among fruits, avocados have potential to fight against various risk factors of CVD such as hypertension, obesity, and type II diabetes mellitus owing to the presence of various functional ingredients such as vitamin B6, polyphenols (flavonoids), fiber, magnesium, potassium, and folate. High levels of these phytochemicals are associated with lowering the levels of homocysteine in plasma. Moreover, beta-sitosterol and other plant sterols present in avocado are associated with lowering the risk of higher cholesterol levels and indirectly mitigating the chances of CVD. Literature revealed that consumption of avocado could improve the plasma lipid profile, biomarkers of antioxidant status, antihemolytic defenses and platelet function in healthy subjects, encouraging further evaluation on a population with higher CVD risk. Cassidy et al. [18] reported that avocados have the potential to treat CVDs due to three major positive effects such as antioxidant, anti-atherosclerotic, and antihypertensive effects.

4.3.5 ANTI-DIABETIC EFFECTS

Diabetes mellitus is one of the most common clinical diseases that is characterized by defects in insulin secretion and insulin action, resulted in disturbance in metabolism of carbohydrates, proteins, and fats. The worldwide burden of type 2 diabetes (T2D), one of the main causes of morbidity and mortality, has increased rapidly in tandem with increases in obesity. Now a day, it is prevalent in almost 150 million people worldwide. Important risk factors for this aberration are sedentary lifestyle, obesity, and consumption of an energy-rich diet. As a result of its increasing prevalence, various therapeutic approaches have been applied in order to treat this disease, unless no particular treatment yet discovered. Emerging management strategies for this disorder are involve insulin and oral anti-diabetic agents, dietary modification, and lifestyle management.

Medicinal therapies are effective but toxic as well, whereas diet therapy is considered as non-toxic strategy. In this context, fruits, and vegetables rich in fiber are usually recommended. In the fruit category, avocados are considered important to cope with type II diabetes mellitus due to BAC such as flavonoids, ACNs, and other phenolic compounds. These avocados play an important role in the treatment of type II diabetes by fighting with related complications due to their antioxidant capacity. A lot of literature exists to defend the functional importance of different forms of avocado against diabetes mellitus and its related complications. Avocado fruit has low sugar content as compared to lipid content. Hence, avocado fruit is generally recommended for people suffering from diabetes because it is a high-energy food.

Chiang and Tseng [19] probed the effect of avocado on diabetes mellitus and its risk factors and found that free radicals and ox-low-density lipoprotein-induced proliferation was reduced significantly in rat aortic smooth muscle cells. Wolfe et al. [28] claimed that avocados are considered best for their contribution in cellular antioxidant activity among all fruits. Yi, Fischer, Krewer and Akoh [23] worked on the treatment of diabetic Mellitus through avocados and concluded that avocado reduce oxidative stress and atherosclerotic lesion owing to a functional ingredient, i.e., ellagic acid. In another study, Pinto et al. [27] explored the anti-hyperglycemic and anti-hypertensive effects of avocado and found that ellagic acid present in avocado can inhibit the activities of carbohydrate and lipid-related enzymes such as α-amylase, α-glucosidase, and angiotensin I-converting enzyme. Moreover, Moazen et al. [26] examined that freeze dried avocado is inversely proportional to type II diabetes owing to the presence of phenolic compounds, mainly anthocyanin and flavonoids. Many other researches were conducted to check the anti-diabetic effect of avocado and found positive results.

4.3.6 ANTI-METABOLIC SYNDROME ACTIVITY

Similar to CVD and obesity, metabolic syndrome, also known as insulin resistance syndrome or syndrome X, is a worldwide prevalent chronic aberration with multiple complications. Risk factors associated with this syndrome are central or visceral obesity, hypertension, high serum triglycerides, insulin resistance, and changed ratio from low to high-density cholesterol levels, increased biomarkers of inflammation and lipid oxidation [28]. As the incidence of complications associated with this syndrome have risen day by day, so there is a need to properly combat these complications through different ways: medicine, diet, life style modification. Now a day, the dietary approach to this syndrome has captured greater attention due to increasing awareness in public regarding its positive effects. In the food category, the trend of fruits consumption for the management of metabolic syndrome is increasing day by day.

Among fruits, avocados are important fruits to combat risk factors of metabolic syndrome such as hypertension, impaired fasting glucose, dyslipidemia, and CVDs [16, 17]. Avocado is one of the most important forms of fruits and its consumption can lowers blood glucose levels owing to high anthocyanin content [28]. Moreover, avocado can overcome the main features of metabolic syndrome such as hypertension and hyperglycemia by deactivating α-glucosidase which is carbohydrate digestive enzyme, α-amylase which is lipid-related enzymes [16].

Similarly, many researches were done to probe the anti-metabolic effect of avocado and it was noticed that avocado supplementation can improve postprandial metabolism and lowered the postprandial hyperglycemia that are the prominent features of metabolic syndrome. Moreover, postprandial need of insulin can be reduced through the regular intake of avocados, which in turn lowers the risk of type II diabetes and metabolic syndrome and is suggested for persons with higher risk of type II diabetes and metabolic syndrome [12]. Main functional ingredient in avocado for anti-metabolic syndrome effect is anthocyanin. Some researchers reported that ellagitannin is also found effective in this sense. So, it is concluded that anthocyanin is the only compound which is effective against features and risk factors of metabolic syndrome [13].

4.4 SUMMARY

This review highlights introduction, production, cultivation, nutritional composition and paying special attention to the therapeutic potential of the avocado. The present review also provides the up-to-date knowledge of the subject regarding the advantages of using avocado. There have been a lot of benefits of the avocado as has been described in various published papers. There is an emerging interest across the world on avocado consumption due to its unique bioactive properties (anti-obesity, anti-inflammatory, anti-allergic, hepatoprotective, cardioprotective, and anti-cancer) and organoleptic characteristics. There is dire to conduct deep research including *in-vivo* and *in-vitro* studies for probing the effectiveness of avocado against various diseases.

KEYWORDS

- **avocado**
- **bioactive compounds**
- **functional foods**
- **health claims**
- **monounsaturated fatty acids**
- **nutritional composition**
- **World Health Organization**

REFERENCES

1. Ameer, K., (2016). Avocado as a major dietary source of antioxidants and its preventive role in neurodegenerative diseases. *Advances in Neurobiology, 12*, 337–354.
2. Anderson, R. A., Evans, M. L., Ellis, G. R., Graham, J., Morris, K., Jackson, S. K., Lewis, M. J., et al., (2001). The relationships between post-prandial lipaemia, endothelial function and oxidative stress in healthy individuals and patients with type 2 diabetes. *Atherosclerosis, 154*, 475–483.
3. Berasategi, I., Barriuso, B., Ansorena, D., & Astiasarán, I., (2012). Stability of avocado oil during heating: Comparative study to olive oil. *Food Chemistry, 1*, 439–446.
4. Bhuyan, D. J., Alsherbiny, M. A., Perera, S., Low, M., Basu, A., Devi, O. A., Barooah, M. S., et al., (2019). The odyssey of bioactive compounds in avocado (*Persea americana*) and their health benefits. *Antioxidants (Basel), 8*(10), 426.
5. Bonora, E., (2002). Postprandial peaks as a risk factor for cardiovascular disease: Epidemiological perspectives. *International Journal of Clinical Practice and Supply, 129*, 5–11.
6. Brownlee, M., (2001). Biochemistry and molecular cell biology of diabetic complications. *Nature, 414*, 813–820.
7. Carvajal-Zarrabal, O., Nolasco-Hipolito, C., Aguilar-Uscanga, M., Melo-Santiesteban, G., Hayward-Jones, P., & Barradas-Dermitz, D., (2014). Avocado oil supplementation modifies cardiovascular risk profile markers in a rat model of sucrose-induced metabolic changes. *Dis. Markers, 2014*, 386–425.
8. Carvajal-Zarrabal, O., Nolasco-Hipolito, C., Aguilar-Uscanga, M., Melo-Santiesteban, G., Hayward-Jones, P., & Barradas-Dermitz, D., (2014). Effect of dietary intake of avocado oil and olive oil on biochemical markers of liver function in sucrose-fed rats. *Biomedical Research International, 2014*, 595479.
9. Ceriello, A., (2000). The post-prandial state and cardiovascular disease: Relevance to diabetes mellitus. *Diabetes Metab. Res. Reviews, 16*, 125–132.

10. Ceriello, A., Esposito, K., Piconi, L., Ihnat, M. A., Thorpe, J. E., Testa, R., Boemi, M., & Giugliano, D., (2008). Oscillating glucose is more deleterious to endothelial function and oxidative stress than mean glucose in normal and type 2 diabetic patients. *Diabetes, 57*, 1349–1354.
11. Das, U. N., (2006). Pyruvate is an endogenous anti-inflammatory and anti-oxidant molecule. *Medical Science Monitor, 12*(5), RA79–RA84.
12. Di Stefano, V., Avellone, G., Bongiorno, D., Indelicato, S., Massenti, R., & Lo Bianco, R., (2017). Quantitative evaluation of the phenolic profile in fruits of six avocado (*Persea americana*) cultivars by ultra-high-performance liquid chromatography-heated electrospray-mass spectrometry. *International Journal of Food Properties, 20*, 1302–1312.
13. Dong, M., Yang, G., Liu, H., Liu, X., Lin, S., Sun, D., & Wang, Y., (2014). Aged black garlic extract inhibits HT29 colon cancer cell growth via the PI3K/Akt signaling pathway. *Biomedical Reports, 2*(2), 250–254.
14. Dreher, M. L., & Davenport, A. J., (2013). Hass avocado composition and potential health benefits. *Critical Reviews in Food Science and Nutrition, 53*(7), 738–750.
15. Flores, M., Saravia, C., Vergara, C. E., Felipe, A., Valdes, H., & Ortiz-Viedma, J., (2019). Avocado oil: Characteristics, properties, and applications. *Molecules, 24*(11), 2172.
16. Fulgoni, V. L., Dreher, M., & Davenport, A. J., (2013). Avocado consumption is associated with better diet quality and nutrient intake, and lower metabolic syndrome risk in US adults: Results from the national health and nutrition examination survey (NHANES) 2001–2008. *Nutrition Journal, 12*(1).
17. Gao, C. M., Takezaki, T., Ding, J. H., Li, M. S., & Tajima, K., (1999). Protective effect of allium vegetables against both esophageal and stomach cancer: A simultaneous case-referent study of a high-epidemic area in Jiangsu province, China. *Japanese Journal of Cancer Research, 90*(6), 614–621.
18. Hanahan, D., & Weinberg, R. A., (2011). Hallmarks of cancer: The next generation. *Cell, 144*(5), 646–674.
19. Heskey, C., Keiji, O., & Sabate, J., (2019). Avocado intake, and longitudinal weight and body mass index changes in an adult cohort. *Nutrients, 11*(3), 691.
20. Jeong, Y. Y., Ryu, J. H., Shin, J. H., Kang, M. J., Kang, J. R., Han, J., & Kang, D., (2016). Comparison of anti-oxidant and anti-inflammatory effects between fresh and aged black garlic extracts. *Molecules, 21*(4), 430.
21. Kim, D. G., Kang, M. J., Hong, S. S., Choi, Y. H., & Shin, J. H., (2017). Anti-inflammatory effects of functionally active compounds isolated from aged black garlic. *Phytotherapy Research, 31*(1), 53–61.
22. Liu, C. T., Su, H. M., Lii, C. K., & Sheen, L. Y., (2009). Effect of supplementation with garlic oil on activity of Th1 and Th2 lymphocytes from rats. *Planta Medica, 75*(03), 205–210.
23. O'Keefe, J. H., Gheewala, N. M., & O'Keefe, J. O., (2008). Dietary strategies for improving post-prandial glucose, lipids, inflammation, and cardiovascular health. *Journal of American College of Cardiology, 51*, 249–255.
24. Osawa, T. (1994) Novel Natural Antioxidants for Utilization in Food and Biological Systems. Postharvest Biochemistry of Plant Food-Materials in the Tropics. Japan Scientific Press, Tokyo, 241–251.

25. Saha, B., (2006). Postprandial plasma glucose level less than the fasting level in otherwise healthy individuals during routine screening. *Indian Journal of Clinical Biochemistry, 21*, 67–71.
26. Setiawan, V. W., Yu, G. P., Lu, Q. Y., Lu, M. L., Yu, S. Z., Mu, L., Zhang, J. G., et al., (2005). Allium vegetables and stomach cancer risk in China. *Asian Pacific Journal of Cancer Prevention: APJCP, 6*(3), 387.
27. Shahidi, F., Janitha, P. K., & Wanasundara, P. D., (1992). Phenolic antioxidants. *Critical Reviews in Food Science and Nutrition, 32*(1), 67–103.
28. Slyper, A. H., (1992). A fresh look at the atherogenic remnant hypothesis. *Lancet, 340*, 289–291.
29. Tan, C., Tan, S., & Tan, S., (2017). Influence of geographical origins on the physicochemical properties of *Hass* avocado oil. *Journal of American Oil Chem. Society*, *94*, 1431–1437.
30. Wang, L., Tao, L., Hao, L., Stanley, T. H., Huang, K. H., Lambert, J. D., & Kris-Etherton, P. M., (2020). A moderate-fat diet with one avocado per day increases plasma antioxidants and decreases the oxidation of small, dense LDL in adults with overweight and obesity: A randomized controlled trial. *Journal of Nutrition, 150*(2), 276–284.
31. Wang, X., Jiao, F., Wang, Q. W., Wang, J., Yang, K., Hu, R. R., Liu, H. C., et al., (2012). Aged black garlic extract induces inhibition of gastric cancer cell growth *in vitro* and *in vivo*. *Molecular Medicine Reports, 5*(1), 66–72.
32. Wong, M., Requejo-Jackman, C., & Woolf, A., (2010). What is unrefined, extra virgin cold-pressed avocado oil? *Information, 21*, 189–260.

CHAPTER 5

HEALTH BENEFITS OF WATERMELON (*CITRULLUS LANATUS*)

AIMAN IJAZ, TABUSSAM TUFAIL, FARHAN SAEED, MUHAMMAD AFZAAL, MUHAMMAD ZIA SHAHID, and HAFIZ ANSAR RASUL SULERIA

ABSTRACT

Watermelon *(Citrullus lanatus)* belongs to cucurbit family *(Cucurbitaceae)*. Watermelon's Citrullus species *(Cucurbitaceae)* have been cultivating since antique times and are innate to Africa. Organic and significant source of citrulline which is non-essential amino acid is watermelon (*Citrullus vulgaris* Schrad.). In nitric oxide (NO) system in humans Citrulline play a vital role in vasodilatation and contain potential antioxidant. Flesh of watermelons which are wild is watery, but pale in color, bland or bitter in taste and usually hard in texture. Biological affairs revision include; antioxidant, anti-inflammatory, antimicrobial, anti-plasmodial, anti-prostatic, antigiardial activity, Hyperplasia activity, anti-oxidant, analgesic properties, antidiabetic, antisecretory, laxative, hepatoprotective, and antiulcerogenis activities. In vision of its widespread biological and pharmacological activities, by tradition its therapeutic potential stated such as, antidiarrheal, and antihypertensive. Watermelon (*Citrallus lunatus*) is known to have bioactive compounds (BACs) such as triterpenes, alkaloids, sterols, minerals, cucurbitacin, and vitamins. Due to the presence of a large variety of bioactive components it is very popular in the native system of traditional medicine.

5.1 INTRODUCTION

The current nutritional reference that increased consumption of a food abundant in natural antioxidants, created attention in exchanging energy-increasing snack each day with fruits that contain antioxidants [1]. *Citrullus lanatus* usually named as watermelon. Nutrients and phytochemicals are present in watermelon which is reported to be useful for human health. Watermelon contains vitamins E, C, and B plus minerals for example, magnesium, phosphorus, iron, and calcium. Epidemiologic revisions have proven that antioxidants present in Citrullus have anti-hypertensive, anti-inflammatory properties along with defensive result against carbon tetrachloride-induced toxicity [2]. Several bioactive ingredients which are present in Plant-based food play a vital role in performing a number of metabolic functions like development growth and defensive mechanism against physical threat. In these circumstances, Phytochemicals as they recover the human health through different paths are of significance importance [3]. In many parts of the world, it is a popular fruit and prominent for its high-water content and eye-catching look. It is mainly eaten as cool seasonal fruit, greatly valued by the buyers due to its eye-catching color, energizing ability, mild flavor, and great content of water to satisfy the summertime thirst [4].

The fruit has various sizes, shapes, and rind pattern [5]. These days, it is cultivated in humid regions of the world, but it is innate to the Kalahari Desert of Africa. In history, its first yield was documented 5,000 years ago in Egypt that later spread to further parts of the world [6]. Watermelon (*Citrullus lanatus)* is a good source of natural antioxidants as well as ascorbic acid, lycopene, and citrulline. These useful components turn as shelter against prolonged health complications such as cardiovascular disorders and cancers. The four most promising cultivars are icebox, picnic, and seedless and yellow flesh but total 1,200 cultivars of *Citrullus lanatus* are produced worldwide [7]. Watermelon contains 93% of water; therefore, it is called "water" melon. The word melon shows that the fruitlet is big and has a sugary, mushy flesh. Watermelon's scientific name comes from Latin as well as Greek origins. *Citrullus* is a Greek word meaning “citrus” which is allusion of fruit, *lanatus* is a Latin word, and has the sense to be wooly, mentioning the fur on leaves and stems of plant. The seeds have no hydrocyanic acid and are a brilliant source of energy. The oil extracted from seeds has glycosides of oleic, stearic, linoleic, and palmitic acids [8, 9].

This chapter explores the potential health advantages of watermelon (*Citrullus Lanatus)* against different diseases. It contains nutrients and beneficial compounds which help in the prevention of cancer, CVDs, GIT problems as well as oxidative stress. It is a red coloring fruit due to the presence of lycopene which is an important antioxidant.

5.1.1 COMMON NAMES

Citrullus lanatus Dialectal names includes; Egusi watermelon and Egusi melon. Furtherinclude; Watermelon, cooking melon, dessert watermelon, West African watermelon. Its botanical classification has been shown in Table 5.1 [10].

5.1.2 PHYSICAL CHARACTERISTICS

It is a big, widespread yearly plant with rough, pinnately-lobed hairy leaves and yellow flowers. Grown-up for eatable fruit, a special form of berry called pepo botanically. The *Citrullus lanatus* fruitlet has light green or gray vertical streaks with deep green smooth thick external rind. Fruit is red in color from inside with small black seeds entrenched in the middle third of the flesh. Red and orange color of watermelon is due to Carotenoids such as β-carotene and lycopene, respectively. Glucose, sucrose, and fructose are mainly the reason of watermelon sweetness. Around 20–40% of Sucrose and glucose and 30–50% fructose are the total sugars in a ripe watermelon [11, 12].

TABLE 5.1 Botanical Classification of WaterMelon

Kingdom	Plantae-Plant
Sub kingdom	Treachiobionta-vascular plants
Super Division	Spermatophyta-Seed plants
Division	Magnoliophyta-Flowering plant
Class	Magnoliosida-Dicotyledons
Order	Cucurbit ales
Family	Curcurbitaceae
Genus	Citrullus
Species	*Citrullus lanatus*

5.2 HEALTH BENEFITS OF WATERMELON

Watermelon is consumed by people because of its positive health effects on the body and good taste as shown in Table 5.2. It is considered as fruit which is low in fat, do not contain cholesterol and have lesser amount of sodium along with this watermelon is rich inessentials vitamins and minerals or phytochemicals playing crucial role in body [13].

TABLE 5.2 Health Benefits of Watermelon in Different Diseases

Parameter	Disease	Results	References
Citrulline	Obese post-menopausal females	Improved cardiac vascular autonomic capacity in fat postmenstrual females	Wong et al. [17]
Lycopene	Hypertension	enhance plasma agitation	Choudhary et al. [11]
Vitamin C	Stroke	Stroke depletion	Pacier et al. [19] Aborashed [20]
Lycopene	Breast cancer suffers	Serum lycopene related with diminished hazards	Dia et al. [21]
Lycopene	Osteoporosis	Can balance the harming impact of oxidative pressure which causes osteoporosis	Choudhary et al. [11]
Vitamin C	Liver disease	Decline of 58.2% serum alanine aminotransferase and 49.4% of high-affectability C-responsive protein and furthermore limit harm and moderate infection progression	Pacier et al. [19]
Vitamin C	Human metabolism	Least risk of deficiency or antagonistic health impact.	Pacier et al. [19]
β-carotene	CVD mortality	diminished the risk/hazard ration in CVD and coronary heart diseases	Choudhary et al. [11]
Lycopene	↑es cholesterol in macrophage cell line	Lowered cholesterol synthesis	Vacca et al. [17]
Citrulline	Low libido	Improve erectile functions	Sotiriou et al. [23]
Vitamin C	Scurvy and flu	Help in the prevention and treatment of different ailments, scurvy, and a simple cold	Choudhary et al. [11]
Vitamin A	Eye health	Enhances optimal eye functioning	USDA [16]

5.3 WATERMELON UTILIZATION

Different pharmaceutical industries are trying to incorporate it more and more in supplements and other products to increase its utilization because of its benefits. It is used in the industry for the synthesis of different functional foods, sauces, smoothies, sweets, jams, and many other products [14]. Those people, who are concerned about their health, but have no time for preparing food can, consume this novel food which is ready to eat healthy snack. However, packaging of this novel fruit is rare in different countries [15, 16].

5.4 PHYTOCHEMICALS PRESENT IN WATERMELON

Watermelon has many phytochemicals like lycopene, vitamin C, β-carotene, and Total polyphenol parts which contain anti-inflammatory, anticancer, and antioxidant properties. Dietary consumption of substances with antioxidant properties are beneficial for human health Studies has shown that natural phytochemicals like polyphenols, vitamin C, β-carotene, and lycopene intercede their effect through various methods, i.e., is regulation of cell growth, immune system response, and modulation of gene expression [17]. Dietary intake of food which contain high amount of phytochemicals has long-lasting impact on human health. Phytochemical which are present in watermelon are as given in subsections.

5.4.1 LYCOPENE

Lycopene termed as carotenoid is associated with the elevated level of degree of unsaturation. It is basically a red pigment which gives respective color to the fruits like tomatoes, the red portion of guava, watermelon, and bell peppers. Lycopene provides about 21–43% of the total carotenoids and is aggregated in the human tissue [18, 19]. This carotenoid is obtained from bright colored vegetables and fruits. Hence, So, its requirement can easily be met by taking adequate amount of fruits and veggies in the diet. The assessed lycopene content consumption in advanced countries around the world is almost 5–7 mg/day and it is predicted that the amount of lycopene absorbed by the human body is almost 10–30%. Watermelon is considered as a main reservoir of lycopene when compared to juice [20, 21].

5.4.1.1 LYCOPENE DEFICIENCY IN AGING

Plasma lycopene level can turn out to be fundamentally diminished during the process of aging. Decrease levels of lycopene and other carotenoids in elderly persons are accepted to reflect age related changes in the intestinal microbiota, which control bioavailability of carotenoids and polyphenols in the large intestine [22]. There are so many clinical proofs are present which are recommending that lycopene supplementation stop or prevent osteoporosis and frequency of bone fractures, make pulmonary function better, postpone skin aging, and improve physical performance in older patients. Clinical proof uncovers the essential job of lycopene in keeping up prostate health and its capacity to prevent or stop prostate cancer in older males [23, 24].

5.4.1.2 LYCOPENE AND CARDIOVASCULAR DISEASE (CVD)

Cardiovascular disease (CVD) stays a main source of mortality and disability all over the world. The decrease pace of CVD mortality can be clarified, at least in part, by the dietary culture of the Mediterranean areas which incorporate utilization of huge amounts of fruit and vegetables. Epidemiological investigations give undeniable proof supporting the immediate job of lycopene in anticipation of CVD. There are numerous clashing reports on how lycopene management affects the development of CVD and its outcomes [25, 26].

Low serum lycopene and absolute carotenoid levels anticipate all-cause mortality as well as poor outcomes and fast development of CVD in the elder population of the USA. Most of the people, even in advanced countries, have amazingly decrease levels of lycopene and all out carotenoids in the blood which convert into higher stroke risk [27]. Heart related disorders are accounted to be the significant reason for mortality in the people of Europe; it is additionally anticipated that in the coming 30 years, the quantity of casualties because of heart disorders will be almost 23.6 million [28].

5.4.1.3 ANTI-MALIGNANT EFFECT OF LYCOPENE DERIVED FROM WATERMELON

Malignant growth has gotten one of the main sources of death worldwide because of the more prominent number of disease patients. It was evaluated that 9 million new malignant growth cases each year. Now days, various confirmations are accessible demonstrating direct linkages between food active parts and cell genomic with exceptional reference to malignant growth treatment [29, 30]. Lycopene is probably related to the formation of stage I and II enzymes that are fundamental for digestion of cancer-causing agent inside the physiological framework. Stage I enzymes can possibly enact the cancer-causing agent while stage II enzymes are liable for appending polar gathering to the active cancer-causing agent that encourages its discharge. Besides, it is associated with relapse of malignancy by intruding on cancer cell development cycle, apoptosis, hormone guideline and cancer-causing agent metabolism [31].

Some extraordinary researches have been recently demonstrated the novel bioactive properties of lycopene. These carotenoids are used in different therapies to target cancer. In one of the researches, 15 men were given the supplementation of lycopene having dose of 30 mg/day and almost 11 men in the benchmark category were prescribed to follow the guidelines of the National Cancer Institute's (NCI) proposals which guides to take almost five servings of fresh fruits and vegetables every day [32]. These discoveries cleared that lycopene eased back the development of prostate cancer. Diets high in vegetables and fruits help forestall colorectal cancer (CRC). Watermelon utilization may diminish CRC hazard because of its concentration of L-Citrulline and its job in endothelial nitric oxide (NO) formation. Research recommends that elevated NO levels have tumoricidal impacts [33, 34].

Lycopene has been recorded to forestall human cancer cell development by meddling with the development factors receptor signaling and cell cycle progress, explicitly in prostate cancer cells, without known proof of poisonous effects or cell apoptosis. Carotenoids and their subsidiaries being oxidized cooperate within a system of interpretation frameworks enacted by various ligands [35, 36]. Lycopene management noticeably reduce insulin enlargement issue-1 insulin-like growth factor 1 stimulation of both the DNA (deoxyribose nucleic acid) binding ability of the activator 1 (AP-1) transcription role as well as tyrosine phosphorylation of insulin receptor substrate-1 [37]. Researchers too showed so as to the lycopene management of Michigan cancer foundation mammary cancer cells lowered insulin-like growth factor 1-stimulated cell series succession, which be not be an adjunct via moreover necrotic cell death or apoptotic. Lycopene-induced impediment in progress via the G1 as well as S phases has also been recognized in other human cancer cell lines (leukemia and endometrium cancer, lung cancer, and prostate cancer). On the basis of these studies, lycopene stops the creation of the pro-inflammatory cytokine interleukin-eight induced through smoke of cigarette [38].

One more study is evidence for those anti-proliferative outcomes of lycopene on person prostate cancer cells engage the commencement of the PPARγ-LXRα-ATP-binding cassette carrier 1 (ABCA1) track. Lycopene had anti-carcinogenic dealings in liver, mammary organ, skin, and lungs in mouse models, and furthermore prohibited the advancement of distorted crypt foci in rat colon as shown in Figure 5.1 [39].

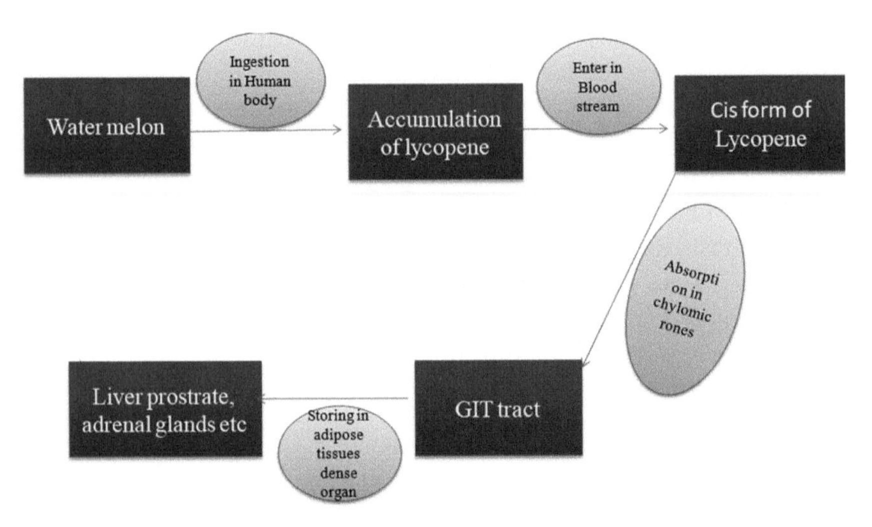

FIGURE 5.1 Absorption of lycopene.

5.4.1.4 ANTI-INFLAMMATORY AND ANTI-SCAVENGING POTENTIAL

Anti-inflammatory foods have been proven to boost immunity and overall wellbeing. Watermelon is called an antioxidant food due to the presence of lycopene in it. Lycopene is an antioxidant to neutralize free radicals and plays a role of determent for various inflammatory procedures. It is also rich in Vitamin C, manganese, and vitamin-B6 (pyridoxine). By consuming foods that are high in Ascorbic acid help out the body to build up immunity in opposition to infective envoy and hunt oxygen-free radicals that are injurious to health. Manganese is used as a co-enzyme for the superoxide dismutase (antioxidant enzyme) by the body. *Citrullus Lanatus* is abundant in Vitamin A, which is an incredible antioxidant, essential to boost immunity and for the betterment of vision [40].

5.4.2 BETA CAROTENE

The β-carotene scavenging potential has been deeply explored and some positive results have been disclosed. B-carotene builds up platelet aggregation, thus expanding the development factor expression, which results in the restoration of walls of blood vessel, along with other remarkable health benefits [41]. It plays a role in the maintenance of many organs such as heart, kidneys as well as cell development. It has also illustrated to be remarkable for the maintenance of the immune system [42]. B-carotene is very helpful for good vision and it is also favorable in the development of epithelial cells and regulates the functioning of the genes which is due to the dioxygenase enzyme, existing in the mucosa of the small intestine. It is also helpful in the conversion of β-carotene into retinol. Moreover, it plays role in defying age naturally by enhancing the diet neutralization of the damaging molecules.

5.4.3 VITAMIN C

Demonstration of biological activity of ascorbic acid by Organic compounds is a general description of Vitamin C. It is a crucial nutrient that the human body cannot synthesize that is why it should be absorbed in the body via diet. Vitamin C, a water-soluble essential nutrient, also called ascorbic acid, is crucial for the biosynthesis of collagen and certain hormones for nutrient improvement due to which it is added to different types of food products for supplementation. Vitamin C inhibits lipid peroxidation that is why it is classified as chain-breaking antioxidant. Therefore, many cohort studies recommend consuming foods that contain vitamin C naturally as compared to synthetic supplements [44].

5.5 HYDRATION AND DIGESTION

To stay hydrated, watermelon is the best choice among other fruits. Watermelons are rich in electrolytes and water content and therefore they are the best nature's gift to quench summer thirst. Watermelon helps keep you regular by encouraging a healthy digestive tract because it contains a lot of fiber [45].

5.5.1 SKIN AND HAIR BENEFITS

Healthy growth of new collagen and elastin cells is promoted by vitamin A and it also keeps the skin and hair moisturized. Healthy collagen growth is also promoted by vitamin C [46].

5.5.2 CITRULLINE

Watermelon contains an abundance of Citrulline and it is Non proteinogenic amino acid. The scientists further announced that citrulline is a beneficial metabolic intermediate in NO cycle and an important precursor for arginine. Citrulline came out as a crucial amino acid, as a result of NO cycle. It is reviewed as a powerful osmolyte and hostile to oxidant against salt pressure and under dry conditions [47].

A cohort study reported that the persons who intake a continuous greater amount of watermelon juice along with every meal had higher level of arginine and ornithine concentrations as compared to the persons who do not intake watermelon juice in their diet [48]. This results that the concentration of arginine can be greater with the consumption of watermelon. Its importance has also been reported to prevent anemia [49].

5.6 ANTIBACTERIAL AND ANTIFUNGAL POTENTIALS OF THE WATERMELON

In numerous nations, the pulp of various fruits, however not the seed and skin, is generally used or expended. This adds to food squanders in the earth. An important case is watermelon, a significant natural product broadly circulated in the tropics that fills in as a refreshment inferable from its high-water content which is almost 92%. To prevent strong waste related dangers to nature, exertion ought to be made to build the usage of food squanders which requires concentrates on their properties and pharmacologic activities [50, 51].

The use of water extract of watermelon seed in traditional herbal medicine to cure catarrhal infection, fever, disorders of the bowel and urinary passage have been reported. Studies on watermelon natural products were accounted for, yet chiefly on the juice/mash and a little on the strip/skin. These justified the investigation planned for contrasting the antibacterial and antifungal possibilities of the unrefined ethanol and watery concentrates of watermelon (*Citrullus lanatus*) skin and seed against these bacterial strains by means of: *Escherichia coli*, *Pseudomonas aeruginosa*, *Klebsiella pneumonia,* *Bacillus pumilus*, *Staphylococcus aureus*, and *Bacillus subtilis* [52].

5.7 RELATIONSHIP BETWEEN WATERMELON AND HYPERTENSION

In hypertensive patient, L-arginine is able to reduce the blood pressure by enhance the bio-availability of this amino acid, which helps as a substrate for endothelial production of the vasodilator NO. The study shows that after consumption of watermelon in the extract form was helpful in decreasing the blood pressure. The scientists proved that this hypotensive effect is due to L-citrulline that exists in watermelon, which can change into L-arginine, which enhanced the production of endothelial NO. Beyond the enhance production in NO, decreased blood pressure through L-arginine supplementation or watermelon has also been accompanied by a reduction in the carotid reflection wave, decrease in wave of amplitude reflection and arterial stiffness, reduce in myocardial oxygen demand during the cold presser test and enhance in the magnitude of reflection waves induced by cold. Although the hypotensive effect of watermelon is well detailed, this data is taken from a single group. Furthermore, it is unclear whether some of the nutrients from watermelon could decrease the activity of sympathetic nervous in hypertensive patients [45].

5.7.1 EFFECT OF WATERMELON EXTRACTS ON BLOOD PRESSURE AND SYMPATHOVAGAL BALANCE IN PRE-HYPERTENSIVE AND HYPERTENSIVE INDIVIDUALS

Studies have been done to assess the result of 6 weeks of watermelon extract in the form of supplementation on blood pressure and sympathovagal balance in prehypertensive and hypertensive patients. It shows the significant result that decrease in blood pressure after taking supplementation with 6 g/day of L-arginine in hypertensive middle-aged female was accompanied by an in-serum nitrate/nitrite level. The effect of L-arginine on Vasodilation has been proved in vascular reactivity studies with humans and rats. Although current evidence shows that relationship between L-arginine/L-citrulline involved in the hypotensive effect attributed to watermelon, the lycopene is enriched in watermelon, it also has carotenoid that has been reported to exert hypotensive effects and cardioprotective [54].

Furthermore, magnesium is also present in watermelon, which also shows that decrease the levels in blood pressure. Sympathetic activity influences vascular function through several mechanisms, including peripheral vasoconstriction, bar reflex dysfunction, vascular remodeling increased the metabolic alterations and blood pressure that compromise endothelial function and structure. The DASH diet included fruits and vegetables, but this is not identifying those which fruits and vegetables would help to reduce the hypotensive effect. In the current study it shows the therapeutic capacity of watermelon. The study show that watermelon can reduce 11.8–15.0 mmHg for systolic and 6.0–7.0 mmHg for diastolic blood pressure [55].

5.8 INHIBITION OF CARBOHYDRATE METABOLIZING ENZYMES

It is reviewed that post-prandial hyperglycemia can be controlled by inhibiting carbohydrate metabolizing enzyme particularly α-glucosidase and α-amylase and, in diabetes type 2. An enzyme called α-glucosidase found in the brush borders of small intestine actually help to delay the release of sugars like sucrose and maltose. This works actually by gradually increasing the surge of glucose after taking the meal which is rich in carbohydrates. It is already observed by different researches on medicinal herbs related to the inhibition of these enzymes. One of the researches on C. Lanatus extracts, when observed in-vitro condition in different quantities has shown the inhibition of these specific enzymes. Seed extracts were used with different concentrations of ethanol. It was observed that 70% ethanol with rind extract has shown maximum inhibition. This study has also shown that watermelon leaf has potential antioxidant property too [56].

5.9 NITRIC ACID SYNTHESIS

Insulin resistance is related to decreased bioavailability of NO. Citrulline, an amino acid present in watermelon works as the precursor of nitric acid and arginine. Research has shown that watermelon juice helps to improve the nitric acid synthesis in Table 5.3.

TABLE 5.3 Mechanism to Improve Glucose Level

Author/Year	Mechanism to Improve Blood Glucose Level	Type of Study	Findings
Chertman et al. [47]	Through structural changes in pancreatic cells	Albino rat study	Decrease of blood glucose level was may be due to the regeneration ability or may be due to proliferation of cells of pancreas, mechanism is still unknown
Jibril et al. [48]	Through inhibiting carbohydrate metabolizing enzymes	Standard lab method	70% ethanol with rind extract has shown maximum inhibition of CHO metabolizing enzyme thus helps to control BGL.
Azizi et al. [49]	Improving nitric acid synthesis	Systematic review	Low bioavailability of nitric acid synthesis is related to insulin resistance

5.10 COMPOSITIONAL CHANGES IN THE GUT MICROBIOME

This study involves the mice which were induced with diabetes type 2 by streptozocin. A specific diet was given to mice which contains 50% or 100% juice of watermelon for almost during night cycle. Laboratory assessments were done by collecting samples of fecal contents, blood, and the other various tissues of those mice. ELISA and clinical analyzer techniques are used for analysis. Data analysis of this study has showed that the blood glucose level of mice these rats were significantly decreased. It has also analyzed with the help of bioinformatics. It also changes the composition of the gut microbiome [56, 57].

5.11 WATERMELON AND OBESITY

Due to aging for one year because nonappearance of a menstrual phase clinically defined as Menopause. Accompany by much physical condition change and is obvious since women grow old. Increasing risk of exclusively atherosclerosis, CVD, inflammation, insulin resistance and increasing the adiposity is linked with Menopause. Considerably high L-arginine and plasma L-citrulline a forerunner to nitrate/ nitrite plasma concentration and NO considerably high by eating watermelon. Additional carotenoids and retinol and ascorbic acid as well as antioxidants wealthy sources with lycopene found in watermelon. By means of menopause failure of estrogen is linked with high middle adiposity and person that increase fat through the usual age development this is exacerbated [57, 58].

In a study, watermelon juice was used for four weeks by hereditarily obese type two diabetic rat and lesser free fatty acids and postprandial blood glucose were noticed. In the current study, daily consumption of watermelon in overweight men and healthy overweight eumenorrheic women convey one gram of L-arginine, L-citrulline reports meant for four weeks thumbs down alteration in insulin and fasting blood sugar level [59]. Though, Insulin resistance rise incidence in postmenopausal women, their premenopausal counterparts compared and by hormone replacement therapy (HRT) that is attenuated. We hypothesized 100% watermelon puree that convey two grams of L-arginine and L-citrulline consuming by obese postmenopausal women and overweight may better the systemic markers that connected by CVD with no changing markers of insulin resistance [60].

In the present study, the fasting L-arginine concentration mean increase was concerning 8% after consuming 0.40 g L-arginine and 1.88 g L-citrulline each day for 6 weeks. Even as avoiding plasma level of L-arginine to increase in obese menopausal women and overweight. In watermelon group be short of changes in plasma L-citrulline concentration might be explained by the blood sample collection 12 hours after consumption of watermelon for the reason that an acute 2 g oral dose of L-citrulline is relatively short the half-life in the blood of humans, concerning for 40 minutes and after consuming watermelon in as little as eight hours returns to baseline [61, 62].

Studies to time have reconstituted freeze-dried watermelon, frozen watermelon puree, watermelon juice, concentrated watermelon juice, and fresh watermelon, to examine the watermelon bioactive components and the health, performance. In the direction of our information, it is the initial learning in the direction of examines the potential physical condition reimbursement of watermelon in obese as well as overweight person by given that the learning contestant ice-covered watermelon puree.

5.12 ANTI-INFLAMMATORY ACTIVITY

Investigations on ulcerogenic properties have been carried out by an unknown methanolic strain of watermelon seed unique lesions of pyloric ligation (PL, 4h ligation) models and on immersion pressure started the wound model in the albino Wister mouse. *Citrullus lanatus* shows a reduction in gastric volume (53.55%), free acidity (57.02%) and absolute sharpness (36.53%) in the case of a PL model. The source of the methanolic seed tends to show an action related to the antiulcer fraction with a significantly higher activity at 800 mg/kg [63].

5.13 ANTIMICROBIAL ACTIVITY

An experimental demonstration of the action of anti-chlorofarmic, hexane, and ethanolic element of leaves, fruits, stems, and seeds from *Citrullus lanatus* var. *Citrullus lanatus* (CL) resistant to microscopic organisms and mold. The antimicrobial was evaluated using a plate cup and disc diffusion

/ diffusion strategy. The results of the study indicate that the chloroform content of fruits exhibits greater antibacterial activity [64]. Nigar had a special weakness in the context of seed chloroform (37 mm) and the ethanolic concept of leaves. It is also mentioned in studies about antibacterial activity. Root origin of *Citrullus lanatus*. They found that seeds differing in comparison with chosen microscopic organisms showed that additives obtained by cool maceration, Soxhlet extraction, as the use of chloroform and methanol could have antibacterial agents especially against *Staphylococcus* sp. and *P. aeruginosa* [65].

5.14 HEPATOPROTECTIVE ACTIVITY

The pathological study on the damaged liver cells in rats evaluating serum hepatic enzyme levels, by giving seeds in carbon tetrachloride of citrus lanatus to check its hepatoprotective activity. Rats were given seed oil of Citrus lanatus; CLSO (125 mg) by comparing it with silymarin oil (100 mg/kg) used as a standard through oral feeding for 10 days. After contrasting the $CC1_4$ promoted liver harm treated groups with serum compared with commonly used medicines it is observed that AST, ALP, and ALT levels are reduced remarkably. Thus, hepatoprotective attribute of *Citrullus lanatus* seed oil is disclosed by seeing its results on liver tissues [53].

5.15 WATERMELON PRESERVATION

Due to its high-water activity that is in between range 0.97–0.99 and high pH (5.2–6.7) watermelon is perishable in nature. As gram positive bacteria are susceptive to pathogenic organisms in lower acidic environment because they are delicate to that environment. Due to its low acidic nature, *Citrullus lanatus* can be a dangerous dormant food in nature. That is why it is very important to conserve juice of watermelon. On consumers demand, many alternate products are formed from juice of watermelon to be used as healthy, nutritious, and favorable food; to increase its use and its shelf life as shown in Table 5.4 [53]. Processing methods having low thermal effect should use to lower the risk of degradations quality parameters of watermelon are easily depleted by heat treatment.

TABLE 5.4 Outcomes on Antioxidants of Watermelon After Processing

Processing Methods	Outcomes on Antioxidants
Electric field with high intensity pulse [1]	Lowers the capacity of antioxidants and vitamin C in watermelon juice and retention of lycopene increases.
Maintaining of lycopene by pulsed electric field [2]	Total antioxidants and polyphenols capacity is reduced and drying loss of vitamin C and lycopene
Thermosonication [13]	Scavenging activity of free radicals is not that much effected, lycopene is increased as phenolic content is not affected and retention of ascorbic acid
Sterilization and pasteurization [14]	Lowering the capacity of lycopene, polyphenols, and vitamin C
Processing in high pressure [19]	Lycopene increases and conserve vitamins and citrulline

5.16 SUMMARY

The modern suggestion about snack that fulfills per day energy should be full of natural antioxidants in our diet such as fruits as they are rich in antioxidants. By consuming watermelon as a snack or fruit, humans can get a very high amount of nutrients as they are a good source of components that are nutritious. Watermelon is a classic summer fruit in functional food category enriched with phytochemicals. Phytochemicals having analgesic, laxative, antigiardial, gastroprotective, antioxidant, hepatoprotective, antifungal, antimicrobial, antibacterial, antiulcer, and anti-inflammatory properties. Moreover, low in sodium, fat free having 40 calories per cup. Components of medicinal properties are also present in it. That is why it could be of great importance in health care services. Essential minerals and amino acids can be obtained from watermelon rind and seeds. On the basis of these attributes of watermelon and its products, one should consume watermelon for health benefits. To prevent yourself future ailments and to spend a healthy life, per day 1 cup/152 g of *Citrullus lanatus* juice snacking could be very beneficial viewing nutritional benefits of *Citrullus lanatus*. Thus, we can conclude that due to various bioactivities of *Citrullus lanatus* present in nature, it is of great importance and beneficial as compared to conventional therapies.

KEYWORDS

- **antioxidants**
- **cardiovascular disease**
- **colorectal cancer**
- **health benefits**
- **micronutrients**
- **watermelon**

REFERENCES

1. Rawson, A., Tiwari, B. K., Patras, A., Brunton, N., Brennan, C., Cullen, P. J., & O'donnell, C., (2015). Effect of thermosonication on bioactive compounds in watermelon juice. *Food Research International, 44*(5), 1168–1173.
2. Ijah, U., Ayodele, H. S., & Aransiola, S. A., (2015). Microbiological and some sensory attributes of watermelon juice and watermelon-orange juice mix. *Journal of Food Resource Science, 4*(3), 49–61.
3. Naz, A., Butt, M. S., Sultan, M. T., Qayyum, M. N., & Niaz, R. S., (2016). Watermelon lycopene and allied health claims. *EXCLI Journal, 13*, 650–660.
4. Romdhane, M. B., Haddar, A., Ghazala, I., Jeddou, K. B., Helbert, C. B., & Ellouz-Chaabouni, S., (2017). Optimization of polysaccharides extraction from watermelon rinds: Structure, functional and biological activities. *Food Chemistry, 2*(16), 355–364.
5. Adunola, A., Chidimma, A. L., Olatunde, D. S., & Peter, O. A., (2015). Antibacterial activity of watermelon (*Citrullus lanatus*) *seed* against selected microorganisms. *African Journal of Biotechnology, 14*(14), 1224–1229.
6. Perkins-Veazie, P., (2017). Cucurbits, watermelon, and benefits to human health. In: *IV International Symposium on Cucurbits* (Vol. 21, No. 871, pp. 25–32).
7. Verma, R., & Tomar, M., (2017). Watermelon: A valuable horticultural crop with nutritional benefits. *Popular Kheti, 5*, 5–9.
8. Erhirhie, E., & Ekene, N., (2016). Medicinal values on *Citrullus lanatus* (watermelon): Pharmacological review. *International Journal of Research in Pharmaceutical and Biomedical Sciences, 4*(4), 1305–1312.
9. Bianchi, G., Rizzolo, A., Grassi, M., Provenzi, L., & Scalzo, R. L., (2018). External maturity indicators, carotenoid and sugar compositions and volatile patterns in 'cuoredolce®'and 'rugby' mini-watermelon (*Citrullus lanatus (Thunb)* Matsumura & Nakai) varieties in relation of ripening degree at harvest. *Postharvest Biology and Technology, 136*, 1–11.
10. Jumde, A. D., Shukla, R. N., & Gousoddin, (2015). Development and chemical analysis of watermelon blends with beetroot juice during storage. *International J. of Science, Engineering and Technology, 3*(4), 960–964.

11. Choudhary, B., Haldhar, S., Maheshwari, S., Bhargava, R., & Sharma, S., (2015). *Phytochemicals and Antioxidants in Watermelon (Citrullus lanatus)* Genotypes *Under Hot Arid Region, 85*(3), 414–417.
12. Sa'id, M. A., (2018). Study in the variability of some nutrient contents of watermelon (*Citrullus lanatus*) before and after ripening consumed within Kano Metropolis, Nigeria. *International Journal of Science and Research (IJSR), 3*(5), 1365–1368.
13. Sivudu, S. N., Umamahesh, K., & Reddy, O., (2017). A comparative study on probiotication of mixed watermelon and tomato juice by using probiotic strains of *Lactobacilli. Int. J. Curr. Microbiol. Appl. Sci., 3*(11), 977–984.
14. Campbell, I., (2017). Macronutrients, minerals, vitamins and energy. *Anaesthesia & Intensive Care Medicine, 18*(3), 141–146.
15. Alam, M., Hoque, M., Morshed, S., Akter, F., & Sharmin, K., (2015). Evaluation of watermelon (*Citrullus lanatus)* juice preserved with chemical preservatives at refrigeration temperature. *Journal of Scientific Research, 5*(2), 407–414.
16. United States National Nutrients database. https://ndb.nal.usda.gov/ndb/foods/show/2393 (accessed on 17 September 2021).
17. Wong, A., Chernykh, O., & Figueroa, A., (2016). Chronic l-citrulline supplementation improves cardiac sympathovagal balance in obese postmenopausal women: A preliminary report. *Autonomic Neuroscience, 1*(198), 50–53.
18. Marie, I., Ducrotte, P., Antonietti, M., Herve, S., & Levesque, H., (2016). Watermelon stomach in systemic sclerosis: Its incidence and management. *Alimentary Pharmacology & Therapeutics, 28*(4), 412–421.
19. Pacier, C., & Martirosyan, D. M., (2017). Vitamin C: Optimal dosages, supplementation and use in disease prevention. *Functional Foods in Health and Disease, 5*(3), 89–107.
20. Abourashed, E. A., (2013). Bioavailability of plant-derived antioxidants. *Antioxidants, 2*(4), 309–325.
21. Dia, M., Wehner, T. C., Perkins-Veazie, P., Hassell, R., Price, D. S., Boyhan, G. E., Olson, S. M., et al., (2016). Stability of fruit quality traits in diverse watermelon cultivars tested in multiple environments. *Horticulture Research, 3*(1), 1–1.
22. Vacca, R. A., Valenti, D., Caccamese, S., Daglia, M., Braidy, N., & Nabavi, S. M., (2016). Plant polyphenols as natural drugs for the management of down syndrome and related disorders. *Neuroscience &Biobehavioral Reviews, 71*, 865–877.
23. Soteriou, G. A., Kyriacou, M. C., Siomos, A. S., & Gerasopoulos, D., (2017). Evolution of watermelon fruit physicochemical and phytochemical composition during ripening as affected by grafting. *Food Chemistry, 165*, 282–289.
24. Doll, S., & Ricou, B., (2016). Severe vitamin C deficiency in a critically ill adult: A case report. *European Journal of Clinical Nutrition, 67*(8), 881, 882.
25. Maoto, M. M., Beswa, D., & Jideani, A. I., (2019). Watermelon as a potential fruit snack. *International Journal of Food Properties, 22*(1), 355–370.
26. Petyaev, I. M., (2016). Lycopene deficiency in aging and cardiovascular disease. *Oxid. Med. Cellular Longevity*.
27. Akhtar, S., Ahmed, A., Randhawa, M. A., Atukorala, S., Arlappa, N., Ismail, T., & Ali, Z., (2018). Prevalence of vitamin A deficiency in South Asia: Causes, outcomes, and possible remedies. *Journal of Health, Population, and Nutrition, 31*(4), 413.

28. Alipanah, N., Varadhan, R., Sun, K., Ferrucci, L., Fried, L. P., & Semba, R. D., (2019). Low serum carotenoids are associated with a decline in walking speed in older women. *JNHA-The Journal of Nutrition, Health and Aging, 13*(3), 170–175.
29. Mariani, S., Lionetto, L., Cavallari, M., Tubaro, A., Rasio, D. D., Nunzio, C., Hong, G. M., et al., (2016). Low prostate concentration of lycopene is associated with development of prostate cancer in patients with high-grade prostatic intraepithelial neoplasia. *International Journal of Molecular Sciences, 15*(1), 1433–1440.
30. Jacques, P. F., Lyass, A., Massaro, J. M., Vasan, R. S., & D'Agostino, Sr. R. B., (2016). Relationship of lycopene intake and consumption of tomato products to incident CVD. *British Journal of Nutrition, 110*(3), 545–551.
31. BabikerAli, A. A., (2016). Effect of watermelon juice consumption on brachial blood pressure among Sudanese hypertensive patients. *Gezira Journal of Health Sciences, 1, 12*(2).
32. Xinli, L. I., & Jiuhong, X. U., (2015). Dietary and circulating lycopene and stroke risk: A meta-analysis of prospective studies. *Scientific Reports, 4*, 5031.
33. Tabiri, B., Agbenorhevi, J. K., Wireko-Manu, F. D., & Ompouma, E. I., (2016). Watermelon seeds as food: Nutrient composition. *Phytochemicals and Antioxidant Activity, 5*(2), 139–144.
34. Paris, H. S., (2015). Origin and emergence of the sweet dessert watermelon, *Citrullus lanatus. Annals of Botany, 116*(2), 133–148.
35. Glenn, K., Klarich, D. S., Kalaba, M., Figueroa, A., Hooshmand, S., Kern, M., & Hong, M. Y., (2018). Effects of watermelon powder and l-arginine supplementation on azoxymethane-induced colon carcinogenesis in rats. *Nutrition and Cancer, 70*(6), 938–945.
36. Dorgan, J. F., Sowell, A., Swanson, C. A., Potischman, N., Miller, R., Schussler, N., & Stephenson, H. E., (2017). Relationships of serum carotenoids, retinol, α-tocopherol, and selenium with breast cancer risk: Results from a prospective study in Columbia, Missouri (United States). *Cancer Causes & Control, 9*(1), 89–97.
37. London, S. J., Stein, E. A., Henderson, I. C., Stampfer, M. J., Wood, W. C., Remine, S., Dmochowski, J. R., et al., (2016). Carotenoids, retinol, and vitamin E and risk of proliferative benign breast disease and breast cancer. *Cancer Causes & Control, 3*(6), 503–512.
38. Gupta, A., Vashist, N., Bajaj, S., Gahlot, V., Yadav, S., Maratha, S., & Rajput, A., (2019). *Can Food Supplements Combat Cancer: An Insight, 9*(1), 32–41.
39. Levy, J., Walfisch, S., Atzmon, A., Hirsch, K., Khanin, M., Linnewiel, K., Morag, Y., et al., (2017). The role of tomato lycopene in cancer prevention. In: *Vegetables, Whole Grains, and Their Derivatives in Cancer Prevention* (pp. 47–66). Springer, Dordrecht.
40. Sun, T., Huang, K., Xu, H., & Ying, Y., (2018). Research advances in nondestructive determination of internal quality in watermelon/melon: A review. *Journal of Food Engineering, 100*(4), 569–577.
41. Barkur, S., Bankapur, A., Chidangil, & Mathur, D., (2017). Effect of infrared light on live blood cells: Role of β-carotene. *Journal of Photochemistry and Photobiology B: Biology, 171*, 104–116.

42. Shao, P., Qiu, Q., Xia, J., Zhu, Y., & Sun, P., (2017). Chemical stability and *in vitro* release properties of β-carotene in emulsions stabilized by ulvafasciata polysaccharide. *International Journal of Biological Macromolecules, 102*, 225–231.
43. World Health Organization. (2016). *Increasing Fruits and Vegetables Consumption to Reduce the Risk of Non-Communicable Diseases*.
44. Rodríguez-Roque, M. J., De Ancos, B., Sánchez-Moreno, C., Cano, M. P., Elez-Martínez, P., & Martín-Belloso, O., (2015). Impact of food matrix and processing on the *in vitro* bio-accessibility of vitamin C, phenolic compounds, and hydrophilic antioxidant activity from fruit juice-based beverages. *Journal of Functional Foods, 14*, 33–43.
45. Figueroa, A., Wong, A., Jaime, S. J., & Gonzales, J. U., (2017). Influence of L-citrulline and watermelon supplementation on vascular function and exercise performance. *Current Opinion in Clinical Nutrition and Metabolic Care, 20*(1), 92–98.
46. Okwechime, I. O., Roberson, S., & Odoi, A., (2015). Prevalence and predictors of pre-diabetes and diabetes among adults 18 years or older in Florida: A multinomial logistic modeling approach. *PLoS One, 10*(12).
47. Chertman, L. S., Neuman, D., & Vendrame, F., (2020). Update on diabetes medical management: Epidemiology and treatment. In: *Behavioral Diabetes* (pp. 261–275). Springer, Cham.
48. Jibril, M. M., Hamid, A. A., Ghazali, H. M., Dek, M. S., Ramli, N. S., Jaafar, A. H., Karrupan, J., & Mohammed, A. S., (2019). Antidiabetic antioxidant and phytochemical profile of yellow-fleshed seeded watermelon (*Citrullus Lanatus*) extracts. *J. Food Nut, 7*. 82–95.
49. Azizi, S., Mahdavi, R., Vaghef-Mehrabany, E., Maleki, V., Karamzad, N., & Ebrahimi-Mameghani, M., (2020). Potential roles of citrulline and watermelon extract on metabolic and inflammatory variables in diabetes mellitus, current evidence and future directions: A systematic review. *Clinical and Experimental Pharmacology and Physiology, 47*(2), 187–198.
50. Wong, S. Y., Wu, L., Lu, P., Ojo, B., Tang, M., Clarke, S., Keirns, B., et al., (2019). Drinking watermelon juice shift the gut microbiome in diabetic mice (P20-025-19). *Current Developments in Nutrition, 3*(Supplement_1), 040–P20.
51. Khoo, H. E., Azlan, A., Kong, K. W., & Ismail, A., (2016). Phytochemicals and medicinal properties of indigenous tropical fruits with potential for commercial development. *Evidence-Based Complementary and Alternative Medicine*.
52. Tchuenchieu, A., Sylvain, S. K., Pop, C., Jean-Justin, E. N., Mudura, E., Etoa, F. X., & Rotar, A., (2018). Low thermal inactivation of *Escherichia coli* ATCC 25922 in pineapple, orange and watermelon juices: Effect of a prior acid-adaptation and of carvacrol supplementation. *Journal of Food Safety, 38*(1), e12415.
53. Madhavi, P., Kamala, V., & Habibur, R., (2016). Hepatoprotective activity of *Citrullus lanatus* seed oil on CCl_4 induced liver damage in rats. *Scholars Academic Journal of Pharmacy, 1*(1), 30–33.
54. Massa, N. M. L., Silva, A. S., Toscano, L. T., Silva, J. D. A. G. R., Persuhn, D. C., & Gonçalves, M. D. C. R., (2016). Watermelon extract reduces blood pressure but does not change sympathovagal balance in prehypertensive and hypertensive subjects. *Blood Pressure, 25*(4), 244–248.

55. Witteman, J. C., Grobbee, D. E., Kok, F. J., Hofman, A., & Valkenburg, H. A., (2017). Increased risk of atherosclerosis in women after the menopause. *BMJ, 298*(6674), 642–644.
56. Keshavarz, H., (2018). *Adipose-Derived Molecular Amplification of Glucose Trafficking in the Bloodstream: A Potential New Role for Leptin*. Michigan State University.
57. Bailey, S. J., Blackwell, J. R., Williams, E., Vanhatalo, A., Wylie, L. J., Winyard, P. G., & Jones, A. M., (2016). Two weeks of watermelon juice supplementation improves nitric oxide bioavailability but not endurance exercise performance in humans. *Nitric Oxide, 59*, 10–20.
58. Wu, G., Collins, J. K., Perkins-Veazie, P., Siddiq, M., Dolan, K. D., Kelly, K. A., Heaps, C. L., & Meininger, C. J., (2007). Dietary supplementation with watermelon pomace juice enhances arginine availability and ameliorates the metabolic syndrome in Zucker diabetic fatty rats. *The Journal of Nutrition, 137*(12), 2680–2685.
59. Lum, T., Connolly, M., Marx, A., Beidler, J., Hooshmand, S., Kern, M., Liu, C., & Hong, M. Y., (2019). Effects of fresh watermelon consumption on the acute satiety response and cardiometabolic risk factors in overweight and obese adults. *Nutrients, 11*(3), 595.
60. Al –Safi, Z. A., & Polotsky, A. J., (2015). Obesity and menopause. *Best Pract. Res. Clin. Obstet. Gynaecol., 29*, 548–553.
61. Shanely, R. A., Nieman, D. C., Perkins-Veazie, P., Henson, D. A., Meaney, M. P., Knab, A. M., & Cialdell-Kam, L., (2016). Comparison of watermelon and carbohydrate beverage on exercise-induced alterations in systemic inflammation, immune dysfunction, and plasma antioxidant capacity. *Nutrients, 8*(8), 518.
62. Figueroa, A., Sanchez -Gonzalez, M. A., Perkins -Veazie, P. M., & Arjmandi, B. H., (2018). Effects of watermelon supplementation on aortic blood pressure and wave reflection in individuals with prehypertension: A pilot study. *American J. .Hypertens, 24*, 40–44.
63. Alok, B., Rajeev, K., Vivek, D., & Niyaz, A., (2016). Evaluation of anti-ulcer activity of *Citrullus lanatus* seed extract in Wistar albino rats. *International Journal of Pharmacy and Pharmaceutical Sciences, 4*(5), 135–139.
64. Loiy, E., & Ahmed, H., (2018). In-vitro Antimicrobial activities of chloroformic, hexane and ethanolic extracts of *Citrullus lanatus* var. citroides. *Journal of Medicinal Plants Research, 5*(8), 1338–1344.
65. Adelani, A. T. A., Ajiba, L. C., Dahunsi, S. O., & Oluyori, A. P., (2017). Antibacterial activity of watermelon (*Citrullus lanatus*) seed against selected microorganisms. *African Journal of Biotechnology, 14*(14), 1224–1229.

CHAPTER 6

ANTHOCYANIN EXTRACTS OF BLUEBERRIES FOR PREVENTION OF CARDIOVASCULAR DISORDERS

TABUSSAM TUFAIL, AIMAN IJAZ, MUHAMMAD UMAIR ARSHAD, SYED AMIR GILANI, SHAHID BASHIR, and HAFIZ ANSAR RASUL SULERIA

ABSTRACT

Berries are the best-known source for polyphenols especially, micronutrients, fiber, and anthocyanins (ACNs). Different varieties of berries are available and used worldwide. It is used as a whole fruit, extract, and powder form. Berries are mostly composed of phenolic compounds, which include flavonols and flavonoids. Some of the phenolic acids and hydrolyzable tannins (ellagitannins) acts as a major BAC in berries. ACNs are the phytopigments belong to one type of flavonoids. They have health benefits on organ and organ systems. ACNs are used as heterosides and are not essential nutrients for the human body but as a functional food that play an important role in different metabolic disorders. In this era cardiovascular diseases (CVDs) are the main cause of morbidity and mortality due to unhealthy life style. The current study focuses on the health benefits of ACNs as an antioxidant in CVDs and related health consequences. Their health advantages are due to their antioxidant, anti-inflammatory properties. ACNs regulate some signaling pathways that contribute to the development and progression of CVDs. There are many natural antioxidants which are present in berries including: vitamin C and E, and Micronutrients.

6.1 INTRODUCTION

Fruit and vegetable consumption is highly recommended in our diet throughout the world. Consumption of fruits like berries can be helpful in combating many diseases as they contain a rich amount of phytochemicals and nutrients. Berries are the highest consumed fruit in human diet and they are very delicious and effective against diseases [1]. Berries are consumed in different forms like fresh, frozen, processed as well as in derived form. They come in the form of yogurt, jams, jellies, and liquid beverages [2]. There are many varieties of berries which are mostly used in diet of humans either in fresh or processed form including: Black raspberry (*Rubus occidentalis*), red raspberry (*Rubus idaeus*), blackberry (*Rubu*s), red raspberry (*Rubus idaeus*), blueberry (*Vaccinium corymbosum*), strawberry (*Fragaria ananassa*) and cranberry (*Vaccinium macrocarpon*) [3]. One of the developing trends in the use of extracts obtained from berries as ingredients for supplements and other functional foods which can also be mixed with other fruits, herbal extracts, and vegetables. Some of the berries (gooseberry, blackberry, and raspberry) extracts are used as free radical inhibitors [4]. There are many benefits of berries in health due to the presence of vitamins, fibers, antioxidants, and minerals, which are in high levels [5]. Anthocyanins (ACNs) are the subclass of plant polyphenols and are shown in Table 6.1.

TABLE 6.1 Subclasses of Polyphenols

Polyphenols	Polyphenols
Hydroxybenzoic acids	Flavonols
Hydroxycinnamic acids	Flavons
Anthocyanins,	Isoflavones
Proanthocyanidins	Stilbenes

It has been shown that various phytochemicals level and antioxidant capacities are present in and throughout the small fruit genus [6]. Furthermore, gathering evidence shows that the genotype is effective on the concentrations of bioactive compounds (BACs) in fruits. Berry extractions used in different supplemental products due to their health advantages properties. Many studies indicate that they have antimutagenic, antiaging, antioxidant, anti-inflammatory properties [7]. Berry bioactive components provide anticancer effect through a variety of mechanisms like they induce metabolic enzymes, change the gene expression, program cell death, and modulate different signaling pathways [8]. Phenolic compounds such as Gallic acid and ellagic acid present in the berries have anticancer effect due to their free radical scavenging activity and apoptosis [9]. Phenolic compounds have a close relationship with human health. Different studies conducted to understand the biological availability as well as metabolism of the phenolic compounds as shown in Figure 6.1.

FIGURE 6.1 Different varieties of berries.

There are reports which point towards the reduced risks of CVD (cardiovascular disease) by the use of fruits and vegetables [10]. There was an objective which points towards the enhanced use of fruits for every person of age 2 years or older up to two servings every day or 75% use every day proposed as National health objective in 2010. However, this percentage has decreased so much that only 32% of adults and only 13% of kids are one who are consuming this much amount of fruits [11]. There is a report which shows that the consumption of fruits is expected to increase by 24–27% between the years of 2000 and 2020 in the US. Consumption of berries increases in per capita in the US market. Black raspberry, red raspberry, cranberry, blueberry, strawberry is most commonly use but chokeberry, mulberry, and black currant are consumed in small amounts [12].

Berries are the fruits with a very high amount of fiber and moisture in them with the least amount of calories. There are many natural antioxidants which are present in berries including: vitamin C and E, and Micronutrients (calcium, alpha, and beta carotene, folic acid, lutein, and selenium). The phytochemicals which are present in berries are polyphenols with enhanced amount of flavonoids like ellagitannins and ACNs. There are almost 400 different ACNs which have already been found. The ACNs are mostly present in fruit skins [13]. There are other fruits as well which contain a good amount of ACNs in them including: cherries and strawberries. The amount of anthocyanin is directly proportional to the intensity of fruit color and it can be from 2 to 4 g/kg which increases with the process of fruit ripening. There are reports which show that the people of the US consume almost 12.5 to 215 mg of ACNs every day [14]. There are reports which also show that ACNs from berries have a low bioavailability and they are removed out of them within 2 to 8 hours after the consumption every time [15]. Different harvesting process affects the polyphones like anthocyanin and micronutrient contents present in the berries as well as their bioactivities. It increases the risk of CVDs [16, 17].

Wild blueberries are a rich source of polyphenols such as ACNs and phenolic acids present in fruit peels and cell voids. The sugar present in them provides stability and water solubility [18]. They not only provide benefits to human health but also protect the plants from environmental stress, i.e., ultraviolet (UV) light, low temperature. About 500 different types of ACNs came to know. Most common present in fruits and vegetables are (pelargonidine, cyanidine, delfinidine, petunidine, peonidine, and malvidine). Different studies indicate that 180–215 mg/day intake of ACNs seems to be effective against different metabolic disorders [19]. They regulate motor functions, blood pressure as well as inflammation [20, 21]. ACNs may also affect lipid accumulation in macrophages [22, 23]. Indeed, blueberry extract can reduce lipid accumulation in cultured macrophages [24]. Supplements of blueberries modulate the activities of antioxidant enzymes; prevent the progression of CVDs [25]. In addition, observations from various epidemiological and clinical studies have shown that polyphenol-rich foods have protective effects on CVDs in regulating cell cholesterol metabolism [26]. In addition, most of these compounds are ingested after rapid separation into phenolic degradation products which are then metabolized [27].

This chapter explores benefits of ACNs flavonoids present in the berries mainly on CVD and their treatment as they are proved to be a good source of combating CVD and many other disorders. CVD are a major issue in the current period, and treating such disorders is a big challenge for the scientists. Plant-based medications have been used for centuries and are a good source of treating many dangerous diseases as well. Berries are a rich source of anthocyanin, vitamins, and polyphenols which are very effective against CVD [28].

6.2 CHEMICAL COMPOSITION OF BERRIES AND BAC

There are different types of berries and their chemical composition also varies with the variety as well as their location of growth, nutrition, harvesting time, and condition of environment. Thus, the quality and content of fruit highly varies from other fruits of the same type. Berries have a very low-fat content in them and there is a high amount of organic acids and dietary fibers [29, 30]. Such compounds are used by industries as ingredients for functional foods as well. Phenolic compounds occur in either free or conjugated forms along with acids, biomolecules, and acids which can be water-soluble as well (Flavonoids, quinones, and phenolic acids) or water-insoluble compounds. The BAC content is berries is mostly composed of phenolic compounds which include flavonols and flavonoids. Some of the phenolic acids and hydrolysable tannins (ellagitannins) acts as a major BAC in berries [31].

All these compounds are the major reason for the health-related benefits of berries as well as their antioxidant properties. Ascorbic acid which is also present in berries could also be a very effective antioxidant as well. This acid is a very important water-soluble vitamin and it has great reducing properties and is highly known for antioxidant activity as it is involved in the neutralization of free radicals and other ROS that are involved in tissue damage and diseases as well which are formed during the metabolism of cell. It is also considered to be an indicator of nutritional quality during processing and storage, as it is known that vitamin C can be retained in foods with minimal changes and losses if other nutrients are retained well. These losses are dependent on the varieties of berries. [45]. However, vitamin C is a strong antioxidant agent with strong reducing power. Many studies have assessed that only a small amount (up to 10%) contributes to the total antioxidant capacity of fruits [32, 33].

6.2.1 *BAC IN BLUEBERRIES*

There are differences in the content of ACNs, TAC (total antioxidant contents), and TPC (total phenolic contents) in different species of blueberries. The relevant BAC in blueberries is mentioned in Table 6.2.

TABLE 6.2 Major Phenolic Compounds in Berries

Berry	Major Phenolic Compounds
Blueberry	• **Flavonols:** o **Myricetin Glycosides:** Myricetin-3-glucoside, Myricetin-3-rhamnoside) Quercetin glycosides (Quercetin-3-galactoside, Quercetin-3-glucoside, Quercetin-3-rutinoside. • **Anthocyanins:** o **Cyanidin Glycosides:** Cyanidin-3-galactoside, Cyanidin-3-glucoside, Cyanidin-3-arabinoside. o **Delphinidin Glycosides:** Delphinidin-3-galactoside, Delphinidin-3-arabinoside, Delphinidin-3-glucoside. o **Malvidin Glycosides:** Malvidin-3-galactoside, Malvidin-3-arabinoside, Malvidin-3-glucoside. o **Petunidin Glycosides:** Petunidin-3-galactoside, Petunidin-3-arabinoside, Petunidin-3-acetylglucoside. o **Peonidin Glycosides:** Peonidin-3-galactoside, Peonidin-3-arabinoside.

The total phenolic content values in blueberries are upwards of 10-times higher or lower depending on the way which is used for the analysis purpose [34]. The phenolic content of berries varies with the location, environment, and harvesting method. The maturity stages increased anthocyanin content [35]. The blueberries were boiled at 85°C for three minutes while the anthocyanin content was allowed to grow by about 7% [36]. However, the anthocyanin content of heat-treated blueberries was reduced by about 30%, osmo dehydrated or air-dried at 70°C.

6.3 ANTIOXIDANT CAPACITY OF BLUEBERRIES

There are many factors which are involved in the antioxidant capability of blueberries. These effects are based on phytochemical complexes out of which the important ones are ACNs, chlorogenic acid, other flavonoids and procyanidins [36]. The most important antioxidant agent is anthocyanin which is the almost 84% of TAC Ascorbic acid which is also found in relative amounts in blueberries is only responsible for almost 10% of antioxidant activity [37]. The best antioxidant activity of blueberries is in early stages of development rather than the ripen stage as compared to other berries. The reason for this is due to the presence of higher levels of flavonols and hydroxycinnamic acids prior to ripening.

This lower antioxidant activity of ripened blueberries is due to lower antioxidant capacity of ACNs as compared to other phenolic compounds like flavonols. The difference in the total phenol content of cultivars and mature phases are related to the obtained changes of the antioxidant activity. The effect of every phenol agent may vary as compared to the combined antioxidant activity [38]. The TAC effect of blueberry processing was not seen for drying with osmotic treatment or blanching [39]. There was an increase in the antioxidant capacity of blueberry fruits after freezing for up to 3 months and then it started to reduce at the end of 6 months [40].

6.4 CARDIOVASCULAR DISORDERS AND RISK FACTORS

6.4.1 DIABETES MELLITUS

It is one of cardiovascular disorder which increases the rate of mortality of the patient by causing many abnormalities in different processes of metabolism, enhanced chance of cardiac arrhythmia, platelet aggregation and endothelial dysfunction. One of a study was based on 120 patients with T2D (type 2 diabetes) out of which 46 of the patients had microalbuminuria and this study showed that deficiency of thiamine was high in 98% of patients with microalbuminuria and 100% without microalbuminuria [41]. A low level of vitamin B1 is seen in the type 1 diabetes mellitus patients which is inversely correlated to blood glucose level. Studies showed that 150 mg daily dosage of B1 for 30 days reduce plasma fasting glucose level [42].

6.4.2 OBESITY

Obesity is one of the disorders which is the reason for multiple diseases including CVDs and T2D as well [43]. One of the studies showed the patients for bariatric surgeries which pointed towards the 15.5% of thiamine deficiency in 303 patients with a 60 kg/m^2 body mass index (BMI) and 29% deficiency in 379 patients with 51.8 kg/m^2 BMI [44]. One of the recent studies on 400 patients with 35 kg/m^2 BMI out of which 16.5% of the patients were the ones who went for thiamine deficiency test and the results showed negative correlation of BMI with neurological and cardiac manifestations [45]. Thiamine deficiency after bariatric surgery has also been reported. In addition to prolonged postoperative vomiting and poor oral intake, altered bowel ecology can lead to SIBO (small intestine bacterial overgrowth), which induces thiamine deficiency [46].

6.4.3 MYOCARDIAL INFARCTION AND CONDUCTION DEFECTS

Myocardial infarction is one the leading death reason throughout the world [47]. One of the animal-based studies showed that in thiamine deficient rats the electrocardiograms showed T-wave changes, bradycardia, and increased ST-segment and such changes were not seen when thiamine was provided to these [48]. The prevalence of thiamine deficiency and thiamine uptake may play a preventive role in MI (myocardial infarction) and increase conduction disturbances in cardiac energy metabolism and minimize necrosis in MI cases.

6.4.3.1 DEPRESSION

CVDs include one of the major factor depressions and there is a delayed recovery as well. One of a Chinese study was based on 1,587 people of ages between 50 and 70 years and their low levels of thiamine was linked with symptoms of depression [49]. There was another study that showed that 74 patients who were malnourished and were in psychic units showed that their deficiency of thiamine was also linked with the symptoms of depression. Cause of Depression is still unknown there is no clear evidence that will justify that low thiamin level effects on depression but metabolic active may be the reasons of depression There is still no clear evidence which proves that the deficiency of thiamine. The malfunction in the neurotransmitter mechanisms is the reason which leads to the cause of depression [50].

6.5 EPIDEMIOLOGICAL OBSERVATIONS: BERRIES IN CARDIOVASCULAR HEALTH

The persuasive substantiates of cardioprotective effect by consuming a large amount of fruits vegetables and micronutrients as well as various phytochemicals is best example of nutritional epidemiology [51]. Dietary regime evidences from dietary models of 52 various countries directed towards a strong inverse association between a prudent diet and a high-risk acute myocardial infarction in fruits and vegetables [52]. Around 2,757-US adults are assessed for selected nutrients and food group entries diagnosed with T2D, 50% of cases have a minimum consumption of fruits and vegetables showed known risk factor for CVD [53, 54]. A 24-hour reminding analyzes, the National Health and Nutrition Examination Survey (NHANES) and this comparison provided that only 40% of Americans consume one or more portions of fruit and vegetables per day in a study conducted between 1999 and 2000 [55]. Such figures indicate that there is a substantial difference between the real amount of consumption of fruit and vegetables and the average number of portions for the US population.

Numerous studies investigated the relationship between strawberry or mulberry flavonoids (anthocyanins) and cardiovascular health. Strawberry intake in fresh, frozen, or canned strawberries was defined as "never" up to 6 servings per day. Basal strawberry consumption research showed that only 7.7% of subjects ate more than two portions of strawberries a week, and 42% recorded 11–3 portions a month. Those who ate more strawberries had a downward trend for CVD 1 (P = 0.06) over 8a follow-up of approximately 11 years. These observational data suggest that fruiting flavonoids have a potential anti-inflammatory role, which may contribute to the reduction of CVD risk [56].

6.5.1 BERRIES AND CARDIOVASCULAR HEALTH: INTERVENTION STUDIES

The effects of acai berries, blackcurrants, blueberries, strawberries, cranberries, raspberries, strawberries, and wolfberries has been investigated in healthy persons who are at risk of CVDs by involving extensive intervention [57]. The most renowned consequences of clinical research postulated an enhancement in antioxidant capacity of plasma or urine, and reduction in low density lipoprotein (LDL). The reduction in plasma glucose or total cholesterol is followed by increased high density lipoprotein (HDL)-cholesterol in berries intervene cases. Subsequently, high concentrations of plasma glucose, lipid or lipid oxidation are contributed toward coronary artery disease (CAD) [58], these figures expound the hypothetical position of edible fruits in refining their effects and minimize these risk factors. Over 20 studies, only nine are scrutinized for prandial status measurements that exhibit drastically reduction in postprandial oxidative stress, particularly lipid peroxidation, in strawberry consumption [58]. Hence, diet-based therapy of fruits has beneficial effects against oxidative damages associated with CVDs [59].

Moreover, blueberry, and black currant extracts, chokeberry juice, cranberry extracts and freeze-dried strawberries are some specific fruits that are effective for people 10 with metabolic risk factor including type 1 diabetes and T2D, dyslipidemia or metabolic syndrome and have positive impact plasma glucose or lipid methods [60]. In diet-based therapy of fruit 4 to 12 weeks, conventional fruit products or purified anthocyanin extracts are used and indicating that both delivery modes were effective. As a consequence, an increase in plasma antioxidant capacity and reduction in peroxidation of lipids is associated with the consumption of berries has been examined in smokers with a high risk of developing CVD [61]. The data depicted that consecutive research on cranberry juice supplementation is an effective remedy for advanced hypertension or hypertension treatment. Moreover, massive clinical investigation did not predict significant effect in the intervention of fruits to any important biomarker, apart from reduction in adhesion molecule is observed following cranberry juice supplementation in healthy volunteers [62]. This will increase the demand for extensive research or clinical studies to check the effect of combination of fruits on inflammatory biomarkers such as C-reactive protein (CRP) or interleukins (ILs) in persons with diabetes mellitus like syndromes under adhesion molecules or via-inflammatory conditions [63].

6.6 MECHANISMS: BERRIES, ENDOTHELIAL FUNCTION, AND ATHEROSCLEROSIS

Oxidative stress mean imbalance between antioxidant enzymes and reactive oxygen species (ROS). Free radicals production directly contributes to the progression of heart diseases [64, 65]. Various evidences suggest that berry ACNs significantly reduce free radical production due to the scavenging activity and play a role in vivo amplification of CVD in cellular and animal models. Youdi metal reported elderberry ACNs also play an important role in suppressing the cytotoxicity caused by chemical inducers of oxidative stress [66]. It has been shown that ACNs derived from blackberry extract protect against oxidative damage caused by peroxynitis in human umbilical vein endothelial cells (HUVEC) [67].

Mulberry ACNs also exhibited antioxidant and anti-atherogenic effects by lowering the oxidation of LDL as well as the development of foam cells of believe in-vitro model of atherosclerosis [68]. In the United States, ACNs derived from commonly consumed fruits such as blueberries have been reported to stimulate the response of inflammatory mediators in human microvascular endothelial cells of TNF-α [69]. In an 8-week study, De Furiaetal. Male C57B1/6j mice showed that their inflammatory gene expression was weakened and fed to a non-supplemented group with a high fat diet supplemented with blueberry powder as shown in Figure 6.2.

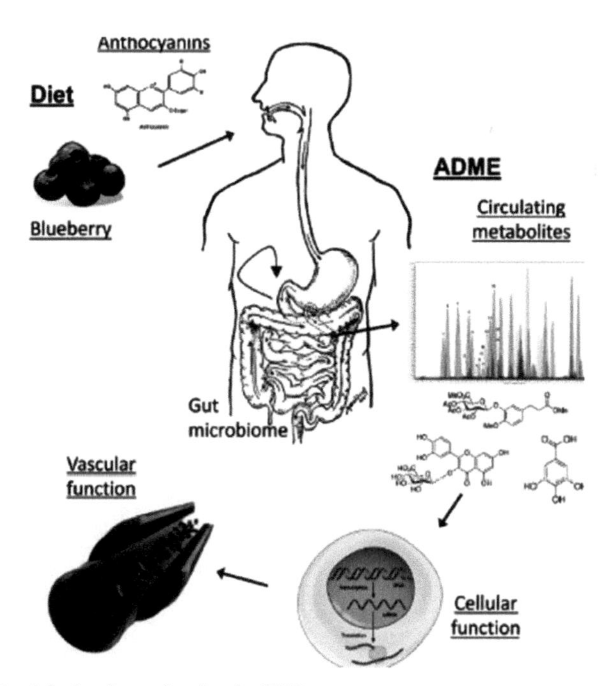

FIGURE 6.2 Mechanism of action in CVDs.

Presence of antioxidants compound indicated that berries provide protective effects against metabolic disorders like insulin resistance, hyperglycemia, and CVDs. A study conducted on a rat model of prediabetes and hyperlipidemia, showed that the activity of intestinal mucosal disaccharides (maltase and sucrose) decreased after dietary supplementation with chokeberry fruit extract for 4 weeks. Different vitro and animal studies demonstrated that fruits provide protective effects in improving glucose and lipid-related abnormalities that lead toward the CVDs. When induced by inducible nitricoxidostasis (iNOS) activation, nitric oxide (NO) causes vascular permeability, increased vascular permeability, a strong oxidizing agent and leads to the formation of peroxynitrite. They also demonstrated that different fractions of berries show inhibitory effects on the synthesis of NO in cell lines [70]. While the potential for iNOS expression increases, the pre-inflammatory sequence plays an important role in maintaining cardiovascular homeostasis and positively modulating blood pressure by NO production by endothelial nitric oxide synthase (eNOS) and reducing endothelial dysfunction as shown in Table 6.3 [71, 72].

TABLE 6.3 Health Protective Mechanisms of Berries

Studies	Anthocyanins	Mechanisms
Pergola et al. [70]	Anthocyanin fraction (blackberry extract)	• Inhibit Biosynthesis of LPS-induced murine monocytes/macrophages J774 cell line • Inhibit inducible NOS protein expression • Reduce the mismatch response in macrophages and stop the formation of foam cells
Pergola et al. [70]	Chokeberry juice	Effective in lipid metabolism in cellular and animals models of dyslipidemia
Xia et al. [62]	Anthocyanins powder	Regulates cholesterol distribution by interfering with the uptake of tumor necrosis factor receptor-dependent factors (TRAF)-2 in lipid rafts, thereby inducing CD40-induced pro-signal signaling

TABLE 6.3 *(Continued)*

Studies	Anthocyanins	Mechanisms
Pergola et al. [70]	Anthocyanins	eNOS is regulated by cyanidine-3-glucoside in bovine artery endothelial cells, and that anthocyanin treatment (cyanidin and delfinidine) in human umbilical vein endothelial cells increases protein levels of eNOS
Xia et al. [62]	Anthocyanins from berries and strawberries	Anthocyanin treatment of human umbilical vein endothelial cells regulates cholesterol distribution by interfering with the uptake of tumor necrosis factor receptor-dependent factors (TRAF)-2 in lipid rafts, thereby inducing CD40-induced pro-signal signaling
Pergola et al. [70]	Supplementation of chokeberry fruit	Activity of intestinal mucosal disaccharides (maltase and sucrose) decreased after dietary supplementation with chokeberry fruit extract for 4 weeks

6.7 SUMMARY

Production of ROS are the major contributor to the progression of some diseases. Antioxidants scavenge the free radicals and provide additional health benefits. Consumption of fruits and vegetables is inversely associated with the oxidative stress. Phytochemicals are the bioactive components released from plants having medicinal properties. Berries are a good example of phytochemicals and belongs to the flavonoid class of the phytochemicals. The etiology of heart diseases is very complex and different studies indicated that flavonoids have heart-healthy properties. ACNs also exhibited antioxidant and anti-atherogenic effects by lowering the oxidation of LDL as well as the development of foam cells. ACNs derived from commonly consumed fruits such as blueberries stimulate the response of inflammatory mediators in human microvascular endothelial cells of TNF-α. Berry bioactive components provide anticancer effect through a variety of mechanisms like they induce metabolic enzymes, change the gene expression, program cell death, and modulate different signaling pathways. Black raspberry, red raspberry, cranberry, blueberry, and strawberry are the most commonly use but chokeberry, mulberry, and black currant are consumed in small amounts. Berries are also rich source of ellagic acid, tannin acid, micronutrients, and fiber that provide some additional health benefits beyond the basic needs.

KEYWORDS

- **anthocyanins**
- **antioxidants**
- **berries**
- **cardiovascular diseases**
- **flavonoids**
- **inducible nitricoxidostasis**
- **reactive oxygen species**

REFERENCES

1. Bazzano, L. A., He, J., Ogden, L. G., Loria, C. M., Vupputuri, S., Myers, L., & Whelton, P. K., (2002). Fruit and vegetable intake and risk of cardiovascular disease in US adults: The first national health and nutrition examination survey epidemiologic follow-up study. *The American Journal of Clinical Nutrition, 76*(1), 93–99.
2. Peleteiro, B., Padrão, P., Castro, C., Ferro, A., Morais, S., & Lunet, N., (2016). Worldwide burden of gastric cancer in 2012 that could have been prevented by increasing fruit and vegetable intake and predictions for 2025. *Br. J. Nutr., 115*, 851–859.
3. Bello, S., Neri, M., Riezzo, I., Othman, M. S., Turillazzi, E., & Fineschi, V., (2011). Cardiac beriberi: Morphological findings in two fatal cases. *Diagnostic Pathology, 6*(1), 8.
4. Wang, S. Y., & Jiao, H., (2000). Scavenging capacity of berry crops on superoxide radicals, hydrogen peroxide, hydroxyl radicals, and singlet oxygen. *Journal of Agricultural and Food Chemistry, 48*(11), 5677–5684.
5. Corley, J., Kyle, J. A., Starr, J. M., McNeill, G., & Deary, I. J., (2015). Dietary factors and biomarkers of systemic inflammation in older people: The Lothian birth cohort 1936. *Br. J. Nutr., 114*, 1088–1098.
6. Manach, C., Scalbert, A., Morand, C., Rémésy, C., & Jiménez, L., (2004). Polyphenols: food sources and bioavailability. *The American Journal of Clinical Nutrition, 79*(5), 727–747.
7. Meyskens, F. L., & Szabo, E., (2005). Diet and cancer: The disconnect between epidemiology and randomized clinical trials. *Cancer Epidemiology and Prevention Biomarkers, 14*(6), 1366–1369.
8. Kim, Y. J., (2007). Anti-melanogenic and antioxidant properties of gallic acid. *Biol. Pharm. Bull., 30*, 1052–1055.
9. Xue, H., Aziz, R. M., Sun, N., Cassady, J. M., Kamendulis, L. M.,; Xu, Y., & Klaunig, J. E., (2001). Inhibition of cellular transformation by berry extracts. *Carcinogenesis, 22*(2), 351–356.

10. Chun, O. K., Chung, S. J., Claycombe, K. J., & Song, W. O., (2008). Serum C-reactive protein concentrations are inversely associated with dietary flavonoid intake in US adults. *The Journal of Nutrition, 138*(4), 753–760.
11. Bakowska-Barczak, A. M., Marianchuk, M., & Kolodziejczyk, P., (2007). Survey of bioactive components in western Canadian berries. *Canadian Journal Of Physiology and Pharmacology, 85*(11), 1139–1152.
12. Wojdyło, A., Oszmiański, J., & Bober, I., (2008). The effect of addition of chokeberry, flowering quince fruits and rhubarb juice to strawberry jams on their polyphenol content, antioxidant activity and color. *European Food Research and Technology, 227*(4), 1043–1051.
13. Hollands, W., Brett, G. M., Radreau, P., Saha, S., Teucher, B., Bennett, R. N., & Kroon, P. A., (2008). Processing blackcurrants dramatically reduces the content and does not enhance the urinary yield of anthocyanins in human subjects. *Food Chemistry, 108*(3), 869–878.
14. Manach, C., Scalbert, A., Morand, C., Rémésy, C., & Jiménez, L., (2004). Polyphenols: food sources and bioavailability. *The American Journal of Clinical Nutrition, 79*(5), 727–747.
15. Kroon, P. A., Clifford, M. N., Crozier, A., Day, A. J., Donovan, J. L., Manach, C., & Williamson, G., (2004). How should we assess the effects of exposure to dietary polyphenols *in vitro*. *The American Journal of Clinical Nutrition, 80*(1), 15–21.
16. Srivastava, A., Akoh, C. C., Yi, W., Fischer, J., & Krewer, G., (2007). Effect of storage conditions on the biological activity of phenolic compounds of blueberry extract packed in glass bottles. *Journal of Agricultural and Food Chemistry, 55*(7), 2705–2713.
17. Wojdyło, A., Figiel, A., & Oszmiański, J., (2009). Effect of drying methods with the application of vacuum microwaves on the bioactive compounds, color, and antioxidant activity of strawberry fruits. *Journal of Agricultural and Food Chemistry, 57*(4), 1337–1343.
18. Nicoue, E. E., Savard, S., & Belkacemi, K., (2007). Anthocyanins in wild blueberries of Quebec: Extraction and identification. *Journal of Agricultural and Food Chemistry, 55*(14), 5626–5635.
19. Tsuda, T., (2012). Dietary anthocyanin-rich plants: Biochemical basis and recent progress in health benefits studies. *Molecular Nutrition & Food Research, 56*(1), 159–170.
20. Speciale, A., Cimino, F., Saija, A., Canali, R., & Virgili, F., (2014). Bioavailability and molecular activities of anthocyanins as modulators of endothelial function. *Genes & Nutrition, 9*(4), 404.
21. Jennings, A., Welch, A. A., Fairweather-Tait, S. J., Kay, C., Minihane, A. M., Chowienczyk, P., & Cassidy, A., (2012). Higher anthocyanin intake is associated with lower arterial stiffness and central blood pressure in women. *The American Journal of Clinical Nutrition, 96*(4), 781–788.
22. Jia, Y., Hoang, M. H., Jun, H. J., Lee, J. H., & Lee, S. J., (2013). Cyanidin, a natural flavonoid, is an agonistic ligand for liver X receptor alpha and beta and reduces cellular lipid accumulation in macrophages and hepatocytes. *Bioorganic & Medicinal Chemistry Letters, 23*(14), 4185–4190.

23. Packard, R. R., & Libby, P., (2008). Inflammation in atherosclerosis: From vascular biology to biomarker discovery and risk prediction. *Clinical Chemistry, 54*(1), 24–38.
24. Chen, J., Uto, T., Tanigawa, S., Kumamoto, T., Fujii, M., & Hou, D. X., (2008). Expression profiling of genes targeted by bilberry (Vacciniummyrtillus) in macrophages through DNA microarray. *Nutrition and Cancer, 60*(S1), 43–50.
25. Wu, T., Tang, Q., Gao, Z., Yu, Z., Song, H., Zheng, X., & Chen, W., (2013). Blueberry and mulberry juice prevent obesity development in C57BL/6 mice. *PLoS One, 8*(10), e77585.
26. Zanotti, I., Dall'Asta, M., Mena, P., Mele, L., Bruni, R., Ray, S., & Del, R. D., (2015). Atheroprotective effects of (poly) phenols: A focus on cell cholesterol metabolism. *Food & Function, 6*(1), 13–31.
27. Faria, A., Fernandes, I., Norberto, S., Mateus, N., & Calhau, C., (2014). Interplay between anthocyanins and gut microbiota. *Journal of Agricultural and Food Chemistry, 62*(29), 6898–6902.
28. Gaziano, J. M., Buring, J. E., Breslow, J. L., Goldhaber, S. Z., Rosner, B., VanDenburgh, M., & Hennekens, C. H., (1993). Moderate alcohol intake, increased levels of high-density lipoprotein and its subfractions, and decreased risk of myocardial infarction. *New England Journal of Medicine, 29*(25), 1829–1834.
29. Kowalenko, C. G., (2005). Accumulation and distribution of micronutrients in Willamette red raspberry plants. *Canadian Journal of Plant Science, 85*(1), 179–191.
30. Nile, S. H., & Park, S. W., (2014). Edible berries: Bioactive components and their effect on human health. *Nutrition, 30*(2), 134–144.
31. Koyuncu, M. A., & Dilmaçünal, T., (2010). Determination of vitamin C and organic acid changes in strawberry by HPLC during cold storage. *Notulae Botanicae Horti Agrobotanici Cluj-Napoca, 38*(3), 95–98.
32. Battino, M., Beekwilder, J., Denoyes-Rothan, B., Laimer, M., McDougall, G. J., & Mezzetti, B., (2009). Bioactive compounds in berries relevant to human health. *Nutrition Reviews, 67*(suppl_1), S145–S150.
33. Giampieri, F., Tulipani, S., Alvarez-Suarez, J. M., Quiles, J. L., Mezzetti, B., & Battino, M., (2012). The strawberry: Composition, nutritional quality, and impact on human health. *Nutrition, 28*(1), 9–19.
34. Stevenson, D., & Scalzo, J., (2012). Anthocyanin composition and content of blueberries from around the world. *Journal of Berry Research, 2*(4), 179–189.
35. Forney, C. F., Kalt, W., Jordan, M. A., Vinqvist-Tymchuk, M. R., & Fillmore, S. A., (2012). Blueberry and cranberry fruit composition during development. *Journal of Berry Research, 2*(3), 169–177.
36. Giovanelli, G., Brambilla, A., Rizzolo, A., & Sinelli, N., (2012). Effects of blanching pre-treatment and sugar composition of the osmotic solution on Physico-chemical, morphological and antioxidant characteristics of osmodehydrated blueberries (*Vaccinium corymbosum* L.). *Food Research International, 49*(1), 263–271.
37. Barberis, A., Spissu, Y., Fadda, A., Azara, E., Bazzu, G., Marceddu, S., & Serra, P. A., (2015). Simultaneous amperometric detection of ascorbic acid and antioxidant capacity in orange, blueberry and kiwi juice, by a telemetric system coupled with a fullerene-or nanotubes-modified ascorbate subtractive biosensor. *Biosensors and Bioelectronics, 67*, 214–223.

38. Pertuzatti, P. B., Barcia, M. T., Rodrigues, D., Da Cruz, P. N., Hermosín-Gutiérrez, I., Smith, R., & Godoy, H. T., (2014). Antioxidant activity of hydrophilic and lipophilic extracts of Brazilian blueberries. *Food Chemistry, 164*, 81–88.
39. Lohachoompol, V., Srzednicki, G., & Craske, J., (2004). The change of total anthocyanins in blueberries and their antioxidant effect after drying and freezing. *BioMed Research International, 2004*(5), 248–252.
40. Reque, P. M., Steffens, R. S., Jablonski, A., Flôres, S. H., Rios, A. D. O., & De Jong, E. V., (2014). Cold storage of blueberry (*Vaccinium* spp.) fruits and juice: Anthocyanin stability and antioxidant activity. *Journal of Food Composition and Analysis, 33*(1), 111–116.
41. Nix, W. A., Zirwes, R., Bangert, V., Kaiser, R. P., Schilling, M., Hostalek, U., & Obeid, R., (2015). Vitamin B status in patients with type 2 diabetes mellitus with and without incipient nephropathy. *Diabetes Research and Clinical Practice, 107*(1), 157–165.
42. González-Ortiz, M., Martínez-Abundis, E., Robles-Cervantes, J. A., Ramírez-Ramírez, V., & Ramos-Zavala, M. G., (2011). Effect of thiamine administration on metabolic profile, cytokines and inflammatory markers in drug-naïve patients with type 2 diabetes. *European Journal of Nutrition, 50*(2), 145–149.
43. Carrodeguas, L., Kaidar-Person, O., Szomstein, S., Antozzi, P., & Rosenthal, R., (2005). Preoperative thiamine deficiency in obese population undergoing laparoscopic bariatric surgery. *Surgery for Obesity and Related Diseases, 1*(6), 517–522.
44. Flancbaum, L., Belsley, S., Drake, V., Colarusso, T., & Tayler, E., (2006). Preoperative nutritional status of patients undergoing Roux-en-Y gastric bypass for morbid obesity. *Journal of Gastrointestinal Surgery, 10*(7), 1033–1037.
45. Nath, A., Tran, T., Shope, T. R., & Koch, T. R., (2017). Prevalence of clinical thiamine deficiency in individuals with medically complicated obesity. *Nutrition Research, 37*, 29–36.
46. Shah, H. N., Bal, B. S., Finelli, F. C., & Koch, T. R., (2013). Constipation in patients with thiamine deficiency after Roux-en-Y gastric bypass surgery. *Digestion, 88*(2), 119–124.
47. Finegold, J. A., Asaria, P., & Francis, D. P., (2013). Mortality from ischaemic heart disease by country, region, and age: Statistics from World Health Organization and United Nations. *International Journal of Cardiology, 168*(2), 934–945.
48. Shin, B. H., Choi, S. H., Cho, E. Y., Shin, M. J., Hwang, K. C., Cho, H. K., & Jang, Y., (2004). Thiamine attenuates hypoxia-induced cell death in cultured neonatal rat cardiomyocytes. *Mol. Cells, 18*(2), 133–140.
49. Zhang, G., Ding, H., Chen, H., Ye, X., Li, H., Lin, X., & Ke, Z., (2012). Thiamine nutritional status and depressive symptoms are inversely associated among older Chinese adults. *The Journal of Nutrition, 143*(1), 53–58.
50. Liu, S., Manson, J. E., Lee, I. M., Cole, S. R., Hennekens, C. H., Willett, W. C., & Buring, J. E., (2000). Fruit and vegetable intake and risk of cardiovascular disease: The women's health study. *The American Journal of Clinical Nutrition, 72*(4), 922–928.
51. Bazzano, L. A., (2005). *Dietary Intake of Fruit and Vegetables and Risk of Diabetes Mellitus and Cardiovascular Diseases*, (p. 66). Geneva: WHO.

52. Iqbal, R., Anand, S., Ounpuu, S., Islam, S., Zhang, X., Rangarajan, S., Chifamba, Jet al., (2008). Interheart study investigators, dietary patterns and the risk of acute myocardial infarction in 52 countries. *Circulation, 118*(19), 1929–1937.
53. Vitolins, M. Z., Anderson, A. M., Delahanty, L., Raynor, H., Miller, G. D., Mobley, C., & Mayer-Davis, E., (2009). Action for health in diabetes (Look AHEAD) trial: baseline evaluation of selected nutrients and food group intake. *Journal of the American Dietetic Association, 109*(8), 1367–1375.
54. Tamers, S. L., Agurs-Collins, T., Dodd, K. W., & Nebeling, L., (2009). US and France adult fruit and vegetable consumption patterns: An international comparison. *European Journal of Clinical Nutrition, 63*(1), 11.
55. Guenther, P. M., Dodd, K. W., Reedy, J., & Krebs-Smith, S. M., (2006). Most Americans eat much less than recommended amounts of fruits and vegetables. *Journal of the American Dietetic Association, 106*(9), 1371–1379.
56. Casagrande, S. S., Wang, Y., Anderson, C., & Gary, T. L., (2007). Have Americans increased their fruit and vegetable intake; The trends between 1988 and 2002. *American Journal of Preventive Medicine, 32*(4), 257–263.
57. Ruel, G., Pomerleau, S., Couture, P., Lemieux, S., Lamarche, B., & Couillard, C., (2006). Favorable impact of low-calorie cranberry juice consumption on plasma HDL-cholesterol concentrations in men. *British Journal of Nutrition, 96*(2), 357–364.
58. Gupta, S., Sodhi, S., & Mahajan, V., (2009). Correlation of antioxidants with lipid peroxidation and lipid profile in patients suffering from coronary artery disease. *Expert Opinion on Therapeutic Targets, 13*(8), 889–894.
59. Suriñach, J. M., Alvarez, L. R., Coll, R., Carmona, J. A., Sanclemente, C., Aguilar, E., & FRENA Investigators, (2009). Differences in cardiovascular mortality in smokers, past-smokers and non-smokers: Findings from the FRENA registry. *European Journal of Internal Medicine, 20*(5), 522–526.
60. Basu, A., Wilkinson, M., Penugonda, K., Simmons, B., Betts, N. M., & Lyons, T. J., (2009). Freeze-dried strawberry powder improves lipid profile and lipid peroxidation in women with metabolic syndrome: Baseline and post-intervention effects. *Nutrition Journal, 8*(1), 43.
61. Wu, X., Kang, J., Xie, C., Burris, R., Ferguson, M. E., Badger, T. M., & Nagarajan, S., (2010). Dietary blueberries attenuate atherosclerosis in apolipoprotein E-deficient mice by upregulating antioxidant enzyme expression. *The Journal of Nutrition, 140*(9), 1628–1632.
62. Qin, Y., Xia, M., Ma, J., Hao, Y., Liu, J., Mou, H., & Ling, W., (2009). Anthocyanin supplementation improves serum LDL-and HDL-cholesterol concentrations associated with the inhibition of cholesteryl ester transfer protein in dyslipidemic subjects. *The American Journal of Clinical Nutrition, 90*(3), 485–492.
63. Lee, S. H., Park, S. A., Ko, S. H., Yim, H. W., Ahn, Y. B., Yoon, K. H., & Kwon, H. S., (2010). Insulin resistance and inflammation may have an additional role in the link between cystatin C and cardiovascular disease in type 2 diabetes mellitus patients. *Metabolism, 59*(2), 241–246.
64. Libby, P., (2007). Inflammatory mechanisms: The molecular basis of inflammation and disease. *Nutrition Reviews, 65*(suppl_3), S140–S146.
65. Real, J. T., Martínez-Hervás, S., Tormos, M. C., Domenech, E., Pallardó, F. V., Sáez-Tormo, G., & García-García, A. B., (2010). Increased oxidative stress levels and

normal antioxidant enzyme activity in circulating mononuclear cells from patients of familial hyper cholesterolemia. *Metabolism, 59*(2), 293–298.
66. Youdim, K. A., Martin, A., & Joseph, J. A., (2000). Incorporation of the elderberry anthocyanins by endothelial cells increases protection against oxidative stress. *Free Radical Biology and Medicine, 29*(1), 51–60.
67. Serraino, I., Dugo, L., Dugo, P., Mondello, L., Mazzon, E., Dugo, G., & Cuzzocrea, S., (2003). Protective effects of cyanidin-3-O-glucoside from blackberry extract against peroxynitrite-induced endothelial dysfunction and vascular failure. *Life Sciences, 73*(9), 1097–1114.
68. Liu, L. K., Lee, H. J., Shih, Y. W., Chyau, C. C., & Wang, C. J., (2008). Mulberry anthocyanin extracts inhibit LDL oxidation and macrophage-derived foam cell formation induced by oxidative LDL. *Journal of Food Science, 73*(6), H113–H121.
69. Youdim, K. A., McDonald, J., Kalt, W., & joseph, J. A., (2002). Potential role of dietary flavonoids in reducing microvascular endothelium vulnerability to oxidative and inflammatory insults. *The Journal of Nutritional Biochemistry, 13*(5), 282–288.
70. Pergola, C., Rossi, A., Dugo, P., Cuzzocrea, S., & Sautebin, L., (2006). Inhibition of nitric oxide biosynthesis by anthocyanin fraction of blackberry extract. *Nitric Oxide, 15*(1), 30–39.
71. Xu, J. W., Ikeda, K., & Yamori, Y., (2004). Upregulation of endothelial nitric oxide synthase by cyanidin-3-glucoside, a typical anthocyanin pigment. *Hypertension, 44*(2), 217–222.
72. Lazzè, M. C., Pizzala, R., Perucca, P., Cazzalini, O., Savio, M., Forti, L., & Bianchi, L., (2006). Anthocyanidins decrease endothelin-1 production and increase endothelial nitric oxide synthase in human endothelial cells. *Molecular Nutrition & Food Research, 50*(1), 44–51.

CHAPTER 7

FUNCTIONAL AND NUTRACEUTICAL SIGNIFICANCE OF MACRO AND MICRO ALGAE

IAHTISHAM-UL-HAQ, AQSA AKRAM, and
HAFIZ ANSAR RASUL SULERIA

ABSTRACT

Algae possess various important features which give it industrial significance. These serve as a source of some novel compounds that are potential to be utilized in human disease prevention and health maintenance. Several unique bioactive compounds (BACs) like different types of proteins (amino acids, lecithins, peptides, and phycobilin-proteins), polysaccharides, polyphenols, and many others which are not present in common food sources make algae effective to modulate chronic diseases. Health claims associated with algal bioactives include anticancer, antiviral, and anticoagulant properties and also abilities like modulating intestinal health and risk factors for diabetes and obesity. Many *in vitro* and animal studies have shown evidences for the beneficial effects of seaweeds and their components in terms of human diseases and health status. This chapter critically evaluates the industrial importance of the macro- and micro-algae and their BACs draw attention towards planning and implementing future studies.

7.1 INTRODUCTION

Algae comprise of a very diverse group of organisms with an extensive genetic and metabolic range [10, 129]. Structurally, algal species ranged from unicellular to multicellular and with colonial and filamentous structures that resemble higher plants like kelp. There are two major classes of algae, i.e., microalgae and macroalgae with the former comprising of members which cannot be seen without a microscope while the latter can be seen with unaided eye. Most of the algae are autotrophic as these are efficient in photosynthesis, enabling them to convert light energy and simple inorganic nutrients to a variety of complex biomolecules. Furthermore, few are heterotrophic species similar to yeast, fungi, and bacteria which are able to grow without light on organic sources like sugar. Occasionally some species show mixotrophy, i.e., they can switch modes, although they perform better in one mode than the other. Metabolically, they yield in proteins, carbohydrates, pigments, lipids, toxins, and numerous bioactive compounds (BACs) [129].

Due to their ecological diversity and adaptability, these are one of the major organisms with industrial significance. The commercial products from algae include foods, their additives, nutraceuticals, and feeds [144]. Among macroalgae, *Nori, Wakame, Kombu,* and *Dulse* are well-known culinary products. For nutritional supplementation of both humans and animals, widely used microalgae are the *Spirulina* and *Chlorella.* Several microalgae are grown commercially as aquaculture feed due to their high nutritional profile alongside energy value. The hydrocolloids, including carrageenan, agars, and alginates are the major products obtained after processing algal biomass, which are used as gelling agents in various food and healthcare products. Amongst the pigments, β-carotene, astaxanthin, and phycobiliproteins are extracted from different algae. The antioxidant properties of these pigments make them to be generally utilized as food colorants and nutraceuticals. Also, they serve as an additive in the animal feed [128, 129]. As these also produce vital bioactives, their use in the nutraceutical industry is widespread and increasing day by day [70]. The detailed discussion with respect to their industrial significance and disease prevention is enclosed herein.

7.2 ALGAE WITH COMMERCIAL SIGNIFICANCE

Algae gained enormous interest among all marine sources due to their simple structures for reproduction and photosynthetic nature. They exist from unicellular-microscopic to multicellular-macroscopic organisms known as microalgae and macroalgae or aquatic plants, respectively. Although the total number of algal species remain unidentified, but an estimated range lies between 1 and 10 million [71, 107]. For biomass and oil production, microalgae use energy from light and fix carbon dioxide with higher efficiency via photosynthesis than other plants. In fact, the oil production rate of microalgae can be 10 to 100 times higher than that from terrestrial oil crops thereby making these a promising biofuel production source with an added advantage of not being competitive with food production. In terms of adaptability, algae adapt well to the changing conditions of the environment compared to other marine organisms by producing various secondary metabolites that are biologically active and cannot be found in other organisms. Moreover, most of them can easily be cultivated, and some BACs can be regulated either by controlling the conditions for cultivation or by using approaches like genetic engineering. Algae can also be used as aquaculture food, as purifier for wastewater and as extractor for high added value foods and pharmaceutical products [71]. It has further been reported that algae are potentially used in biomedical products [169]. Some of the common algal products and their functional properties are outlined in Table 7.1.

TABLE 7.1 Common Algal Products, Their Functional Ingredients and Properties

Common Name	Scientific Name	Functional Ingredient(s)	Functional Properties	References
Wakame	*Undaria pinatifida*	Fucoidan Lactoferrin	Pancreatic cancer treatment Obesity and diabetes risk reduction Prebiotic Anti-inflammatory	[47, 165]
Kombu or Haidai	*Laminaria japonica*	Alginate	Excipients in pharmaceuticals	[133]
Chlorella spp.	*Arthrospira platensis*	Lutein Astaxanthin	Anti-angiogenic effect against pancreatic cancer	[103, 133]
Green algae	*Fucus vesiculosis*	Mucopolysaccharides	Anti-cellulitis	[115]

TABLE 7.1 *(Continued)*

Common Name	Scientific Name	Functional Ingredient(s)	Functional Properties	References
Brown seaweeds	*Macrocystis pyrifera*	Alginate	Emulsifying Gelling Stabilizer	[27, 92]
	Ascophyllum nodosum	Phlorotannin	Anti-diabetic Neuroprotective	
	Laminaria digitata	Phenols chiefly phlorotannins	Anti-oxidant Prevent CVD Anti-biotic Anti-HIV	
Hiziki	*Hizikia fusiformis*	Fucoxanthin	Prebiotics Plasma glucose level regulation	[27, 56]
Kelp	*Saccharina japonica*			
Sea ferns	*Bryopsis pinnata*	Kahalalide F.	Cytotoxic	[92]
Red macroalgae	*E. spinosum* *Eucheuma cottonii* *Chondrus crispus*	Carrageenan (polymers of sulfonated galactose)	Anti-viral	[159]

7.2.1 MACROALGAE

Macroalgae or aquatic plants can be seen with unaided eyes due to their macroscopic structures [59]. These are used as major algal foods, generally harvested either from cultivated and managed population or from wild populations. Among global macroalgae producers, Indonesia, Philippines, Malaysia, and China are prominent countries [68]. Drying the whole algal biomass often carried out via sun exposure, spray, or oven dryers, and incorporated into other foods which are then traded as a sheet, powder, tablet, or capsule. Sorting, cleaning, or processing of biomass are the only processing [59, 71]. Many species of macroalgae are consumed as foods but those with major commercial markets include Nori, Wakame, Kombu, and Dulse as discussed below.

Nori is made from either blade or leaf of Porphyra; red algae (Figure 7.1). It is being collected since ages however, its active cultivation was started in 1640 AD. It stands as a dominant algal product in the market with a revenue of more than a billion dollars annually [128]. Majorly, Nori is used in the preparation of sushi. Further, toasted nori sheets are also most popular products [8]. Underwater farming is the essential method for Nori cultivation and is similar to other food products made from macroalgae and hydrocolloid species. Seed-like propagules also known as Concho spores are seeded onto webs and hung in oceanic areas that are sheltered, looped between poles, or attached to surface buoys. Many devices have recently been developed that are designed to keep them raised out of water and allows it to be more mechanically harvested and processed [129, 155].

FIGURE 7.1 Some common algae and their commercial products.

Kombu, a popular algal product, is either served as a vegetable or alongside meat and fish. It is also added in the soups. The Kombu is obtained from *Laminaria japonica* and brown algal species related to it. Algae are dried and boiled after getting collected. It is claimed that Kombu trade yield in 600 million dollars annually. For human consumption, many other macroalgae are also utilized that include *Dulse*, derived from the macroalgae *Palmaria palmata* and is widely consumed in European coastal areas [129]. Another product Wakame is derived from brown algae *Undaria pinnatifida*; majorly cultivated in China, Japan, and Korea. Being sold in multiple forms, the boiled and salted wakame products are most popular earning about $600 million dollars annually. It is also a useful ingredient within noodles, soups, and salads [59].

7.2.2 MICROALGAE

A global increase in population has raised concerns for searching alternative protein sources so that future food shortage can be addressed. Microalgae have been identified as promising candidates as they contain high protein content. It is worth mentioning that proteins derived from algae are of superior quality than many conventional plants [151]. Microalgae also provide essential fatty acids especially introducing polyunsaturated fatty acids (PUFA) to the food chain. Majority of essential vitamins can be obtained from microalgae alongside nutraceutically effective pigments such as carotenoids, chlorophyll, and phycobiliproteins [151, 169]. Although microalgae and its extracts are productive in certain niche food applications, the algae production on large scale has not been appeared yet in order to solve food crisis and protein shortage around the world due to certain concerns like high cost, undesirable sensory traits, and slimy pond-scum perception [110].

Historically, the leading commercial product made from microalgae are the nutritional supplements. Both cell biomass and extracts are marketed as nutraceuticals and functional food ingredients and consumed as a nutritional supplement, or in combination with other products [10, 59]. Chlorella and Spirulina (*Arthrospira platensis*) are dominant in the market and are used predominantly in healthcare food products [128].

The filamentous cyanobacteria *Arthrospira platensis* and other members of same species, commonly known as Spirulina, hold a longer history of human consumption. Natural alkaline lakes were the original source for obtaining it [59]. However, now it is also produced in commercial sites yielding tons of dry product each year [151]. Due to its filamentous morphology, cultivation of Spirulina is relatively easy which makes it readily harvested through filtration. Conditions for cultivation such as alkalinity and salinity prevent the development of contamination and also the progression of other major algal species [59, 129, 155]. Commercially, it is produced in huge open raceway ponds. Harvesting from ponds is followed by filtration and drying via solar exposure or by the use of oven and spray dryers. Its products are mainly marketed in Asia, Europe, and North America in the form of powder, tablets, and capsules [107, 151]. It primary sales are either as a health food or additive within the animal feed. In animal and human diets, massive literature is available that examines the nutritional value of Spirulina [59]. However, safety tests for possible algal contamination must be conducted before consumption. Spirulina biomass is claimed to offer various health-promoting effects, e.g., improving immunity, protection against viruses and promoting growth of beneficial intestinal flora [151]. However, in terms of clinical studies, confirmations have not yet been made.

Chlorella belongs to green microalga, primarily marketed as a health promoting food product. It was first cultivated commercially by Nihon Chlorella (Taiwan) however, numerous firms are now producing it commercially [151]. Because of its growth under moderate conditions, its successful cultivation in open reactors depends on its rapid growth rate to beat contamination. Heterotrophically, it also grows on glucose or acetate to produce valuable compounds [172]. After harvesting biomass through centrifugation, it is dried into powder for further utilization. Many health-promoting effects have been decorated to chlorella, including stimulation of immunity, reduction in blood lipids, antioxidant properties, effectiveness on stomach ulcers, wounds, and constipation [150, 151, 169]. However, various studies were unable to demonstrate conclusions and are often introduced by the companies that produce Chlorella mainly as a marketing strategy.

7.2.3 ALGAL CONSTITUENTS WITH COMMERCIAL SIGNIFICANCE

Algae are regarded as an excellent nutritional source for animals and humans as they sufficiently constitute essential macro- and micro-nutrients like protein, lipids, carbohydrates, vitamins and minerals. Out of all algal products, macroalgal food products that are primarily sold and produced in the East, possess the largest value in the market [59]. Microalgae could not capture much market despite its potentials due to high production cost associated. However, the dominant markets for microalgae where premium prices can be commanded by algal biomass are nutraceuticals and healthcare food items [63].

From macroalgae, various functional ingredients like polysaccharides are extracted that serve as hydrocolloids sharing a large market around the globe. Red and brown species of macroalgae are major sources of these hydrocolloids. A series of extraction and purification steps in hot solvent are involved in production of hydrocolloids providing thickening and gelling properties in food products. For instance, agar is primarily extracted from red algae like *Gracilaria*, *Gelidium*, *Pterocladia*, *Acanthopeltis*, and *Ahnfeltia*. Agar can further be converted to agarose in a series of fractionation and purification steps. These are widely used in laboratories as a gel forming agent. Furthermore, alginates; salts of alginic acid and polymers of D-mannuronic acid and L-gluronic acid, are extracted majorly from brown algae like *Laminaria*, *Macrocystis*, and *Ascophyllum* and particularly used as thickening and gel-forming agents in food, paper, and biomedical products. Carrageenan; complex polysaccharides made of polymers of sulfonated galactose, are extracted from red macroalgae like *Eucheuma cottonii*, *E. spinosum* and *Chondrus crispus*. Foods, cosmetics, pharmaceuticals, and other products are gelled, thickened, suspended, and stabilized using carrageenan [128, 129].

Algae majorly produce chlorophyll, carotenoids, phycocyanin (PC), and phycobiliproteins as major pigments which provide added efficiency to use light energy and protection against free radicals and solar radiation [43, 46, 128]. In industries like food, cosmetics, pharmaceuticals, textile, and printing there is a greater demand for natural colorants obtained from sustainable sources for utilization [43]. Few of the natural colorants possess tinctorial and persistence properties which are essential traits required to be used in industry [140]. Commercially used pigments from algae include β-carotene from *Dunaliella*, astaxanthin from *Haematococcus* and the phycobiliproteins from cyanobacteria and some red algae [59].

Carotenoids with color ranging from brown-red to orange-yellow mainly include β-carotene and astaxanthin, and sparely include lutein, zeaxanthin, and lycopene [151]. Green halotolerant alga *Dunaliella* naturally produces β-carotene. Hypersaline open ponds are the habitats where *Dunaliella* and related species are generally grown. After the accumulation of biomass, it is exposed to environments with high salt and light. β-carotene accumulates due to exposure to such stress conditions making up to 14% of the dry weight [100, 129, 151]. Carotenoids are Vitamin A precursors and have protective properties against oxidation and inflammation [128]. Although, synthetically produced carotenoids are cheaper but safer natural sources are consumer's first choice nowadays [70, 151].

Commercially, the astaxanthin is obtained from *Haematococcus pluvialis*. Up to 3% dry weight of astaxanthin can be accumulated with this alga [128]. It is primarily marketed for coloration purposes as a feed additive in aquaculture especially salmon feed [119, 128, 151]. Both open and closed systems can also be combined, which might not only initiate the generation of unialgal biomass but also offer stress conditions in pigment accumulation during the final cultivation stages. Synthetic manufacturing of Astaxanthin is also possible. Natural production cost cannot be compared with the synthetic form, but still can be marketed as cosmetics and nutraceuticals because the consumers prefer natural products [70]. In the aquaculture market, consumers stay unbiased as they are not informed well about nurturing practices, however, in some applications, enhanced performance is not offered by natural astaxanthin, e.g., the natural form is greatly deposited in carp tissues, chicken, and red sea bream [119, 151].

Many hydrophilic proteins that are also light-harvesting proteins found in cyanobacteria, rhodophytes, cryptomonads, and glaucophytes are unique algae known as phycobiliproteins [46, 140]. They are able to absorb the light range of 495–650 nm while chlorophyll and carotenoids are unable to absorb this light absorption range [10]. According to their spectral properties, phycobiliproteins are classified into three major groups PC, phycoerythrin (PE), and allophycocyanin (APC). On the outer surface of thylakoid membranes, phycobilisomes are the complexes formed by the organization of different phycobiliproteins [140]. In many cyanobacteria, phycobiliproteins are among the most abundant proteins [46]. Factors that can influence their cellular content might be the culture conditions, mainly light and nutrients. In *Spirulina platensis*, glucose, and acetate improve cell growth and PC production [29]. PE production in *Porphyridium* largely depends on the chloride and nitrate concentrations, small amounts of sulfate and phosphate within the culture medium also influence [82].

In order to extract PC as an edible pigment, water-soluble components are extracted into buffer generally from dry biomass after which solids are removed. Then with the chromatographic techniques, further separation and purification is done to prepare pure phycobiliproteins such as PE so that they can be utilized in fluorescent applications [140]. Fluorescent properties of Phycobiliproteins enable them to be utilized as fluorescent dyes and markers. Other phycobiliproteins also fluoresce at different wavelengths and yield a spectrum of colors to detect multicolor. In this regard, PE, and APC are most commonly used that are able to conjugate with molecules specific to certain substrates, e.g., antibodies, receptions, streptavidin, and biotin, yielding fluorescent tags or probes that will bind to a specific protein, tissue, or cell type. Fluorescent phycobiliproteins are also employed in fluorescent microscopy, diagnostics, cell sorting, immunolabeling, and immunohistochemistry [46, 140]. In cancer tumor photodynamic therapy, their usage as a light sensitive agent has also been proposed [114]. Their applications are also expanding specially in the fields of cosmetics, nutrition, and pharmacy [128]. Phycobiliproteins just like other pigments also exhibit pharmaceutical activities. Many studies indicated that they possess properties that are not just immunomodulatory but also protective against oxidation, inflammation, neural damages, liver damages, viruses, and tumors [46, 140].

The nutritional importance of PUFAs, including the Ω-3 fatty acids eicosatetraenoic acid (EPA) and docosahexaenoic acid (DHA) as well as arachidonic acid (AA), are eminent. These are essential for brain development, flexibility, and selectively permeable properties of cellular membranes. Furthermore, these are equally beneficial to the cardiovascular system and in both human and animal subjects form important nutraceutical and pharmaceutical targets. Overall health benefits associated with these fatty acids include development of brain health, cardiovascular disease (CVD) prevention, stroke, asthma, as well as prevention from rheumatoid arthritis (RA) [76, 105, 106, 170]. The competitive inhibition of compounds produced from Ω-6 fatty acids, and the direct production of beneficial compounds (eicosanoids) from the Ω-3 fatty acids ultimately result in the beneficial effects of Ω-3 fatty acids [148]. Plants, microalgae, and some bacteria exclusively synthesize PUFA's *de novo*. Through elongation and desaturation, animals convert one form of PUFA to another, but very few of them can synthesize these fatty acids [20]. Fat molecules of microalgae, such as phytoflagellates and dinoflagellates, green, and red microalgae, and in diatoms are the longer chained PUFAs containing 20 or more carbon atoms [130]. In deep-sea cold-water microalgae, commonly found PUFA's are EPA and DHA attributed to survival strategy of these species at cold temperatures [20, 160]. Algal biotechnology is advantageous in producing PUFA's which can easily be extracted or purified at low operational costs along with increased product content, quality, and safety. Relatively large amounts of EPA can be accumulated by using Fungi, especially of the order Mucorales, and bacteria of the genera Shewanella, Alteromonas, Flexibacter, and Vibrio [177], however, the fermentation ability of bacteria and fungi to economically compete with traditional sources of Ω-3 fatty acids is limited by low productivities and excessively long times for fermentation [13]. Accumulation of total fatty acid as EPA or DHA is accomplished especially by microalgal species such as *Skeletonema costatum* and *Chlorella minutissima*. Additionally, long-chain Ω-3 fatty acid production is reported to be 1–2 folds more from the microalgal fermentation systems than production reported from fungal or bacterial systems [59].

7.2.4 ALGAL BIOACTIVE COMPOUNDS (BACs)

The potential to utilize a range of algal bioproducts and their complex chemical components as feed, pharmaceuticals, nutraceuticals, research industries is of great interest [59]. Bhatnagar and Kim [16] defined BACs as secondary metabolites which are produced by particular species of microalgae that represent bioactivity. This bioactivity involves protecting against other microbes or normalizing certain abnormal physiological states in humans [16]. Many of secondary metabolites produced by algae possess protective properties against cancers, fungi, bacteria, viruses as well as toxins.

Algae are considered an imperative source of biologically active moieties owing to their wide spectrum of activities as well as the advantage of possessing environment friendly cultivation methods [109]. Algae contain an abundance of vital nutrients and bioactive entities like tocopherols, carotenoids, terpenoids, polyphenols, diterpenes, sulfated polysaccharides and phlorotannins [70]. These BACs have antioxidant, tumor-suppressing, antimicrobial, antiviral, and anti-inflammatory properties [104, 126, 171]. Hence, efficient extraction and purification is necessary to retain their beneficial properties [137].

Natural pigments, phlorotannins, diterpenes, and sulfated polysaccharides from algae have gained significant reputation. Being a photosynthetic organism, algae majorly contain chlorophylls pigment [31]. Chlorophyll is converted to pheophorbide, pheophytin, and pyropheophytin when processed that demonstrate reasonable cancer-preventive roles [32, 33]. The pheophytin has been attributed for more cytotoxic properties than pheophorbide [96]. It has also been noticed that the pigment contents are dependent on season such as fucoxanthin, violaxanthin, chlorophyll A and carotene in brown algae (Ascophyllum). Seasonal variations are also demonstrated by *Fucus serratus* showing lowered levels during autumn. Furthermore, chlorophyll concentrations are also influenced with the light exposure as harbored algae showed more pigment as compared to openly grown algae [33].

Sulfated polysaccharides from algae have been endorsed for their prominent pharmacological functions as anticancer and immunomodulators. A range of polysaccharides such as fucoidans, alginates, and laminarans are derived from algae [70]. Fucoidans and laminarans are mainly water-soluble whereas, alkali-soluble polysaccharides comprise of high molecular alginic acids. The brown algae have laminarans as their storage glucans. There are two types of extracellular acid polysaccharides, namely alginic acid and sulfated fucans in algae where later is further classified into glycorunogalactofucans, xylofucoglycuronans, and fucoidans [60]. Sulfated polysaccharides from marine algae have claimed to possess anticoagulant, antiviral, antioxidant, and antitumor properties [12, 60, 173].

Phlorotannins are present in marine algae and possess a vital position as bioactives. Biologically, polyketide or acetate-malonate pathway yields phlorotannins which are chemically derived from phloroglucinol (1,3,5-trihydroxybenzene) monomers' polymerization. It is highly hydrophilic nature and range between 126 Da and 650 kDa in molecular size [131]. Both phenoxy and phenyl units are present in these phloroglucinol-based polyphenols. Based on linkage, phlorotannins are classified into fuhalols and phlorethols (having ether linkage), eckols (with a dibenzodioxin linkage), fucols (containing phenyl linkage) and fucophloroethols (having both ether and phenyl linkage). Marine brown algae (*Ecklonia cava*) are claimed to be the richest source of phlorotannins in which eckol, dieckol, phloroglucinol, phlorofucofuroeckol A, 7-phloroeckol and dioxinodehydroeckol possess significant biological activities [108]. Studies have shown that phlorotannins have s anti-cancer, anti-HIV, bactericidal, antioxidant, and enzyme inhibitory functions [44, 94].

Diterpenes are also among an important class of vital bioactive moieties from algal sources. Different diterpenoids are present in algae like xenicanes, dolabellanes, sesquiterpenoids, and hydroazulenoids. Species of Dictyotaceae family produce these compounds as their main secondary metabolites that possess high algicidal, anti-viral, and cytotoxic properties [95]. All BACs of algae have been shown in Figure 7.2.

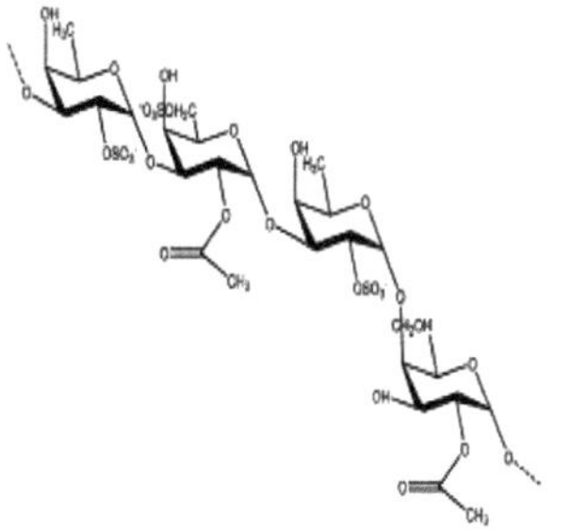

Fucoidan from *Undaria pinnatifida*

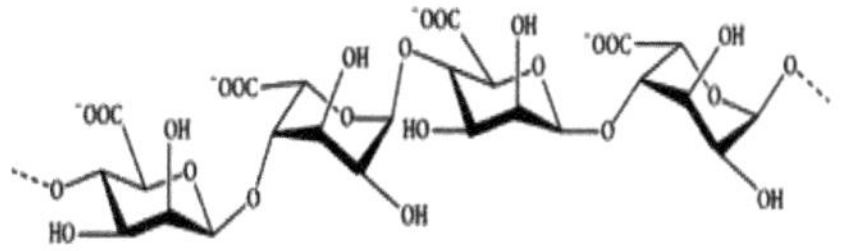

Alginate from *Lamonaria japonica*

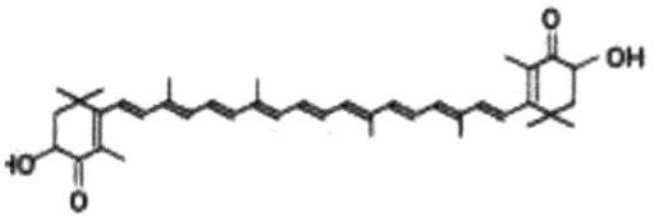

Astaxanthin

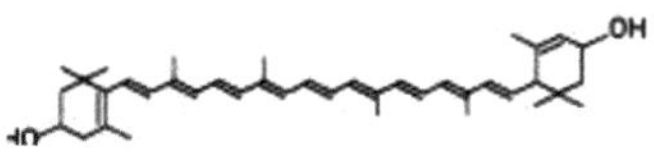

Lutein

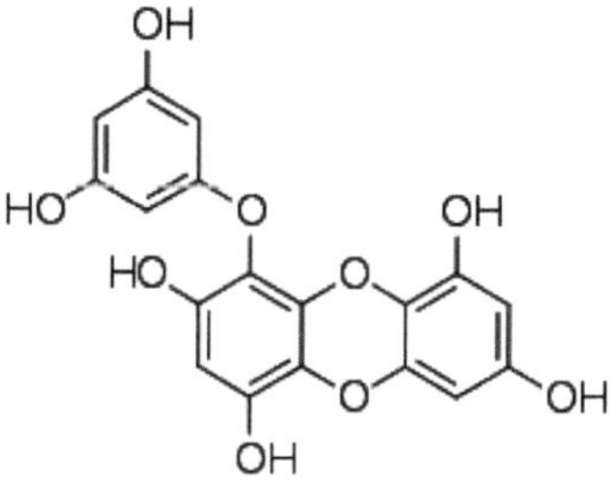

Phlorotannin

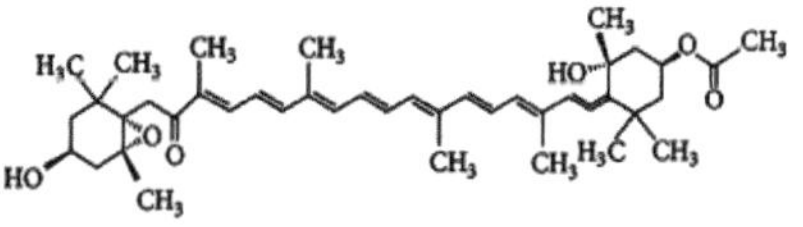

Fucoxanthin

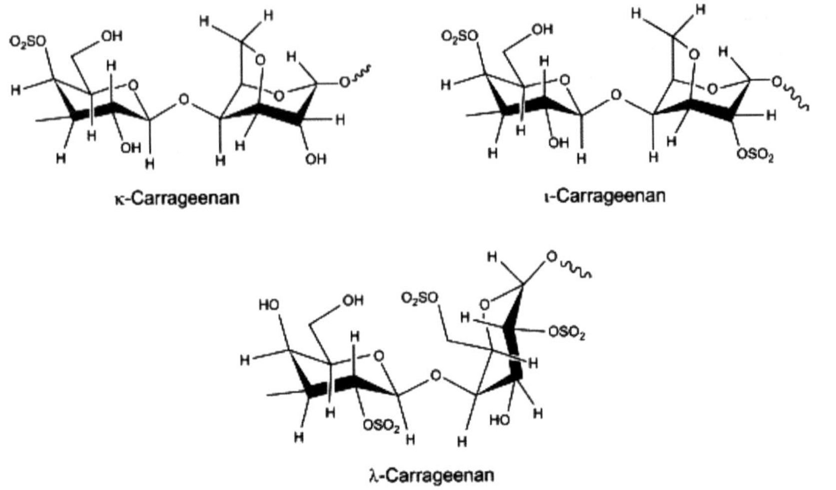

κ-Carrageenan

ι-Carrageenan

λ-Carrageenan

Carrageenans

FIGURE 7.2 Chemical constituents from algal sources.

7.3 NUTRACEUTICAL WORTH OF ALGAL BIOACTIVE COMPOUNDS (BACS)

Nutraceuticals from different algae and their allied health claims are summarized in Table 7.2. A short summary of functions of such compounds with their associated health benefits is discussed as follows.

7.3.1 ANTIOXIDANT

Oxidative stress is caused by the imbalance between production and neutralization of free radicals that further involves the onset of various degenerative disorders. BACs are considered important to reduce the oxidative stress in the human body [69]. In this respect, phlorotannins from brown algae are considered promising to suppress oxidative stress by controlling the generation and progression of reactive oxygen species (ROS) due to their strong antioxidant activities [94].

It has been reported that eckstolonol extracted from *E. stolonifera* have significant *in vitro* antioxidant activities [78]. In addition, phlorotannins from *H. fusiform* have proven as vital radical scavengers as well [149]. It has further been reported that eckol from *E. cava* effectively reduces lung cell damage caused by oxidative stress [81]. Likewise, bioactives from *E. cava*, *E. bicyclis*, *E. kurome* and *H. fusiformis* have also been claimed to possess antioxidant properties [94, 95]. Eckol, dieckol, 8,8'-dieckol and phlorofucofuroeckol A are promising antioxidants to control phospholipid peroxidation [145]. Particularly, phloroglucinol, and triphlorethol-A are promising algal compounds having ameliorative effect against oxidative cell damages [79–81].

7.3.2 CARDIOVASCULAR HEALTH

Regular intake of seaweed, soy, and fish has been attributed to longer and healthier life expectancy of people. Similarly, it has been found that such diet is also correlated with the lowered risk of CVDs. Since the Japanese diet is full of seaweeds, hence individuals show less cardiac disease prevalence [21].

Limited human data is available to support the fact that seaweed components possess protective properties against inflammations, coagulation, and adhesion, however, the association of low CVD risk with algal consumption is supported through many animal models [37]. Functional foods based on seaweeds have been identified to possess such potential properties that facilitate cardiac health, while it has also been recognized that before making any health claim, it is necessary to consider the compositional variations that could result due to environmental and seasonal changes [18]. The potential health benefits of associated mechanisms have been explored through CVD rat models where fucoxanthin from Wakame is found to be responsible for reduction in blood pressure and risk factors for stroke [73]. Whereas, low levels of cholesterol, low density lipoproteins (LDLs), triglycerides, and reduced lipid peroxidation alongside erythrocyte glutathione peroxidase were observed with tropical seaweeds [105].

Research on carrageenan derived from *Chondrus crispus* unveiled its anticoagulant properties, which favors to inhibit the aggregation of platelets [97], however further investigation via dietary interventional is required. The anticoagulant properties of fucoidan from brown seaweeds (*Laminaria* and *Ascophyllum*) are vital in preventing CVDs [15]. The same species also produce Laminarin which has been reported to exhibit protective properties against oxidation, inflammation, and coagulation [21]. Consequently, in preventing heart diseases, both fucoidan and laminarin serve as useful mediators. Antioxidant compounds like PCs and carotenoids have antiatherogenic properties [35, 135]. The rich microalgal source of these antioxidant compounds is *Spirulina platensis* that also promising in the inhibition of oxidative stress and programmed cell death in cardiomyocytes [83]. Astaxanthins, Ω-3 PUFA, EPA, and DHA from algal sources have also been found to deliver cardioprotection via control of oxidative stress, immunomodulation, and antiatherosclerosis effects. As most of the research is conducted on animal-based models, there is a need to further explore the potential properties of marine BACs in humans so that their cardioprotective effects can be better elucidated.

7.3.3 ANTICANCER

Natural anticancer bioactives are considered promising tool to overcome cancer progression due to their capability of preventing or curing oncogenic events [95]. In this context, sulfated polysaccharides from algae are found to represent significant potential to cure cancer insurgence. These polysaccharides control oxidative stress by scavenging free radicals, ultimately lowering the incidence of cancer progression in human body [173].

PUFAs, EPA, carotenoids, and lipopeptides are promising anticancer BACs from algae. PUFAs are used for the treatment of cancer because these compounds either directly act as anticancer or they alleviate the effects of radiation and chemotherapy. The damaging effects caused by oxidative stress are reduced by EPA due to its anti-inflammatory nature [168]. It has also been suggested that direct effects of these compounds as anticancer specifically work against tumors by inhibiting angiogenesis and metastasis [59]. Antioxidant effects of carotenoids reduce oxidative damage at cellular levels. *Dunaliella salina* is a good natural source of β-carotene for humans capable of reducing cancer and other degenerative diseases [14]. Curacin-A and dolastatin 10 have also been shown to serve as potential anticancer drugs in preclinical and clinical trials [57]. Most of similar biomolecules are found in *Nostocales*, and some members of genera *Lyngbya*, *Oscillatoria*, and *Symploca* [59].

Sulfated polysaccharides deliver anti-metastasis activity via blocking interaction between cancer cells and basement membrane. Nevertheless, actual mechanisms are yet to be unveiled, including adhesion and blockage of tumor cells propagation. However, studies have shown that oral intake of seaweeds significantly lowers *in vivo* carcinogenesis. The sulfated polysaccharides extracted from *Ecklonia cava* have represented vital anti-proliferative properties and stimulated apoptosis in human leukemic monocyte lymphoma cell line (U-937) [12].

Phlorotannins from brown algae are also useful to ameliorate cancer insurgence. Animal studies have shown that *Laminaria* lowers the risk of intestinal and mammary cancer development via antioxidant and cell proliferation reduction mechanisms [178]. Likewise, a phloroglucinol derivative namely dioxinodehydroeckol, persuades apoptosis in MCF-7 cell lines and lowers breast cancer risk [87]. Moreover, eckol, dieckol, phlorofucofurofuroeckol, and fucodiphloroethol G extracted from *E. cava* potentially exhibit cytotoxic effects on HT-29, HT-1080, A549 and HeLa human cancer cell lines [95].

Fucoidans from brown algae have been reported as potential anticancer agents attributed to their molecular weight (Mw) and sulfate content. Broadly, depolymerization, and fucoidans with lowered Mw show improved anticancer properties [60, 102]. The sulfated polysaccharides derived from seaweeds acquire protective effects against oxidative damage that may prevent cancer hence, making them suitable as promising chemo-preventing agents [173].

7.3.4 ANTIVIRAL

Globally, human immunodeficiency virus (HIV) is considered amongst the biggest threats to human life. The people suffering from the infection of HIV receive a combination of antiretroviral therapy (ART) for its treatment/management [91]. Scientists have been seeking novel BACs to prepared drugs for treating HIV infection with minor side effects comparative to already existing drugs [11]. On the basis of epidemiological evidences, initial research on the protective properties of algae against viruses has been stimulated, showing a link between low infection rates and high algae consumption [159]. In East Asia and Africa, the annual incidence of HIV/AIDS (acquired immunodeficiency syndrome) is increasing annually, which suggests that there may be involvement of other factors than just use of intravenous drug and sexual behavior [159]. Variations have been reported within Africa and low rates were associated with daily consumption of *Spirulina* algae [158, 159]. It has been shown that blue-green algae *Lyngbya lagerheimeii* and *Pormidium tenue* have potential to protect human lymphoblastoid T-cells against the cytopathic effects of HIV infection [59, 61]. In this respect, marine brown algae possess a significant amount of innovative anti-HIV constituents for the production of drugs [24, 94].

Seaweed species contain sulfated polysaccharides, which appear to possess antiviral properties, and it has been identified through *in vitro* as well as *in vivo* animal studies that carrageenan, fucoidans, and sulfated thamnogalactans work against viruses like herpes and HIV. The ability of these polysaccharides to block the entrance of virus to cell is their principal mechanism of action. Fucoidans have inhibiting ability for the growth of a variety of viruses [6, 163]. Antiviral activity of carrageenan work against type 1 and 2 herpes simplex, HIV-1 and human rhinovirus [58]. It has been reported that carrageenan potentially works as vaginal microbiocide [150, 180].

Anti-viral polysaccharides from algae are considered safe due to their very low cytotoxic effects on mammalian cells. This property is of great importance when designing antiviral drugs. Amongst algal polysaccharides, fucoidan possess significant antiviral activities against human cytomegalovirus (HCMV) and HIV [60]. *Adenocystis utricularis* yields galactofuran that acquires strong inhibitory effects against herpes simplex virus (HSV) 1 and 2 and showed no cytotoxic effects [126]. Likewise, aqueous polysaccharide fraction from *Sargassum patens* also represented strong activity against HSV-1 and 2 with low cytotoxic effects [182]. Similarly, anti-viral activities of *U. pinnatifida* derived galactofucan sulfate extract against HCMV and HSV-1 and 2 have also been reported [66].

The brown algae especially *Ecklonia cava* is potentially important due to phlorotannin compounds [94]. *In vitro* studies have shown inhibitory effect of phlorotannins 8,8′-bieckol and 8,4‴-dieckol on HIV-1 protease and reverse transcriptase (RT). Both of these compounds are dimmers of eckol but their relative inhibitory effect is different. For an instance, 8,8′-bieckol exhibited 10-folds higher inhibitory effect than dieckol [5]. This bioactive showed selective inhibitory effect on RT which is comparable to nevirapine [11].

Amongst major phloroglucinol derivative, 6,6′-bieckol also acquire strong inhibitory effects against syncytia formation induced by HIV-1, lytic responses and *in vitro* and in vivo viral p24 antigen production [11]. It also blocked entry of HIV-1 and its RT enzyme activity when exposed to a dose rate of 1.07 μM. Besides, it did not show any cytotoxic effects at such concentrations where HIV-1 replication is completely inhibited. Henceforth, phlorotannins from marine algae have remarkable potential to produce novel therapeutics against HIV with minor side effects [11, 70].

7.3.5 ANTI-OBESITY

In developed as well as developing countries, obesity is spreading with each passing day [141]. Obesity greatly facilitates the development of risk for other diseases like diabetes mellitus, hypertension, and dyslipidemia, and moderately increases the risk of developing bone and CVDs like osteoarthritis and CHD, respectively [65]. Obesity may involve the accumulation of excessive fat-causing an increase in weight of an individual [86]. Critically, obesity is described as the accumulation of fat in adipose tissue which is due to hyperplasia and hypertrophy of fat cells [127]. In this context, delaying or stopping the adipogenesis process can avoid the risk of obesity [70].

Algal bioactives especially fucoxanthin and fucoxanthinol have been seen to inhibit 3T3-L1 preadipocytes differentiation to adipocytes. Mechanistically, these compounds down-regulate adipogenic transcription factors [99]. Another mechanism involving structure specific suppression in adipocyte differentiation has been suggested where neoxanthin and fucoxanthin exhibited significant suppressive responses on adipocyte differentiation [117]. It has been shown that induction of uncoupling protein 1 (UCP1) mediated by fucoxanthin administration showed anti-obesity effects [174]. This protein can bring about uncoupled oxidative phosphorylation causing production of heat instead of ATP (adenosine triphosphate) especially in white adipose tissues (WAT). Hence, most of the heat-producing mechanisms are held in brown adipose tissues (BAT), so the UCP1 is fully expressed in these tissues. Whereas, BAT is not primarily involved in human weight reduction as adults contain less BAT than WAT which contributes to weight gain [162]. Also, WAT is currently known to secrete adipokines which is an important biological mediator [38]. Fucoxanthin can serve as a potential anti-obesity agent in clinical trials as it controls the expression of UCP1 [122]. Combined therapy has also depicted anti-obesity effects of Xanthigen™ comprising of pomegranate oil and fucoxanthin in non-diabetic obese women that offered promising liver-protecting and weight-reducing effects [2].

Seaweeds also provide dietary fiber which is reported to assist in weight loss due to the prolonged rate of gastric emptying which enhances satiety and results in food intake reduction [88, 98]. The appetite attenuating role of seaweed isolates particularly alginate is evident, but only limited studies have investigated the appetite and associated markers' role of marine fiber from whole seaweed. Breakfast meal comprising of bread enriched with 4% seaweed (*Ascophyllum nodosum*) showed lowered intake of energy. Further, various algal BACs positively influenced the lipid metabolism making these a right choice for control of obesity [21].

7.3.6 METABOLIC SYNDROMES

Role of algae has been suggested for the prevention and treatment of metabolic syndrome through researches based on animal models [72]. Various foods consisting of seaweeds have been found to be potentially protective against metabolic syndromes [175]. Dietary interventions including whole dried seaweed *Undaria pinnatifida* showed reduced waist circumference in individuals receiving the treatment as well as blood pressure modulation as compared to placebo [157].

However, conflicting results were observed in a cross-sectional study that showed an association of increased risk of developing metabolic syndrome with the consumption of grilled laver (seaweed) [146]. It is being claimed that this cross-sectional study had possible associated problems like under-reporting and misreporting in the used food frequency questionnaire along with the limitations such as selection bias. Nevertheless, the need for the development of well-designed human studies have been highlighted by this study so that the results from animal studies can be confirmed and specificity for the beneficial effects of seaweeds maybe determined that either they are specie or population-specific [21, 146, 147].

7.3.7 ANTI-ANGIOGENIC

The development of new blood vessels from already present vasculature in response to physiological or pathological circumstances is denoted as angiogenesis [70]. Uncontrolled spread of blood vessels is noted in pathological events including RA, inflammatory diseases and tumor metastasis that further worsens the pathological state [45]. The control of angiogenesis is an effective strategy to hinder cancer growth and other angiogenic diseases alongside lowering their severity [122, 124]. In this respect, algal bioactive provides promising anti-angiogenic effects. For instance, siphonoxanthin extracted from green algae (*Codium fragile*) significantly blocks the propagation of human umbilical vein endothelial cells (HUVEC) at dose rates of 2.5 μM and above. HUVEC tube formation is sufficiently inhibited upon exposure to siphonoxanthin treatment and no tube formation is further identified when the concentrations of the compound are raised to 25 μM suggesting the suppression of angiogenic mediators [53, 54].

In an alternative investigation, it was demonstrated that fucoxanthin significantly hinder tube development and spread of HUVEC at a dose rate of 10 μM. It efficiently reduces the differentiation of endothelial cells from endothelial progenitor cells. Alongside, *in vivo* and *ex vivo* micro vessel outgrowth was also suppressed by fucoxanthin and fucoxanthinol [152]. It can be stated that algal bioactives such as fucoxanthin and siphonoxanthin possess strong anti-angiogenic properties primarily attributed to hydroxyl groups as both of these compounds are structurally similar, having hydroxyl groups at 3 and 3′ positions, respectively. Nevertheless, further investigations in diversified diet compositions may depict a true picture of their efficacy when a routine meal is consumed.

7.3.8 DIABETES MELLITUS

Prolonged obesity often results in type 2 diabetes (T2D) along with impaired glucose tolerance, hypertension, and hyperlipidemia [136]. The majority of animal studies related to diabetes have demonstrated desirable outcomes by using *Undaria* spp. as therapeutic agent resulting in improvements in lipid status, declined inflammatory effects, reduction in weight gain and blood glucose reduction [99, 111]. Although a limited number of studies have been conducted on diabetic individuals, the results of human studies are generally found favorable [21]. In order to interpret the mechanisms responsible for improving diabetic parameters, further investigations are required.

Emerging evidences have linked the consumption of seaweed with the progression of diabetes. Seaweed consumption has shown a weak association with decreased incidence of diabetes [93]. Favorable regulation in insulin levels and sensitivity have been noticed with just a single ingestion of whole dried brown seaweed extracts obtained from *Ascophyllum nodosum* and *Fucus vesiculosis* after a meal rich in carbohydrates [125].

Attenuation in parameters related to diabetes has been demonstrated through many dietary interventional studies conducted on individuals with diabetes, obesity, and impaired glucose tolerance. In an earlier investigation, postprandial reduction in serum insulin, blood glucose and plasma concentrations of c-peptide in diabetic individuals have been noticed after alginate drink consumption [161]. However, it was noticed that consumption of agar along with the conventional diet and routine exercise significantly reduces body mass index (BMI), body weight and total levels of cholesterol [101].

In a 4-week intervention of 48 g/day high fiber intake from whole dried seaweed supplements containing *Saccharina japonica* and *Undaria pinnatifida*, significant reduction in concentration of fasting and postprandial blood glucose and favorable alteration in lipid profile, i.e., raised HDL cholesterol level and lowered triglyceride levels were observed in individuals having diabetes and obesity [85]. Subsequently, as this study was unable to report weight or BMI changes or fiber-rich control use, therefore, it was difficult to determine that blood glucose and lipid level changes were either because of increased seaweed fiber intake or due to seaweed BACs. Overall, prolonged studies conducted on diabetic individuals demonstrated beneficial effects of supplementation of seaweed and seaweed products on weight reduction and diabetes markers including BMI, blood glucose and lipids as well. The conclusions by scientists of these researches focused on the effects that possibly managed the symptoms of insulin independent diabetes mellitus.

7.3.9 DIGESTIVE TRACT HEALTH

A great amount of recognition is being given to gut microflora in humans as they play a vital role in health maintenance and disease prevention [165]. At both systemic and local (intestinal mucosa) levels, interaction of microflora with its host occurs which results in a broad range of physiological, immunological, and metabolic effects [21, 23]. Diet may modulate the effects of bacteria on host besides their direct influence on the host in both beneficial and unfavorable way to health [17, 23, 134]. Emerging evidences have shown that many seaweed-derived fibers positively affect gut health [39, 167] and numerous studies have examined polysaccharides of low-molecular-weight for their prebiotic activity [40, 132].

Seaweed fiber differ from those of carbohydrates from higher plants in terms of chemical, physiochemical, and fermentation characteristics, but available studies are not very clear about its characteristics in the human gut or the particular species of bacteria that it may modulate [40]. Such characteristics have been determined to date by using either animal models or just by *in vitro* assessments. Novel sources of prebiotic carbohydrates include structural components of macroalgae as well as marine-derived products like alginates, laminarin, and agar [116]. Increased probiotic survival have been observed from alginate, xanthum, and carrageenan gum as they provide live bacteria that physically protect against adverse digestive conditions [42].

The combined use of alginate and chitosan is emerging as effective delivery systems [28, 75]. Overall, there is a huge gap in the knowledge of marine BACs and their role in human digestive tract health. Surely, there is a potential to explore and develop these compounds as prebiotics or as materials that augment the probiotic survival. However, insufficient data from human studies causes hindrance to derive definite conclusion for use of algal polysaccharides and possible probiotics' modulation. In the future, potential mode of action and specific compounds of interest might be explored and better understood if researchers may incorporate intelligentially designed human models.

7.3.10 NEUROPROTECTIVE

Neurodegenerative ailments are predicted as a second most prevalent reason for mortality among the elderly people by 2040s surpassing causalities due to cancer [9]. It is postulated that neuroprotection may be achieved by avoiding damage to the neuronal cells, dysfunction, or degeneration of the central nervous system (CNS), apoptosis, and limiting post-injury death or dysfunction in CNS [164]. Protection of CNS dysfunctions and neural damage is recently being accomplished by the use of drugs from natural and synthetic sources. However, synthetic drugs also pose several side effects delivering serious health issues. Hence, drugs or neuroprotective agents from natural sources are being explored due to their professedly safer nature [123]. Consequently, a wide-ranging exploration is being conducted to identify neuroprotective agents from seaweeds [122, 123].

In an earlier investigation, fucoxanthin from *H. fusiformis* have been demonstrated to inhibit the progression of GOT0 cells and expression of N-myc in human neuroblastoma cell line. Though, the complete mechanism is yet to be explored but a dose of 10 μg/mL showed a 38% reduction in the growth rate of these cells [118]. In stroke-prone spontaneously hypertensive rats (SHRSP), it was found that "wakame" has attenuating effect against hypertension [73]. Fucoxanthins were confirmed to possess significant neuroprotective activity by protecting cortical neurons cell damage during oxygen reperfusion and hypoxia [84]. This protective effect can be presumed associate to radical scavenging abilities of fucoxanthins as re-oxygenation and hypoxia release free radicals which are causative of degenerative neural damages [122]. Moreover, pheophytin A isolated from *Sargassum fulvellum* has been shown to protect neurite outgrowths which are cardinal in neuronal development in embryogenesis [74]. Such protective effects of pheophytin A are attributed to its low Mw due to which it acquires easy penetration in cells and effectively promote the growth of neurite [122]. Considering the scientific evidences, algal bioactives can be endorsed for their neuroprotective abilities and potential utilization in clinical therapies.

7.3.11 BONE HEALTH

Risk of decreased bone strength and increased fracture occurs due to reduced bone mass. Also, the deterioration of bone microarchitecture clearly defines osteoporosis as a degenerative disease [64]. To address this disease burden, dietary modifications like increased consumption of food sources rich in calcium can greatly contribute and promote bone health maintenance [19, 156]. It is well established that the most important micronutrients responsible for maintaining bone health are calcium and vitamin D. non-digestible carbohydrates acting as prebiotics are other dietary components that can possibly contribute to enhance absorption and bioavailability of calcium [26, 139].

In this respect, marine algae being rich in minerals, particularly calcium, may help in increasing calcium intake through dietary sources as these contain higher calcium contents as compared to other dietary sources. For instance, *Ascophyllum nodosum* contains 575 mg/100 g of calcium in terms of wet weight while *Laminaria digitata* and *Ulva* spp. contains 364.7 mg/100 mg and 325 mg/100 g, respectively which is comparatively higher calcium than delivered by whole milk (115 mg/100 g) [98]. Contradictory results have been reported by cross-sectional studies conducted in two Asian countries regarding the association between bone mineral density (BMD) and seaweed consumption [143, 147]. According to one study, it was demonstrated that consumption of seaweed as a part of traditional diet is positively correlated with lumbar BMD. Although, after the adjustment of other variables, this change was non-significant [143]. In a larger study, no association was found between consumption of seaweed and risk of osteoporosis or osteopenia using BMD and self-reported food frequency questionnaires [147]. Lower risk of osteoporosis and osteopenia was seen in individuals reporting increased calcium intake. However, their dietary calcium source was not elaborated which probably could have been seaweed. Regardless of these findings, several interventional studies have reported that using different seaweed species can exert beneficial effects on osteoporosis markers like BMD, joint mobility, and pain [50–52]. It has also been noted that BMD increased when *Cystophyllum fusiforme* was consumed with oyster shell powder for absorbable calcium enhancement and symptoms of osteoarthritis were reduced when red algae (*Lithothamnion calcareum*) was supplemented in the diet of individuals [48–52].

Moreover, supplementation of whole dried *Fucus vesiculosis*, *Macrocystis pyrifera* and *Laminaria japonica* in showed promising outcomes probably due to their high contents of fucoidans and polyphenols [113]. Besides the promising results, conclusions must be drawn carefully because of the small size of the study population. Marine derived calcium supplementation is associated with bone health to date, still further studies are needed to contribute to this study. Bone maintenance process is extremely complex as many endogenous factors are responsible for regulating bone absorption and resorption processes. Thyroid hormone produced by the thyroid plays a pivotal role in bone homeostasis and altered production of thyroid is associated with greater risk of osteoporosis. It has been evident through several studies that normal function of the thyroid is interfered with high consumption of seaweeds that contains Iodine, for example, individuals consuming kelp supplements rich in iodine were reported hyperthyroidism therefore, while investigating any type of supplements or extracts derived from marine sources, these effects must be considered. Paradoxically, hypothyroidism can be induced if seaweeds with elevated iodine is consumed. Collectively, all these evidences indicate that iodine status must be considered while undertaking intervention in humans with either whole or extracted seaweed [21].

7.3.12 PHOTOPROTECTION

The use of marine biological sources is increasing for production of cosmeceutical and pharmaceutical compounds. Various compounds from marine organisms are of significant importance to protect against harmful effects of ultraviolet (UV) radiations and photoaging complications [120, 121]. Photo-oxidation in skin cells is a dynamic phenomenon involved in oxidative stress that primarily involves interaction of UV radiations which have strong oxidative properties [25]. In this context, phlorotannins including dieckol and eckol from *E. cava* have shown strong protective effects to lower the damages caused by exposure to the UV spectrum of light. The compounds can lower the intracellular ROS arising from γ-rays exposure. Furthermore, DNA damage and lipid peroxidation in membranes exposed to radiations are curtailed by eckol administration. Based on scientific evidences, it can be stated that phlorotannins possess consequential competencies to lower skin damage arising from oxidative stress induced by radiations. Further, these compounds can profoundly be utilized in pharmaceutical, functional foods and cosmetic industries to alleviate skin damages [94].

It has further been studied that methanolic extract of red seaweed (*Corallina pilulifera*) possess significant photoprotective agents which protects DNA from UV-A induced oxidative stress alongside inhibiting matrix metalloproteinases [121]. Similarly, naturally occurring photoprotective ingredients from algae (*E. cava*, *C. pelulifera* and *P. rosengurttii*) have also been reported including plastoquinones, sargachromenol, fucoxanthins, astaxanthins, dieckol, and phlorotannins [138].

TABLE 7.2 Nutraceutical Worth of Algal Bioactive Compounds

Source	Bioactive Compounds	Functions	References
Codium fragile	Siphonoxanthin	Antiangiogenic	[179]
Dictyota pfaffii	8,10,18-trihydroxy-2,6-dolabelladiene	HSV-1 inhibitor and early protein modulator	[3]
Dictyota sp.	4,18-dihydroxy dictyolactone	Cytotoxic	[34]
Dictyota sp.	Diterpenes	Anti-retroviral	[30]
Ecklonia cava	8,8′-bieckol 8,4‴dieckol 6,6′-bieckol	HIV-1 RT inhibitors	[5, 11]
Ecklonia cava	Dioxinodehydroeckol	Anticancer	[122]
Ecklonia cava *Corallina pilulifera*	Dieckol Eckol	Whitening agent Photoprotection	[22, 121]
Ecklonia kurome	Phlorofucofuroeckol A	Algicidal	[36]
Ecklonia kurome *Eisenia arborea*	Phlorofucofuroeckol-B	Antiallergy	[153, 154]
Fucus evanescens *Adenocystis utrlicularis* *Undria pinnatifida*	Fucoidan Galactofuran Galactofucan sulfate	Anticancer, Anti-metastatic, antitumor, HSV-1 and 2 and HCMV inhibitors	[3, 60, 67, 89, 102]
Fucus vesiculosus	Fucan	Antithrombin, Avian RT inhibitor	[90]
Hizikia fusiformis	Fucoxanthins	Neuroprotective	[55, 181]
Ishige okamurae	Diphlorethohydroxycarmalol	Antidiabetic, whitening agent	[112, 142]

TABLE 7.2 *(Continued)*

Source	Bioactive Compounds	Functions	References
Laminaria japonica	Laminarin	Anti-apoptotic	[77]
Pelvetia siliquosa	Fucosterol	Antidiabetic	[1, 60]
Sargassum fulvellum	Pheophytin a	Neuroprotective	[7, 73, 84]
Sargassum species	Polysaccharide fractions	Anti-HIV, Anti-HSV-1 and 2	[41, 166, 176]
		Cytotoxic	
Sargassum vulgare	Alginic acid	Antitumor	[62]
Undaria pinnatifida	Sulfated polysaccharides	Anti-viral	[4]

7.4 SUMMARY

Marine algae provide sufficient diversity for their functional and nutraceutical use in products ranging from food to cosmeceutical items. Though, functional food and nutraceuticals market is a dynamically developing segment around the globe. However, there are always challenges in designing and developing novel food and cosmetic products due to complexity, expensive technologies and market risks involved in acceptance of these products. Nowadays, consumers are also quite aware of the safety concerns of the products for healthcare so, technical requirements, legislation, and consumer demands must be considered to attain sustainable product market while using seaweeds and other sources for obtaining BACs. Furthermore, carefully designed scientific studies and precisely forecasted market sustainability should be utilized to affirm the product's claims and predict the costumer behavior, respectively, as there is a lot of potential for commercial exploitation of seaweeds in routine products. Sustainable business-led products should be launched alongside introducing novel and less expensive technologies considering the demands of industry and fulfilling the consumer demands.

KEYWORDS

- **algae**
- **anticancer**
- **arachidonic acid**
- **functional foods**
- **gastric health**
- **nutraceuticals**

REFERENCES

1. Abdul, Q. A., Choi, R. J., Jung, H. A., & Choi, J. S., (2016). Health benefit of fucosterol from marine algae: A review. *Journal of the Science of Food and Agriculture, 96*(6), 1856–1866.
2. Abidov, M., Ramazanov, Z., Seifulla, R., & Grachev, S., (2010). The effects of xanthigen™ in the weight management of obese premenopausal women with non-alcoholic fatty liver disease and normal liver fat. *Diabetes, Obesity and Metabolism, 12*(1), 72–81.
3. Abrantes, J. L., Barbosa, J., Cavalcanti, D., Pereira, R. C., Fontes, C. L. F., Teixeira, V. L., Souza, T. L. M., & Paixão, I. C., (2010). The effects of the diterpenes isolated from the Brazilian brown algae *Dictyota pfaffii* and *Dictyota menstrualis* against the herpes simplex type-1 replicative cycle. *Planta Medica, 76*, 339–346.
4. Ahmadi, A., Zorofchian, M. S., Abubakar, S., & Zandi, K., (2015). Antiviral potential of algae polysaccharides isolated from marine sources: A review. *BioMed Research International.*
5. Ahn, M. J., Yoon, K. D., Min, S. Y., Lee, J. S., Kim, J. H., Kim, T. G., Kim, S. H., et al., (2004). Inhibition of HIV-1 reverse transcriptase and protease by phlorotannins from the brown alga *Ecklonia* cava. *Biological and Pharmaceutical Bulletin, 27*(4), 544–547.
6. Aisa, Y., Miyakawa, Y., Nakazato, T., Shibata, H., Saito, K., Ikeda, Y., & Kizaki, M., (2005). Fucoidan induces apoptosis of human HS-sultan cells accompanied by activation of caspase-3 and down-regulation of ERK pathways. *American Journal of Hematology, 78*(1), 7–14.
7. Alghazwi, M., Kan, Y. Q., Zhang, W., Gai, W. P., Garson, M. J., & Smid, S., (2016). Neuroprotective activities of natural products from marine macroalgae during 1999–2015. *Journal of Applied Phycology, 28*(6), 3599–3616.
8. Allen, M. & Sakamoto, R. (2011). 'Sushi reverses course: Consuming American sushi in Tokyo,' *Asia-Pacific Journal: Japan Focus, vol. January,* pp. 1–16.
9. Ansari, J., Siraj, A., & Inamdar, N., (2010). Pharmacotherapeutic approaches of Parkinson's disease. *International Journal of Pharmacology, 6*(5), 584–590.

10. Apt, K. E., & Behrens, P. W., (1999). Commercial developments in microalgal biotechnology. *Journal of Phycology, 35*(2), 215–226.
11. Artan, M., Li, Y., Karadeniz, F., Lee, S. H., Kim, M. M., & Kim, S. K., (2008). Anti-HIV-1 activity of phloroglucinol derivative, 6, 6′-bieckol, from *Ecklonia* cava. *Bioorganic & Medicinal Chemistry, 16*(17), 7921–7926.
12. Athukorala, Y., Ahn, G. N., Jee, Y. H., Kim, G. Y., Kim, S. H., Ha, J. H., Kang, J. S., Lee, K. W., & Jeon, Y. J., (2009). Antiproliferative activity of sulfated polysaccharide isolated from an enzymatic digest of *Ecklonia* cava on the U-937 cell line. *Journal of Applied Phycology, 21*(3), 307–314.
13. Barclay, W., Meager, K., & Abril, J., (1994). Heterotrophic production of long-chain omega-3 fatty acids utilizing algae and algae-like microorganisms. *Journal of Applied Phycology, 6*(2), 123–129.
14. Ben-Amotz, A., (1999). Production of beta-carotene from *Dunaliella. Chemicals from Microalgae,* 196–204.
15. Berteau, O., & Mulloy, B., (2003). Sulfated fucans, fresh perspectives: Structures, functions, and biological properties of sulfated fucans and an overview of enzymes active toward this class of polysaccharide. *Glycobiology, 13*(6), 29–40.
16. Bhatnagar, I., & Kim, S. K., (2010). Immense essence of excellence: Marine microbial bioactive compounds. *Marine Drugs, 8*(10), 2673–2701.
17. Blaut, M., & Clavel, T., (2007). Metabolic diversity of the intestinal microbiota: Implications for health and disease. *The Journal of Nutrition, 137*(3), 751–755.
18. Bocanegra, A., Bastida, S., Benedi, J., Rodenas, S., & Sanchez-Muniz, F. J., (2009). Characteristics and nutritional and cardiovascular-health properties of seaweeds. *Journal of Medicinal Food, 12*(2), 236–258.
19. Bonjour, J. P., Guéguen, L., Palacios, C., Shearer, M. J., & Weaver, C. M., (2009). Minerals and vitamins in bone health: The potential value of dietary enhancement. *British Journal of Nutrition, 101*(11), 1581–1596.
20. Brett, M., & Müller-Navarra, D., (1997). The role of highly unsaturated fatty acids in aquatic food web processes. *Freshwater Biology, 38*(3), 483–499.
21. Brown, E. M., Allsopp, P. J., Magee, P. J., Gill, C. I., Nitecki, S., Strain, C. R., & Mcsorley, E. M., (2014). Seaweed and human health. *Nutrition Reviews, 72*(3), 205–216.
22. Brunt, E., & Burgess, J., (2018). The promise of marine molecules as cosmetic active ingredients. *International Journal of Cosmetic Science, 40*(1), 1–15.
23. Cani, P. D., Neyrinck, A. M., Fava, F., Knauf, C., Burcelin, R. G., Tuohy, K. M., Gibson, G., & Delzenne, N. M., (2007). Selective increases of bifidobacteria in gut microflora improve high-fat-diet-induced diabetes in mice through a mechanism associated with endotoxaemia. *Diabetologia, 50*(11), 2374–2383.
24. Cardoso, S. M., Pereira, O. R., Seca, A. M., Pinto, D. C., & Silva, A., (2015). Seaweeds as preventive agents for cardiovascular diseases: From nutrients to functional foods. *Marine Drugs, 13*(11), 6838–6865.
25. Cardozo, K. H., Marques, L. G., Carvalho, V. M., Carignan, M. O., Pinto, E., Marinho-Soriano, E., & Colepicolo, P., (2011). Analyses of photoprotective compounds in red algae from the Brazilian coast. *Revista Brasileira de Farmacognosia, 21*(2), 202–208.
26. Cashman, K. D., (2007). Calcium and vitamin D. *Dietary Supplements and Health: Novartis Foundation Symposium, 282*, 123–142. Wiley Online Library.

27. Charoensiddhi, S., Abraham, R. E., Su, P., & Zhang, W., (2020). Seaweed and seaweed-derived metabolites as prebiotics. *Advances in Food and Nutrition Research, 91*, 97–122.
28. Chávarri, M., Marañón, I., Ares, R., Ibáñez, F. C., Marzo, F., & Del, C. V. M., (2010). Microencapsulation of a probiotic and prebiotic in alginate-chitosan capsules improves survival in simulated gastro-intestinal conditions. *International Journal of Food Microbiology, 142*(1, 2), 185–189.
29. Chen, F., & Zhang, Y., (1997). High cell density mixotrophic culture of spirulina platensis on glucose for phycocyanin production using a fed-batch system. *Enzyme and Microbial Technology, 20*(3), 221–224.
30. Chen, J., Li, H., Zhao, Z., Xia, X., Li, B., Zhang, J., & Yan, X., (2018). Diterpenes from the marine algae of the genus *Dictyota*. *Marine Drugs, 16*(5), 159.
31. Chen, K., Ríos, J. J., Pérez-Gálvez, A., & Roca, M., (2017). Comprehensive chlorophyll composition in the main edible seaweeds. *Food Chemistry, 228*, 625–633.
32. Chen, K., & Roca, M., (2018). Cooking effects on chlorophyll profile of the main edible seaweeds. *Food Chemistry*.
33. Chen, K., & Roca, M., (2018). *In vitro* bioavailability of chlorophyll pigments from edible seaweeds. *Journal of Functional Foods, 41*, 25–33.
34. Cheng, S., Zhao, M., Sun, Z., Yuan, W., Zhang, S., Xiang, Z., Cai, Y., Dong, J., Huang, K., & Yan, P., (2014). Diterpenes from a Chinese collection of the brown alga *Dictyota plectens*. *Journal of Natural Products, 77*(12), 2685–2693.
35. Cheong, S. H., Kim, M. Y., Sok, D. E., Hwang, S. Y., Kim, J. H., Kim, H. R., Lee, J. H., et al., (2010). Spirulina prevents atherosclerosis by reducing hypercholesterolemia in rabbits fed a high-cholesterol diet. *Journal of Nutritional Science and Vitaminology, 56*(1), 34–40.
36. Chowdhury, M., Sukhan, Z., Kang, J., Ehsan, M., Hannan, M., Shahrin, T., Gatachow, P., et al., (2014). Algicidal activity of the brown seaweed, Ecklonia cava against red tide microalgae. *Proceedings of the 5th International Conference Environ.* (pp. 101–103). Aspects of Bangladesh [ICEAB (2014)].
37. Cumashi, A., Ushakova, N. A., Preobrazhenskaya, M. E., D'incecco, A., Piccoli, A., Totani, L., Tinari, N., et al., (2007). A comparative study of the anti-inflammatory, anticoagulant, antiangiogenic, and antiadhesive activities of nine different fucoidans from brown seaweeds. *Glycobiology, 17*(5), 541–552.
38. Curat, C., Wegner, V., Sengenes, C., Miranville, A., Tonus, C., Busse, R., & Bouloumie, A., (2006). Macrophages in human visceral adipose tissue: Increased accumulation in obesity and a source of resistin and visfatin. *Diabetologia, 49*(4), 744.
39. Devillé, C., Damas, J., Forget, P., Dandrifosse, G., & Peulen, O., (2004). Laminarin in the dietary fibre concept. *Journal of the Science of Food and Agriculture, 84*(9), 1030–1038.
40. Devillé, C., Gharbi, M., Dandrifosse, G., & Peulen, O., (2007). Study on the effects of laminarin, a polysaccharide from seaweed, on gut characteristics. *Journal of the Science of Food and Agriculture, 87*(9), 1717–1725.
41. Dinesh, S., Menon, T., Hanna, L. E., Suresh, V., Sathuvan, M., & Manikannan, M., (2016). *In vitro* anti-HIV-1 activity of fucoidan from *Sargassum* swartzii. *International Journal of Biological Macromolecules, 82*, 83–88.

42. Ding, W., & Shah, N. P., (2009). Effect of various encapsulating materials on the stability of probiotic bacteria. *Journal of Food Science, 74*(2), 100–107.
43. Dufossé, L., Galaup, P., Yaron, A., Arad, S. M., Blanc, P., Murthy, K. N. C., & Ravishankar, G. A., (2005). Microorganisms and microalgae as sources of pigments for food use: A scientific oddity or an industrial reality? *Trends in Food Science & Technology, 16*(9), 389–406.
44. El Gamal, A. A., (2010). Biological importance of marine algae. *Saudi Pharmaceutical Journal, 18*(1), 1–25.
45. Elshabrawy, H. A., Chen, Z., Volin, M. V., Ravella, S., Virupannavar, S., & Shahrara, S., (2015). The pathogenic role of angiogenesis in rheumatoid arthritis. *Angiogenesis, 18*(4), 433–448.
46. Eriksen, N. T., (2008). Production of phycocyanin-a pigment with applications in biology, biotechnology, foods and medicine. *Applied Microbiology and Biotechnology, 80*(1), 1–14.
47. Etman, S. M., Abdallah, O. Y., & Elnaggar, Y. S., (2020). Novel fucoidan based bioactive targeted nanoparticles from *Undaria pinnatifida* for treatment of pancreatic cancer. *International Journal of Biological Macromolecules, 145*, 390–401.
48. Frestedt, J. L., Kuskowski, M. A., & Zenk, J. L., (2009). A natural seaweed derived mineral supplement (Aquamin F) for knee osteoarthritis: A randomized, placebo-controlled pilot study. *Nutrition Journal, 8*(1), 7.
49. Frestedt, J. L., Walsh, M., Kuskowski, M. A., & Zenk, J. L., (2008). A natural mineral supplement provides relief from knee osteoarthritis symptoms: A randomized controlled pilot trial. *Nutrition Journal, 7*(1), 9.
50. Fujita, T., (1997). Osteoporosis: Past, present and future. *Osteoporosis International, 7*(3), 6–9.
51. Fujita, T., Fujii, Y., Goto, B., Miyauchi, A., & Takagi, Y., (2000). Peripheral computed tomography (pQCT) detected short-term effect of AAACa (heated oyster shell with heated algal ingredient HAI): A double-blind comparison with $CaCO_3$ and placebo. *Journal of Bone and Mineral Metabolism, 18*(4), 212–215.
52. Fujita, T., Ohue, T., Fujii, Y., Miyauchi, A., & Takagi, Y., (1996). Heated oyster shell-seaweed calcium (AAA Ca) on osteoporosis. *Calcified Tissue International, 58*(4), 226–230.
53. Ganesan, P., Matsubara, K., Ohkubo, T., Tanaka, Y., Noda, K., Sugawara, T., & Hirata, T., (2010). Anti-angiogenic effect of siphonaxanthin from green alga, *Codium fragile*. *Phytomedicine, 17*(14), 1140–1144.
54. Ganesan, P., Noda, K., Manabe, Y., Ohkubo, T., Tanaka, Y., Maoka, T., Sugawara, T., & Hirata, T., (2011). Siphonaxanthin, a marine carotenoid from green algae, effectively induces apoptosis in human leukemia (HL-60) cells. *Biochimica et Biophysica Acta (BBA)-General Subjects, 1810*(5), 497–503.
55. Garcia-Vaquero, M., & Hayes, M., (2016). Red and green macroalgae for fish and animal feed and human functional food development. *Food Reviews International, 32*(1), 15–45.
56. Ge, H., Zang, Y., Cao, Z., Ye, X., & Chen, J., (2020). Rheological properties, textural and compound preservative of kelp recombination noodles. *LWT, 118*, 108729.
57. Gerwick, W. H., Tan, L. T., & Sitachitta, N., (2001). Nitrogen-containing metabolites from marine cyanobacteria. *The Alkaloids. Chemistry and Biology, 57,* 75–118.

58. Grassauer, A., Weinmuellner, R., Meier, C., Pretsch, A., Prieschl-Grassauer, E., & Unger, H., (2008). Iota-Carrageenan is a potent inhibitor of rhinovirus infection. *Virology Journal, 5*(1), 107.
59. Griffiths, M., Harrison, S. T., Smit, M., & Maharajh, D., (2016). Major commercial products from micro-and macroalgae. *Algae Biotechnology.* Springer.
60. Gupta, S., & Abu-Ghannam, N., (2011). Bioactive potential and possible health effects of edible brown seaweeds. *Trends in Food Science & Technology, 22*(6), 315–326.
61. Gustafson, K. R., Cardellina, J. H., Fuller, R. W., Weislow, O. S., Kiser, R. F., Snader, K. M., Patterson, G. M., & Boyd, M. R., (1989). AIDS-antiviral sulfolipids from cyanobacteria (blue-green algae). *JNCI: Journal of the National Cancer Institute, 81*(16), 1254–1258.
62. Gutiérrez-Rodríguez, A. G., Juárez-Portilla, C., Olivares-Bañuelos, T., & Zepeda, R. C., (2017). Anticancer activity of seaweeds. *Drug Discovery Today.*
63. Harrison, S., Richardson, C., & Griffiths, M., (2013). Analysis of microalgal biorefineries for bioenergy from an environmental and economic perspective focus on algal biodiesel. *Biotechnological Applications of Microalgae: Biodiesel and Value-Added Products.* CRC Press.
64. Harvey, N., Dennison, E., & Cooper, C., (2010). Osteoporosis: Impact on health and economics. *Nature Reviews Rheumatology, 6*(2), 99.
65. Haslam, D., (2006). What's new--tackling childhood obesity. *The Practitioner, 250*(1682), 26.
66. Hemmingson, J. A., Falshaw, R., Furneaux, R., & Thompson, K., (2006). Structure and antiviral activity of the galactofucan sulfates extracted from *Undaria pinnatifida* (Phaeophyta). *Journal of Applied Phycology, 18*(2), 185.
67. Heo, S. J., Ko, S. C., Kang, S. M., Cha, S. H., Lee, S. H., Kang, D. H., Jung, W. K., et al., (2010). Inhibitory effect of diphlorethohydroxycarmalol on melanogenesis and its protective effect against UV-B radiation-induced cell damage. *Food and Chemical Toxicology, 48*(5), 1355–1361.
68. Hurtado, A. Q., Neish, I. C., & Critchley, A. T., (2015). Developments in production technology of *Kappaphycus* in the Philippines: More than four decades of farming. *Journal of Applied Phycology, 27*(5), 1945–1961.
69. Iahtisham-Ul-Haq, Butt, M. S., Shamshad, A., & Suleria, H. A. R., (2018). Heath benefits of anthocyanins in black carrot (*Daucus carota*). In: Goyal, M. R., & Suleria, H. A. R., (eds.), *Human Health Benefits of Plant Bioactive Compounds: Potentials and Prospects.* USA: Apple Academic Press, Inc.
70. Iahtisham-Ul-Haq, Masood, S. B., Amjad, N., Yasmin, I., & Suleria, H. A. R., (2019). Marine-algal bioactive compounds: A comprehensive appraisal. *Handbook of Algal Technologies and Phytochemicals: Two Volume Set, II* (pp. 69–78).
71. Ibañez, E., & Cifuentes, A., (2013). Benefits of using algae as natural sources of functional ingredients. *Journal of the Science of Food and Agriculture, 93*(4), 703–709.
72. Ichimura, M., Kato, S., Tsuneyama, K., Matsutake, S., Kamogawa, M., Hirao, E., Miyata, A., et al., (2013). Phycocyanin prevents hypertension and low serum adiponectin level in a rat model of metabolic syndrome. *Nutrition Research, 33*(5), 397–405.

73. Ikeda, K., Kitamura, A., Machida, H., Watanabe, M., Negishi, H., Hiraoka, J., & Nakano, T., (2003). Effect of *Undaria* pinnatifida (Wakame) on the development of cerebrovascular diseases in stroke-prone spontaneously hypertensive rats. *Clinical and Experimental Pharmacology and Physiology, 30*(1, 2), 44–48.
74. Ina, A., Hayashi, K. I., Nozaki, H., & Kamei, Y., (2007). Pheophytin a, a low molecular weight compound found in the marine brown alga *Sargassum* fulvellum, promotes the differentiation of PC12 cells. *International Journal of Developmental Neuroscience, 25*(1), 63–68.
75. Islam, M. A., Yun, C. H., Choi, Y. J., & Cho, C. S., (2010). Microencapsulation of live probiotic bacteria. *Journal of Microbiology and Biotechnology, 20*(10), 1367–1377.
76. Judé, S., Roger, S., Martel, E., Besson, P., Richard, S., Bougnoux, P., Champeroux, P., & Le Guennec, J. Y., (2006). Dietary long-chain omega-3 fatty acids of marine origin: A comparison of their protective effects on coronary heart disease and breast cancers. *Progress in Biophysics and Molecular Biology, 90*(1–3), 299–325.
77. Kadam, S. U., Tiwari, B. K., & O'donnell, C. P., (2015). Extraction, structure and biofunctional activities of laminarin from brown algae. *International Journal of Food Science & Technology, 50*(1), 24–31.
78. Kang, H. S., Chung, H. Y., Jung, J. H., Son, B. W., & Choi, J. S., (2003). A new phlorotannin from the brown alga *Ecklonia stolonifera. Chemical and Pharmaceutical Bulletin, 51*(8), 1012–1014.
79. Kang, K. A., Lee, K. H., Chae, S., Koh, Y. S., Yoo, B. S., Kim, J. H., Ham, Y. M., et al., (2005). Triphlorethol-A from *Ecklonia* cava protects V79-4 lung fibroblast against hydrogen peroxide induced cell damage. *Free Radical Research, 39*(8), 883–892.
80. Kang, K. A., Lee, K. H., Chae, S., Zhang, R., Jung, M. S., Ham, Y. M., Baik, J. S., et al., (2006). Cytoprotective effect of phloroglucinol on oxidative stress-induced cell damage via catalase activation. *Journal of Cellular Biochemistry, 97*(3), 609–620.
81. Kang, K. A., Lee, K. H., Chae, S., Zhang, R., Jung, M. S., Lee, Y., Kim, S. Y., et al., (2005). Eckol isolated from *Ecklonia* cava attenuates oxidative stress induced cell damage in lung fibroblast cells. *FEBS Letters, 579*(28), 6295–6304.
82. Kathiresan, S., Sarada, R., Bhattacharya, S., & Ravishankar, G., (2007). Culture media optimization for growth and phycoerythrin production from *Porphyridium purpureum. Biotechnology and Bioengineering, 96*(3), 456–463.
83. Khan, M., Varadharaj, S., Shobha, J. C., Naidu, M. U., Parinandi, N. L., Kutala, V. K., & Kuppusamy, P., (2006). C-phycocyanin ameliorates doxorubicin-induced oxidative stress and apoptosis in adult rat cardiomyocytes. *Journal of Cardiovascular Pharmacology, 47*(1), 9–20.
84. Khodosevich, K., & Monyer, H., (2010). Signaling involved in neurite outgrowth of postnatally born subventricular zone neurons *in vitro. BMC Neuroscience, 11*(1), 18.
85. Kim, M. S., Kim, J. Y., Choi, W. H., & Lee, S. S., (2008). Effects of seaweed supplementation on blood glucose concentration, lipid profile, and antioxidant enzyme activities in patients with type 2 diabetes mellitus. *Nutrition Research and Practice, 2*(2), 62–67.
86. Kong, C. S., Kim, J. A., & Kim, S. K., (2009). Anti-obesity effect of sulfated glucosamine by AMPK signal pathway in 3T3-L1 adipocytes. *Food and Chemical Toxicology, 47*(10), 2401–2406.

87. Kong, C. S., Kim, J. A., Yoon, N. Y., & Kim, S. K., (2009). Induction of apoptosis by phloroglucinol derivative from *Ecklonia* cava in MCF-7 human breast cancer cells. *Food and Chemical Toxicology, 47*(7), 1653–1658.
88. Kristensen, M., & Jensen, M. G., (2011). Dietary fibres in the regulation of appetite and food intake. Importance of viscosity. *Appetite, 56*(1), 65–70.
89. Kwak, J. Y., (2014). Fucoidan as a marine anticancer agent in preclinical development. *Marine Drugs, 12*(2), 851–870.
90. Lahrsen, E., Schoenfeld, A. K., & Alban, S., (2018). Size-dependent pharmacological activities of differently degraded fucoidan fractions from *Fucus vesiculosus*. *Carbohydrate Polymers, 189*, 162–168.
91. Langebeek, N., Gisolf, E. H., Reiss, P., Vervoort, S. C., Hafsteinsdóttir, T. B., Richter, C., Sprangers, M. A., & Nieuwkerk, P. T., (2014). Predictors and correlates of adherence to combination antiretroviral therapy (ART) for chronic HIV infection: A meta-analysis. *BMC Medicine, 12*(1), 142.
92. Leandro, A., Pereira, L., & Gonçalves, A., (2020). Diverse applications of marine macroalgae. *Marine Drugs, 18*, 17.
93. Lee, H. Y., Won, J. C., Kang, Y. J., Yoon, S. H., Choi, E. O., Bae, J. Y., Sung, M. H., et al., (2010). Type 2 diabetes in urban and rural districts in Korea: Factors associated with prevalence difference. *Journal of Korean Medical Science, 25*(12), 1777–1783.
94. Li, Y. X., Wijesekara, I., Li, Y., & Kim, S. K., (2011). Phlorotannins as bioactive agents from brown algae. *Process Biochemistry, 46*(12), 2219–2224.
95. Li, Y., Qian, Z. J., Kim, M. M., & Kim, S. K., (2011). Cytotoxic activities of phlorethol and fucophlorethol derivatives isolated from *Laminariaceae Ecklonia* cava. *Journal of Food Biochemistry, 35*(2), 357–369.
96. Lin, C. Y., Lee, C. H., Chang, Y. W., Wang, H. M., Chen, C. Y., & Chen, Y. H., (2014). Pheophytin a inhibits inflammation via suppression of LPS-induced nitric oxide synthase-2, prostaglandin E2, and interleukin-1β of macrophages. *International Journal of Molecular Sciences, 15*(12), 22819–22834.
97. Liu, L., Heinrich, M., Myers, S., & Dworjanyn, S. A., (2012). Towards a better understanding of medicinal uses of the brown seaweed sargassum in traditional Chinese medicine: A phytochemical and pharmacological review. *Journal of Ethnopharmacology, 142*(3), 591–619.
98. Macartain, P., Gill, C. I., Brooks, M., Campbell, R., & Rowland, I. R., (2007). Nutritional value of edible seaweeds. *Nutrition Reviews, 65*(12), 535–543.
99. Maeda, H., Hosokawa, M., Sashima, T., Murakami-Funayama, K., & Miyashita, K., (2009). Anti-obesity and anti-diabetic effects of fucoxanthin on diet-induced obesity conditions in a murine model. *Molecular Medicine Reports, 2*(6), 897–902.
100. Maeda, H., Tsukui, T., Sashima, T., Hosokawa, M., & Miyashita, K., (2008). Seaweed carotenoid, fucoxanthin, as a multi-functional nutrient. *Asia Pacific Journal of Clinical Nutrition, 17*.
101. Maeda, H., Yamamoto, R., Hirao, K., & Tochikubo, O., (2005). Effects of agar (kanten) diet on obese patients with impaired glucose tolerance and type 2 diabetes. *Diabetes, Obesity and Metabolism, 7*(1), 40–46.
102. Malyarenko, O. S., & Ermakova, S. P., (2017). Fucoidans: Anticancer activity and molecular mechanisms of action. *Seaweed Polysaccharides*. Elsevier.

103. Marková, I., Koníčková, R., Vaňková, K., Leníček, M., Kolář, M., Strnad, H., Hradilová, M., et al., (2020). Anti-angiogenic effects of the blue-green alga *Arthrospira platensis* on pancreatic cancer. *Journal of Cellular and Molecular Medicine.*
104. Martínez-Hernández, G. B., Castillejo, N., Carrión-Monteagudo, M. D. M., Artés, F., & Artés-Hernández, F., (2017). Nutritional and bioactive compounds of commercialized algae powders used as food supplements. *Food Science and Technology International.*
105. Matanjun, P., Mohamed, S., Muhammad, K., & Mustapha, N. M., (2010). Comparison of cardiovascular protective effects of tropical seaweeds, *Kappaphycus alvarezii*, *Caulerpa lentillifera*, and *Sargassum polycystum*, on high-cholesterol/high-fat diet in rats. *Journal of Medicinal Food, 13*(4), 792–800.
106. Menotti, A., Lanti, M., Kromhout, D., Blackburn, H., Jacobs, D., Nissinen, A., Dontas, A., et al., (2008). Homogeneity in the relationship of serum cholesterol to coronary deaths across different cultures: 40-year follow-up of the seven countries study. *European Journal of Cardiovascular Prevention & Rehabilitation, 15*(6), 719–725.
107. Metting, F., (1996). Biodiversity and application of microalgae. *Journal of Industrial Microbiology, 17*(5, 6), 477–489.
108. Michalak, I., & Chojnacka, K., (2015). Algae as production systems of bioactive compounds. *Engineering in Life Sciences, 15*(2), 160–176.
109. Michikawa, T., Inoue, M., Shimazu, T., Sawada, N., Iwasaki, M., Sasazuki, S., Yamaji, T., et al., (2012). Seaweed consumption and the risk of thyroid cancer in women: The Japan public health center-based prospective study. *European Journal of Cancer Prevention, 21*(3), 254–260.
110. Milledge, J. J., (2011). Commercial application of microalgae other than as biofuels: A brief review. *Reviews in Environmental Science and Bio/Technology, 10*(1), 31–41.
111. Murata, M., Ishihara, K., & Saito, H., (1999). Hepatic fatty acid oxidation enzyme activities are stimulated in rats fed the brown seaweed, *Undaria pinnatifida* (wakame). *The Journal of Nutrition, 129*(1), 146–151.
112. Murray, M., Dordevic, A. L., Bonham, M. P., & Ryan, L., (2017). Do marine algal polyphenols have antidiabetic, antihyperlipidemic or anti-inflammatory effects in humans? A systematic review. *Critical Reviews in Food Science and Nutrition,* 1–16.
113. Myers, S. P., O'connor, J., Fitton, J. H., Brooks, L., Rolfe, M., Connellan, P., Wohlmuth, H., et al., (2010). A combined phase I and II open-label study on the effects of a seaweed extract nutrient complex on osteoarthritis. *Biologics: Targets & Therapy, 4*, 33.
114. Niu, J. F., Wang, G. C., Lin, X. Z., & Zhou, B. C., (2007). Large-scale recovery of C-phycocyanin from Spirulina platensis using expanded bed adsorption chromatography. *Journal of Chromatography B, 850*(1, 2), 267–276.
115. Nova, P., Pimenta-Martins, A., Laranjeira, S. J., Silva, A. M., Gomes, A. M., & Freitas, A. C., (2020). Health benefits and bioavailability of marine resources components that contribute to health-what's new? *Critical Reviews in Food Science and Nutrition,* 1–13.

116. O'sullivan, L., Murphy, B., Mcloughlin, P., Duggan, P., Lawlor, P. G., Hughes, H., & Gardiner, G. E., (2010). Prebiotics from marine macroalgae for human and animal health applications. *Marine Drugs, 8*(7), 2038–2064.
117. Okada, T., Nakai, M., Maeda, H., Hosokawa, M., Sashima, T., & Miyashita, K., (2008). Suppressive effect of neoxanthin on the differentiation of 3T3-L1 adipose cells. *Journal of Oleo Science, 57*(6), 345–351.
118. Okuzumi, J., Nishino, H., Murakoshi, M., Iwashima, A., Tanaka, Y., Yamane, T., Fujita, Y., & Takahashi, T., (1990). Inhibitory effects of fucoxanthin, a natural carotenoid, on N-myc expression and cell cycle progression in human malignant tumor cells. *Cancer Letters, 55*(1), 75–81.
119. Olaizola, M., (2003). Commercial development of microalgal biotechnology: From the test tube to the marketplace. *Biomolecular Engineering, 20*(4–6), 459–466.
120. Pallela, R., (2014). Antioxidants from marine organisms and skincare. *Systems Biology of Free Radicals and Antioxidants.* Springer.
121. Pallela, R., Na-Young, Y., & Kim, S. K., (2010). Anti-photoaging and photoprotective compounds derived from marine organisms. *Marine Drugs, 8*(4), 1189–1202.
122. Pangestuti, R., & Kim, S. K., (2011). Biological activities and health benefit effects of natural pigments derived from marine algae. *Journal of Functional Foods, 3*(4), 255–266.
123. Pangestuti, R., & Kim, S. K., (2011). Neuroprotective effects of marine algae. *Marine Drugs, 9*(5), 803–818.
124. Pangestuti, R., & Kim, S. K., (2015). Seaweeds-derived bioactive materials for the prevention and treatment of female's cancer. *Handbook of Anticancer Drugs from Marine Origin.* Springer.
125. Paradis, M. E., Couture, P., & Lamarche, B., (2011). A randomized crossover placebo-controlled trial investigating the effect of brown seaweed (*Ascophyllum nodosum* and *Fucus vesiculosus*) on post-challenge plasma glucose and insulin levels in men and women. *Applied Physiology, Nutrition, and Metabolism, 36*(6), 913–919.
126. Ponce, N. M., Pujol, C. A., Damonte, E. B., Flores, M. L., & Stortz, C. A., (2003). Fucoidans from the brown seaweed adenocystis utricularis: Extraction methods, antiviral activity and structural studies. *Carbohydrate Research, 338*(2), 153–165.
127. Pozza, C., & Isidori, A. M., (2018). What's behind the obesity epidemic. *Imaging in Bariatric Surgery.* Springer.
128. Pulz, O., & Gross, W., (2004). Valuable products from biotechnology of microalgae. *Applied Microbiology and Biotechnology, 65*(6), 635–648.
129. Radmer, R. J., (1996). Algal diversity and commercial algal products. *Bioscience, 46*(4), 263–270.
130. Radwan, S. S., (1991). Sources of C 20-polyunsaturated fatty acids for biotechnological use. *Applied Microbiology and Biotechnology, 35*(4), 421–430.
131. Ragan, M., & Glombitza, K., (1986). *Handbook of Physiological Methods*. Cambridge University Press: Cambridge, UK.
132. Ramnani, P., Chitarrari, R., Tuohy, K., Grant, J., Hotchkiss, S., Philp, K., Campbell, R., et al., (2012). *In vitro* fermentation and prebiotic potential of novel low molecular weight polysaccharides derived from agar and alginate seaweeds. *Anaerobe, 18*(1), 1–6.

133. Rani, K., Sandal, N., & Sahoo, P., (2018). A comprehensive review on chlorella-its composition, health benefits, market and regulatory scenario. *The Pharma Innovation Journal, 7*(7), 584–589.
134. Rastall, R. A., Gibson, G. R., Gill, H. S., Guarner, F., Klaenhammer, T. R., Pot, B., Reid, G., et al., (2005). Modulation of the microbial ecology of the human colon by probiotics, prebiotics and synbiotics to enhance human health: An overview of enabling science and potential applications. *FEMS Microbiology Ecology, 52*(2), 145–152.
135. Riss, J., Décordé, K., Sutra, T., Delage, M., Baccou, J. C., Jouy, N., Brune, J. P., et al., (2007). Phycobiliprotein C-phycocyanin from spirulina platensis is powerfully responsible for reducing oxidative stress and NADPH oxidase expression induced by an atherogenic diet in hamsters. *Journal of Agricultural and Food Chemistry, 55*(19), 7962–7967.
136. Roberts, A. W., (2010). Cardiovascular risk and prevention in diabetes mellitus. *Clinical Medicine, 10*(5), 495–499.
137. Roohinejad, S., Nikmaram, N., Brahim, M., Koubaa, M., Khelfa, A., & Greiner, R., (2018). Potential of novel technologies for aqueous extraction of plant bioactives. *Water Extraction of Bioactive Compounds*. Elsevier.
138. Saewan, N., & Jimtaisong, A., (2015). Natural products as photoprotection. *Journal of Cosmetic Dermatology, 14*(1), 47–63.
139. Scholz-Ahrens, K. E., Ade, P., Marten, B., Weber, P., Timm, W., Açil, Y., GlüEr, C. C., & Schrezenmeir, J. R., (2007). Prebiotics, probiotics, and synbiotics affect mineral absorption, bone mineral content, and bone structure. *The Journal of Nutrition, 137*(3), 838S–846S.
140. Sekar, S., & Chandramohan, M., (2008). Phycobiliproteins as a commodity: Trends in applied research, patents and commercialization. *Journal of Applied Phycology, 20*(2), 113–136.
141. Selassie, M., & Sinha, A. C., (2011). The epidemiology and aetiology of obesity: A global challenge. *Best Practice & Research Clinical Anaesthesiology, 25*(1), 1–9.
142. Senevirathne, M., & Kim, S. K., (2011). 14 brown algae-derived compounds as potential cosmeceuticals. *Marine Cosmeceuticals: Trends and Prospects,* 179.
143. Shaw, C. K., (1993). An epidemiologic study of osteoporosis in Taiwan. *Annals of Epidemiology, 3*(3), 264–271.
144. Shehzad, A., Zahid, A., Latif, A., Amir, R. M., & Suleria, H. A. R., (2019). Marine foods: Nutritional significance and their industrial applications. *Technological Processes for Marine Foods, From Water to Fork: Bioactive Compounds, Industrial Applications, and Genomics,* 289.
145. Shibata, T., Ishimaru, K., Kawaguchi, S., Yoshikawa, H., & Hama, Y., (2008). Antioxidant activities of phlorotannins isolated from Japanese *Laminariaceae*. *Journal of Applied Phycology, 20*(5), 705.
146. Shin, A., Lim, S. Y., Sung, J., Shin, H. R., & Kim, J., (2009). Dietary intake, eating habits, and metabolic syndrome in Korean men. *Journal of the American Dietetic Association, 109*(4), 633–640.
147. Shin, C. S., Choi, H. J., Kim, M. J., Kim, J. T., Yu, S. H., Koo, B. K., Cho, H. Y., et al., (2010). Prevalence and risk factors of osteoporosis in Korea: A community-based cohort study with lumbar spine and hip bone mineral density. *Bone, 47*(2), 378–387.

148. Simopoulos, A. P., (2002). Omega-3 fatty acids in inflammation and autoimmune diseases. *Journal of the American College of Nutrition, 21*(6), 495–505.
149. Siriwardhana, N., Lee, K. W., & Jeon, Y. J., (2005). Radical scavenging potential of hydrophilic phlorotannins of hizikia fusiformis. *Algae, 20*(1), 69–75.
150. Spieler, R., (2002). Seaweed compound's anti-HIV efficacy will be tested in southern Africa. *The Lancet, 359*(9318), 1675.
151. Spolaore, P., Joannis-Cassan, C., Duran, E., & Isambert, A., (2006). Commercial applications of microalgae. *Journal of Bioscience and Bioengineering, 101*(2), 87–96.
152. Sugawara, T., Matsubara, K., Akagi, R., Mori, M., & Hirata, T., (2006). Antiangiogenic activity of brown algae fucoxanthin and its deacetylated product, fucoxanthinol. *Journal of Agricultural and Food Chemistry, 54*(26), 9805–9810.
153. Sugiura, Y., Matsuda, K., Yamada, Y., Nishikawa, M., Shioya, K., Katsuzaki, H., Imai, K., & Amano, H., (2007). Anti-allergic phlorotannins from the edible brown alga, *Eisenia arborea. Food Science and Technology Research, 13*(1), 54–60.
154. Sugiura, Y., Nagayama, K., Kinoshita, Y., Tanaka, R., & Matsushita, T., (2015). The anti-allergic effect of the ethyl acetate fraction from an *Ecklonia kurome* extract. *Food and Agricultural Immunology, 26*(2), 181–193.
155. Suryanarayan, S., Neish, I. C., Nori, S., & Vadassery, N., (2018). Cultivation and conversion of tropical red seaweed into food and feed ingredients, agricultural biostimulants, renewable chemicals, and biofuel. *Blue Biotechnology: Production and Use of Marine Molecules, 1*, 241–264.
156. Tang, B. M., Eslick, G. D., Nowson, C., Smith, C., & Bensoussan, A., (2007). Use of calcium or calcium in combination with vitamin D supplementation to prevent fractures and bone loss in people aged 50 years and older: A meta-analysis. *The Lancet, 370*(9588), 657–666.
157. Teas, J., Baldeón, M. E., Chiriboga, D. E., Davis, J. R., Sarriés, A. J., & Braverman, L. E., (2009). Could dietary seaweed reverse the metabolic syndrome? *Asia Pacific Journal of Clinical Nutrition, 18*(2), 145.
158. Teas, J., Hebert, J. R., Fitton, J. H., & Zimba, P. V., (2004). Algae-a poor man's HAART? *Medical Hypotheses, 62*(4), 507–510.
159. Teas, J., & Irhimeh, M., (2012). Dietary algae and HIV/AIDS: Proof of concept clinical data. *Journal of Applied Phycology, 24*(3), 575–582.
160. Thompson, Jr. G., (1996). Jr. Lipids and membrane function in green algae. *Biochim. Biophys. Acta, 1302*(1), 17–45.
161. Torsdottir, I., Alpsten, M., Holm, G., Sandberg, A. S., & Tölli, J., (1991). A small dose of soluble alginate-fiber affects postprandial glycemia and gastric emptying in humans with diabetes. *The Journal of Nutrition, 121*(6), 795–799.
162. Trayhurn, P., & Wood, I., (2005). *Signaling Role of Adipose Tissue: Adipokines and Inflammation in Obesity.* Portland Press Limited.
163. Trinchero, J., Ponce, N. M., Córdoba, O. L., Flores, M. L., Pampuro, S., Stortz, C. A., Salomón, H., & Turk, G., (2009). Antiretroviral activity of fucoidans extracted from the brown seaweed adenocystis utricularis. *Phytotherapy Research: An International Journal Devoted to Pharmacological and Toxicological Evaluation of Natural Product Derivatives, 23*(5), 707–712.
164. Tucci, P., & Bagetta, G., (2008). *How to Study Neuroprotection?* Nature Publishing Group.

165. Tuohy, K. M., Probert, H. M., Smejkal, C. W., & Gibson, G. R., (2003). Using probiotics and prebiotics to improve gut health. *Drug Discovery Today, 8*(15), 692–700.
166. Vaseghi, G., Sharifi, M., Dana, N., Ghasemi, A., & Yegdaneh, A., (2018). Cytotoxicity of *Sargassum angustifolium* partitions against breast and cervical cancer cell lines. *Advanced Biomedical Research, 7*.
167. Vaugelade, P., Hoebler, C., Bernard, F., Guillon, F., Lahaye, M., Duee, P. H., & Darcy-Vrillon, B., (2000). Non-starch polysaccharides extracted from seaweed can modulate intestinal absorption of glucose and insulin response in the pig. *Reproduction Nutrition Development, 40*(1), 33–47.
168. Vaughan, V., Hassing, M., & Lewandowski, P., (2013). Marine polyunsaturated fatty acids and cancer therapy. *British Journal of Cancer, 108*(3), 486.
169. Wang, H. M. D., Li, X. C., Lee, D. J., & Chang, J. S., (2017). Potential biomedical applications of marine algae. *Bioresource Technology, 244*(1407–1415.
170. Ward, O. P., & Singh, A., (2005). Omega-3/6 fatty acids: Alternative sources of production. *Process Biochemistry, 40*(12), 3627–3652.
171. Wells, M. L., Potin, P., Craigie, J. S., Raven, J. A., Merchant, S. S., Helliwell, K. E., Smith, A. G., et al., (2017). Algae as nutritional and functional food sources: revisiting our understanding. *Journal of Applied Phycology, 29*(2), 949–982.
172. Wen, Z. Y., & Chen, F., (2003). Heterotrophic production of eicosapentaenoic acid by microalgae. *Biotechnology Advances, 21*(4), 273–294.
173. Wijesekara, I., Pangestuti, R., & Kim, S. K., (2011). Biological activities and potential health benefits of sulfated polysaccharides derived from marine algae. *Carbohydrate Polymers, 84*(1), 14–21.
174. Woo, M. N., Jeon, S. M., Shin, Y. C., Lee, M. K., Kang, M., & Choi, M. S., (2009). Anti-obese property of fucoxanthin is partly mediated by altering lipid-regulating enzymes and uncoupling proteins of visceral adipose tissue in mice. *Molecular Nutrition & Food Research, 53*(12), 1603–1611.
175. Yeh, C. J., Chang, H. Y., & Pan, W. H., (2011). Time trend of obesity, the metabolic syndrome and related dietary pattern in Taiwan: From NAHSIT 1993-1996 to NAHSIT 2005-2008. *Asia Pacific Journal of Clinical Nutrition, 20*(2), 292.
176. Yende, S. R., Harle, U. N., & Chaugule, B. B., (2014). Therapeutic potential and health benefits of *Sargassum* species. *Pharmacognosy Reviews, 8*(15), 1.
177. Yongmanitchai, W., (1989). Omega-3 fatty acids: Alternative sources of production. *Process Biochemistry, 24*, 117–125.
178. Yuan, Y. V., & Walsh, N. A., (2006). Antioxidant and antiproliferative activities of extracts from a variety of edible seaweeds. *Food and Chemical Toxicology, 44*(7), 1144–1150.
179. Yue, P. Y., Leung, H., Li, A. J., Chan, T. N., Lum, T., Chung, Y., Sung, Y., et al., (2017). Angio suppressive properties of marine-derived compounds—a mini-review. *Environmental Science and Pollution Research, 24*(10), 8990–9001.
180. Zeitlin, L., & Whaley, K. J., (2002). Microbicides for preventing transmission of genital herpes. *Herpes: The Journal of the IHMF, 9*(1), 4–9.

181. Zhang, L., Wang, H., Fan, Y., Gao, Y., Li, X., Hu, Z., Ding, K., Wang, Y., & Wang, X., (2017). Fucoxanthin provides neuroprotection in models of traumatic brain injury via the Nrf2-ARE and Nrf2-autophagy pathways. *Scientific Reports, 7*, 746–763.
182. Zhu, W., Ooi, V., Chan, P., & Ang, Jr. P., (2003). Inhibitory effect of extracts of marine algae from Hong Kong against herpes simplex viruses. *Proceedings of the 17th International Seaweed Symposium* (pp. 159–164). Oxford University Press, Oxford.

PART II

Health Claims of Bioactive Compounds from Medicinal Plants

CHAPTER 8

MEDICINAL PLANTS TO HERBAL DRUGS: DEVELOPMENT OF CRUDE DRUGS

VINITHA SASEENDRA BABU and
PUTHUPARAMBIL MADHAVAN RADHAMANY

ABSTRACT

Plants provide to be the richest sources of medicinal value. Time long past, medicinal plants have been used by traditional practitioners for the treatment of several ailments and hence have been popularized. The botanical constituents from medicinal plants can help in discovering new medicinal entities. But, before being used as a crude drug, clinical trials are necessary to determine its availability, safety, efficacy as well as stability. The trials should be done with careful planning so as to ensure the health of participants before being released and given to patients. Hence, this chapter tries to explain several methodologies adopted to elucidate a crude drug from plants. This includes standardization of herbal drugs, extraction techniques as well as its characterization through spectrophotometric and non-spectrophotometric methods. Standardization is the first key step in the development of drugs from any medicinal plant. This includes determining the morphology, anatomy, and so on of plants. Extraction refers to the process of obtaining certain components from plants. Several methods of extraction include the traditional maceration to the advanced supercritical fluid extraction. Finally, characterization of the extractants obtained through thin layer chromatography (TLC), high-performance chromatography (HPTLC), nuclear magnetic resonance imaging (NMR), and Fourier transform infra-red (FTIR) are discussed.

8.1 INTRODUCTION

Medicinal herbs are the most resource base of just about all indigenous preventive medicine. The long-established system of drugs is still being practiced. The resurrection of phytomedicines is due to increasing evidence of the human health hazards associated with many of the synthetic medicines. Since ages, mankind has used extracts from different plants to cure many diseases and thus relieve him from physical agony [10]. Herbal plants are rich in formulations that can be used as active ingredients for the development of drugs that are either pharmacopeial, non-pharmacopeial, or synthetic. Aside from that, these plants play a critical role within the development of human civilizations around the whole world. Moreover, many plants are also considered as a very important source of nutrition and as a result of which they are recommended for his or her therapeutic values.

About 20,000 plant species are used for medicinal purposes as declared by World Health Organization (WHO). The importance of using medicinal plants is ascribed to a variety of reasons, including inexpensiveness and short supply of Western medicine to the under developed countries. Further, they have also developed trust in herbal medicine due to the positive results obtained when applying herbs [11]. One of the many reasons that prompted indigenous people from such communities to rely heavily on plant-based traditional herbs is of the high cost of conventional medicines. Therefore, plants are of infinite value to less affluent populations.

Traditional medicines utilize biological resources and the indigenous knowledge of traditional plant groups, the latter being conveyed verbally from generation to generation. This is closely linked to the conservation of biodiversity and the related intellectual property rights of indigenous people. Even though the bygone decades have witnessed an increase in the use of traditional medical system, the research data in this field is scanty. As a result of this, the WHO has published three volumes of adumbration on selected medicinal plants. Recently, WHO declared that nearly 80% of people around the world rely for their health care on herbal plants. Accordingly, around 22,000 promising plants can be used as medicine. Population expansion, inadequate supply of medication, henpecked cost of treatments, concomitant effects of many synthetic drugs and development of defiance to currently used drugs for communicable diseases have led to increased emphasis on the employment of plant materials as a source of medicines for a large kind of human ailments [2].

As stated by WHO, medicinal plants can be described as the "diverse health practices, approaches, knowledge, and belief incorporating plant-, animal-, and/or mineral-based medicines, spiritual therapies, manual techniques and exercises applied singularly or in combination to maintain well-being, as well as to treat, diagnose, or prevent illness." It is clear, however, that there is a need to validate the information through an organized infrastructure for it to be used as an effective therapeutic means, either in conjunction with existing therapies, or as a tool in novel drug discovery. A brief summary of the overall process of standardization, extraction, characterization, and isolation of botanicals is depicted (Figure 8.1). Hence, understanding of the target species, standardization, extraction, isolation, and characterization of metabolites from the botanicals for drug development and adoption of conservation strategies are necessary to retain the target species in nature.

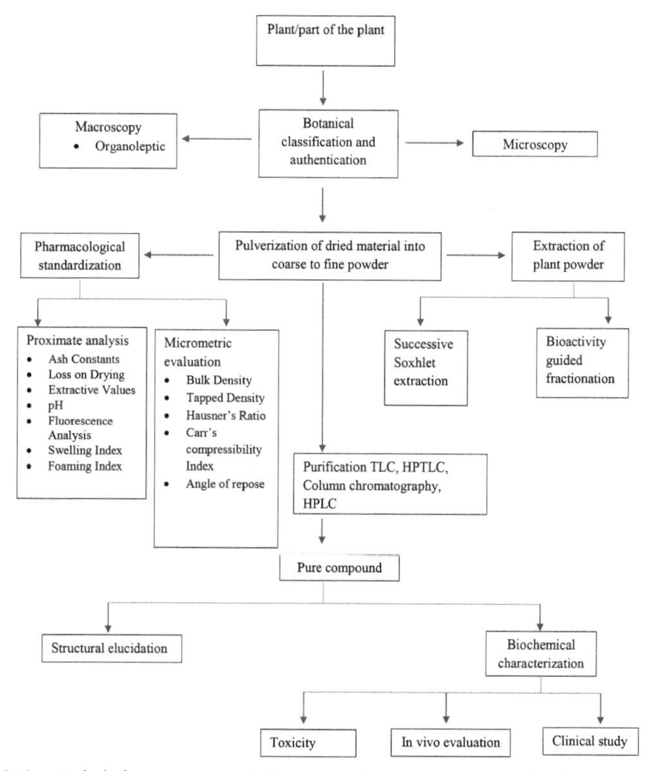

FIGURE 8.1 A brief summary of the overall process of standardization, extraction, characterization, and isolation of botanicals.

8.2 STANDARDIZATION OF HERBALS

Standardization of herbals refers to a set of parameters which can attribute to intrinsic characteristics in order to ensure the quality of crude drugs, so that they can be safely handled. With respect to plants, these standardization parameters at first consists of botanical identification and authentication by an eminent taxonomist, along with allied literatures. This can be further guaranteed through macroscopical and microscopical techniques. Within macroscopy, determination of sensory parameters like color, odor, texture, and surface characteristics are considered. The microscopical methods uses, preparation of anatomical sections to understand the plethora of cells and tissues.

Once, the authentication is done, the plant material is pulverized into fine powder to conduct pharmacognostical standardization and extraction. Pharmacognostical standardization is essential as it helps in the determination of the proximate elements which include ash values, loss on drying, swelling index, foaming index extractive values, heavy metal contamination, fluorescence analysis and so on. This is critical as it helps in knowing the amount of various contaminants in the plant powder. On the other hand, micrometric evaluation helps in determining the flow properties of the plant sample in order to develop as a crude drug. It includes bulk density, tapped density, Hausner's ratio, Carr's compressibility index, angle of repose, and so on. These are very important for the development of herbals into crude drugs [4].

8.3 EXTRACTION

Extraction, as the term is used pharmaceutically, involves the separation of medicinally active portions of plant or animal tissues from the inactive or inert components by using selective solvents. The products so obtained from plants are relatively impure liquids, semisolids or powders intended only for oral or external use [1, 5, 6]. Thus, standardization of extraction procedures contributes significantly to the final quality of the herbal drug. Different extraction methods employed may vary from the traditional maceration, infusion, digestion, decoction, and percolation methods to the advanced methods like Soxhlet extraction, sonication, supercritical fluid extraction and photonics process. In the process of maceration, the intact tissue gets converted into a suspension of cells that ultimately releases products in the form of pulp. This can be used for the development of dairy goods like yogurt and puddings. This process helps in improving the flavor and texture of fresh herbals. In the process of infusion, the chemical compounds can be obtained by keeping the plant material either in polar or non-polar solvents for a long period of time. For infusion, the flowers or leaves of the plant are chiefly used.

It is in fact, a process of mixing together to make it stronger. The methodology employed in digestion is similar to maceration except that, in digestion, warming is required for extracting the components. For that, slight temperature is provided that, the temperature will not alter the chemical composition. Decoction is similar to infusion except that, it is usually made with bark or root of a plant. Generally, herbal decoctions have to be immediately used after making them or it has to be consumed afresh within 24 hours of preparation. Percolation is an extraction technique where a solvent is passed through the herbal material drop by drop for a long period of time to obtain the extractant. Soxhlet is a laboratory apparatus used for the extraction of a liquid from a solid substance. Here, the herbals are added once in powder form.

This technique is generally employed, when the plant material possesses limited availability, and also when the compound to be isolated has less solubility in its desired solvent. In sonication, ultrasonic sound waves are employed to agitate the components in the sample, thereby extracting the desired product. Here, high power ultrasounds are applied to a slurry of plant tissues. It helps in the removal of gases from liquids. In supercritical fluid extraction, the separation of components can be done by applying using supercritical fluids like carbon dioxide. The extraction is usually done to separate solids but liquid components can also be separated. The separation technique is innovative and ecologically sound. As an alternative to the supercritical carbon dioxide, photonics extraction methods have been developed. Here, a novel non-toxic solvent based on hydrocarbons is used. This method is employed to extract oils and flavors from plant extracts. The substances obtained so can be directly used without any further treatments [3].

8.4 CHROMATOGRAPHIC TECHNIQUES

Once the extraction process is completed, the identification of the extractant is the next crucial step in drug discovery. This can be done with the help of chromatographic and spectroscopic techniques. Chromatographic techniques like column chromatography, thin-layer chromatography (TLC), high performance TLC and high-performance liquid chromatography and spectroscopic methods like ultraviolet (UV), infrared, mass spectroscopy, nuclear magnetic resonance and Fourier transform infrared spectroscopy serve this purpose. Marker compounds may be utilized for identification of herbal materials, for setting the specifications for raw materials, for standardizing herbal preparations during all aspects of manufacturing processes, and for obtaining stability profiles.

Chromatography refers to a method of separation of components from a mixture. Several chromatographic methods wide application as far as drug development is concerned. The desired extractant once obtained after successful phytochemical investigations are first subjected to column chromatography. This method is done to obtain a single compound from a mixture of compounds. The main principle behind column chromatography is the adsorption of components of an extractant through a stationary phase and its separation into individual components. The stationary phase may be either liquid, solid or ions based on the nature of chromatography. The fractions obtained from the column are checked for individual components through another chromatographic technique called TLC. This is mainly employed to separate non-volatile components from the mixture.

It is usually performed in an inert adsorbent material like silica, cellulose, and so on. An upgraded version of TLC is high performance TLC where accurate quantitative measurements can be made. In general, the principle of adsorption is used to separate components. Capillary action is the chief cause for the flow of components in high performance TLC. High performance liquid chromatography is a sophisticated method for separation, identification, purification, and quantification of components from a mixture [9]. The principle behind this technique is the ability of some compounds to travel at a faster rate than the others in a column. This is an advanced version of column chromatography, where separation takes place due to the pressure applied to the solvent.

8.5 SPECTROSCOPIC TECHNIQUES

The UV spectroscopic methods are employed chiefly in liquid formulations. Very often, solids, and gases are also considered. Here, quantitative, and qualitative determination of samples are usually done. It is very versatile and one of the most popular spectroscopic methods. The principle behind this technique is the difference in measurement of absorption of a ray of UV light once it passes through a sample. It can also detect functional groups, check impurities, and can identify drugs with and without a chromophore [7]. On the other hand, infrared spectroscopy uses infrared radiation as a source of light. As is for UV, the difference in absorption of light along with the emission and reflection is also measured in infrared spectroscopy. It helps in the qualitative determination of the sample. The mass spectroscopy is used to measure the ratios of mass with a charge of the ions. As the name indicates, this is done to determine the mass of the compound as well as elucidate the chemical identity and structure of the molecules.

Nuclear magnetic resonance spectroscopy is a versatile technology that has been adopted to derive the structure of any chemical molecule [8]. This technique is based on the observation of magnetic field around atomic nuclei. It helps in determination of the amount and purity of any crude drug. Organic molecules in solutions, crystals as well as non-crystalline materials are chiefly employed for this purpose. Fourier transform infrared spectroscopy is used to identify organic as well as inorganic materials in sample. The principle behind this technique is the difference in absorption of light between the bonds of elements. This technique is chiefly used to determine the functional groups of a compound. It can also detect the presence of metal contamination in any sample.

8.6 SUMMARY

The process of obtaining a pure compound from plants is tedious as it involves standardization, extraction, isolation, characterization, along with clinical studies. A thorough knowledge in these areas can help to devise new therapeutic drugs that can combat infectious diseases. As modern allopathic medicines can cause several side effects like memory loss, gastrointestinal (GI) troubles, and so on, herbal compounds are gaining much attention these days. So, the potential of medicinal plants can also be explored with the help of indigenous plant healers. Coupled with their knowledge along with the sophisticated instrumentation facilities of the contemporary world, can help to develop new medications.

KEYWORDS

- **extraction**
- **isolation**
- **medicinal plants**
- **metabolites**
- **phytomedicines**
- **thin-layer chromatography**

REFERENCES

1. Abubakar, A. R., & Haque, M., (2020). Preparation of medicinal plants: Basic extraction and fractionation procedures for experimental purposes. *Journal of Pharmacy and Bioallied Sciences, 1*(2), 1–10.
2. Dirar, A. I., Alsaadi, D. H., Wada, M., Mohamed, M. A., Watanabe, T., & Devkota, H. P., (2019). Effects of extraction solvents on total phenolic and flavonoid contents and biological activities of extracts from Sudanese medicinal plants. *South African Journal of Botany, 6*(1), 261–267.
3. Kalantzopoulos, G. N., Lundvall, F., Checchia, S., Lind, A., Wragg, D. S., Fjellvåg, H., & Arstad, B., (2018). In situ flow MAS NMR spectroscopy and synchrotron PDF analyses of the local response of the Bronsted acidic site in SAPO-34 during hydration at elevated temperatures. *Chem. Phys. Chem., 1*(9), 519–528.
4. Khan, S. U., Anjum, S. I., Ansari, M. J., Khan, M. H., Kamal, S., Rahman, K., Shoaib, M., et al., (2019). Antimicrobial potentials of medicinal plant's extract and their derived silver nanoparticles: A focus on honey bee pathogen. *Saudi Journal of Biological Sciences, 6*(3), 1815–1834.
5. Mohotti, S., Rajendran, S., Muhammad, T., Strömstedt, A. A., Adhikari, A., Burman, R., & Gunasekera, S., (2020). Screening for bioactive secondary metabolites in Sri Lankan medicinal plants by microfractionation and targeted isolation of antimicrobial flavonoids from *Derris scandens. Journal of Ethnopharmacology, 2*(6), 112158.
6. Pedan, V., Weber, C., Do, T., Fischer, N., Reich, E., & Rohn, S., (2018). HPTLC fingerprint profile analysis of cocoa proanthocyanidins depending on origin and genotype. *Food Chemistry, 2*(7), 277–287.

7. Sobolev, A. P., Thomas, F., Donarski, J., Ingallina, C., Circi, S., Marincola, F. C., & Mannina, L., (2019). Use of NMR applications to tackle future food fraud issues. *Trends in Food Science & Technology, 2*(4), 167–175.
8. Tahir, M. A., & Abbasi, M. S., (2020). FTIR spectroscopic analysis of *Mentha spicata* L. (Garden Mint). *Asian Journal of Chemical Sciences, 1*(4), 1–5.
9. Tiwari, B., Shirsat, M. K., & Kulkarni, A., (2020). Analytical method development and validation for the determination of brinzolamide by RP-HPLC. *Journal of Drug Delivery and Therapeutics, 10*(6), 92–96.
10. Xiao, Y., Qunying, Tu, F., & Yan, Y., (2020). The complete chloroplast genome of *Mahonia oiwakensis* (Berberidaceae), a traditional Chinese medicinal plant. *Mitochondrial DNA Part B, 5*(4), 692–694.
11. Dong-Ping, X., Ya, L., Xiao, M., Tong, Z., Yue, Z., Jie, Z., et al., (2017). Natural antioxidants in foods and medicinal plants: Extraction, assessment, and resources. *International Journal of Molecular Sciences, 1*(8), 96–113.

CHAPTER 9

CITRULLUS COLOCYNTHIS: PHYTOCHEMICAL AND MEDICINAL PERSPECTIVES

ANEES AHMED KHALIL, QURAT UL AIN SHAHID,
UBAID UR RAHMAN, ANUM ISHAQ, and
RANA MUHAMMAD AADIL

ABSTRACT

Citrullus colocynthis, commonly known as bitter apple, is a desert plant that is native to Asia and the Mediterranean basin. The plant has large and fleshy roots, vine-like stem, angular leaves, yellowish flowers, and spherical fruits. The extracts obtained from different parts of the plant contain a vast range of health-promoting phytogenic compounds having potent bio-functionalities. Phytochemical screening of these bioactive compounds (BACs) has revealed that these substances are useful in combating several lifestyle-related metabolic disorders due to their antioxidant potential. Accordingly, the current chapter is meant to provide an overview for highlighting the health-promoting potentiality of these bioactive components. This chapter mainly focuses on exploring the anti-oxidative, anti-inflammatory, anti-cancerous, anti-diabetic, anti-microbial, and hepatoprotective perspectives of bitter apple extracts. Based on the present compilation, future directions can be extracted regarding the application of this beneficial plant as a source of developing innovative pharmaceuticals and functional food products.

9.1 INTRODUCTION

Plants have potent biologically active components having a vast range of functional and medicinal activities. Due to the technological advancements, these chemicals have found wide applications of industrial importance. These diversified phytochemicals are found in varied parts of plant such as leaves, seeds, bark, flowers, fruits, seeds, and roots, etc. The health-promoting aspects of plant materials are primarily due to the development of several secondary products in plants during the growth stages. Additionally, the medicinal perspectives of plant-based active components are dependent on particular species and types of plants due to the taxonomical differences [22].

Citrullus colocynthis L. Schrad is a member of *the Cucurbitaceae* family. It is widely grown in the desert areas and has known for its functional and therapeutic applications to combat various lifestyle-related disorders [34]. The well-known members of *Cucurbitaceae* family include bitter apple, cucumber, pumpkin, gourd, and melon, etc. In Pakistan, about 17 genera and 32 species of this family have been grown out of which 25 plants of *Citrullus* have been reported to possess medicinal value. *Citrullus* plants are generally grown in Pakistan (Punjab and Sindh), Southern, and Central parts of India, West Asia, Arabia, and Tropical Africa.

A bitter apple plant generally produces 15–30 fruits with diameter ranging from 7 to 10 cm. The fruit is covered with green skin having yellow stripes. The plant contains angular leaves which are located on an alternate manner on the long petioles. The length of leaves is ranged from 5–10 cm and each leaf contain 3 to 7 lobes. The seeds of *Citrullus* plant are usually of 6 mm size with ovoid shape, smooth texture, and compressed appearance. The color of seeds is varied from dark brown to light yellowish-orange. Additionally, Bitter Apple plant has large roots with long, angular, slender, and tough stem similar to vines. The stems generally spread on the surface but have the ability to climb the trees and other plants through tendrils. The flowers of the plant are of yellow color and present singly at axils. Moreover, the plants are monoecious in nature, i.e., stamens and pistils are found on the same plant at different flowers [15].

This chapter summarizes the potential medicinal properties and major bioactive components present in different parts of *Citrullus colocynthis*.

9.2 PHYTOCHEMISTRY

Various parts of *Citrullus colocynthis* plant (stem, leaves, fruit, roots, and seeds) have been used as fresh, in the form of extracts or after drying. Different parts of the plants are claimed to have significant antidiabetic, anti-inflammatory, anti-hyperlipidemic, antioxidant, antimicrobial laxative, analgesic, and hair-growth-promoting properties that can be used effectively for pharmaceutical purposes [16].

The unripe fruit of bitter apple contains pulp which can be used for its purgative action for the cleaning of intestinal tract after drying. Additionally, dried pulp and leaves have also shown their applications for the treatment of various types of cancers [19]. Furthermore, fruit extract of the plants can be used for pain reduction and improving the nerve functionality in patients suffering from diabetic neuropathy. The extracts from different parts of bitter apple have also been used for the treatment of many types of skin ailments such as scrap, boils, warts, acne, and pimples. Roots of the plants can also be employed for the treatment of jaundice/icterus, rheumatoid arthritis (RA), and several urologic diseases [19].

9.3 BIOACTIVE COMPONENTS PRESENT IN *CITRULLUS COLOCYNTHIS*

Fruits of *Citrullus colocynthis* are reported to have 17 different compounds which are classified as epoxy compounds, alcohols, hydrocarbons, ketones, and acids [13]. Kernels of *Colocynthis* about 52% essential oil, 28.4% proteins, 8.2% carbohydrates, 3.6% ash and 2.7% crude fiber. Whereas, leaves of *Citrullus colocynthis* possesses 3.61% ash, 7.95% moisture content, 12.02% crude protein, and 46.66% Nitrogen free extract [27]. Regarding the quality of proteins, the plant possesses significant quantities of different essential amino acids such as methionine, arginine, and tryptophan. These plants are also good sources of niacin, vitamin B1 and vitamin B2, and different minerals such as calcium, magnesium, potassium, zinc (Zn), manganese, phosphorous, and iron. Additionally, the most abundantly present flavonoid present in the fruit, roots, leaves, and stems is quercetin [22].

The seed oil of *Citrullus colocynthis* contains considerable proportions of unsaturated fatty acids primarily palmitic acid (8.1–17.3%) and stearic acid (6.1–10.5%). Additionally, oil contains linoleic acid and oleic acid which are the primary monounsaturated fatty acids having potential health impacts. The presence of higher quantity (50.6–60.1%) of linoleic acid in oil extracted from the seed portion of bitter apple makes the oil restoratively profitable. Seeds of colocynthis are rich in protein (8.25%) with significant proportions of lysine, leucine, and methionine [11].

Various physicochemical attributes of *Citrullus colocynthis* seed oil also support the fact that it is a good source of various bioactive compounds (BACs) having considerable health-promoting aspects. The seed oil has an iodine value of 114.46 g I/100 g, density value of 905.3: Kg/m^3 (at 15°C), saponification value of 204.44 mg KOH/g, acid value of 0.98 mg KOH/g, caloric value of 39.37 MJ/kg and average molecular weight (Mw) of 874 g. Additionally, seed oil also possesses significant quantities of free fatty acids (0.49%), phenolic compounds (0.74%) and flavonoid contents (0.13%) [9]. Table 9.1 shows the quantities of different BACs present in different parts of the plant.

TABLE 9.1 Bioactive Compounds in Different Parts of *Citrullus colocynthis* Plant

Compound	Whole Fruit (mg/100 g)	Leaves (mg/100 g)	Roots (mg/100 g)
Catechin	16.4	95.4	65.3
Myricetin	2.9	381.7	25.8
Quercetin	1.3	579.9	7.0
Kaempferol	2.3	5.9	6.7
Gallic acid	2.7	11.0	3.7
Caffeic acid	2.9	4.9	2.7
Vanillic acid	2.7	13.7	1.5
p-coumaric acid	1.2	12.9	0.9

Source: Ref. [14].

9.4 MEDICINAL PERSPECTIVES

9.4.1 *ANTI-INFLAMMATORY PROPERTIES OF CITRULLUS COLOCYNTHIS*

Inflammation is considered as one of the most important and common symptoms of several lifestyle-related diseases. Inflammation results in the response of damaged cells in the body which generate chemical signals. As a result of these signals, various complications start to onset, which include infections, production of toxic substances in the damaged tissues, physical damage of the cells, edema, and pain. Several studies have revealed that inflammatory reactions in the body are connected to the development of various metabolic disorders such as RA and cardiovascular diseases (CVDs). Several anti-inflammatory drugs have been applied for the treatment of different kinds of inflammation, however, the extensive use of these medicines can impose serious side effects. Contrarily, various herbal substances from plant origin can be successfully used to develop novel anti-inflammatory drugs with minimal or no side effects [26].

Plant-based essential oils or extracts are generally taken as the most attractive sources to develop new drugs and have shown excellent results in treating inflammatory and other immune-related ailments. Extracts of *Citrullus colocynthis* are also being extensively used for the treatment of many types of inflammations as a folk medicine. For instance, Marzouk et al. [21] assessed the anti-inflammatory potential of *Citrullus colocynthis* seed and fruit extracts for the treatment of paw edema in rats which was intentionally induced by injecting carrageenan to the sub-plantar region. Results of the investigation stated that different doses of *C. colocynthis* (C.C.) extracts (1 mg/kg and 4 mg/kg) caused a significant reduction in the severity of edema during the entire observation period. They also observed that the level of reduction was varied for the seed and fruit extracts. It was also observed from the findings that immature fruits showed better activity to reduce edema as compared to the seeds [21].

In another study, the anti-inflammatory activity of C.C. extracts was investigated by Rajamanickam et al. [28] Outcomes of this trial indicated that extracts of C.C. is quite effective in controlling paw edema in experimental rats. Additionally, root extracts of C.C. have also shown their potential for the treatment of pro-inflammatory osteoarthritis. It has been investigated that C.C. extract can reduce the expression of TNF-α, COX-2, TNF-α, INOS, inflammatory cytokine, and NO and PGE2-production [4].

9.4.2 ANTI-OXIDATIVE POTENTIAL OF CITRULLUS COLOCYNTHIS

Generation of free radicals is responsible for oxidation leading to the onset of many lifestyle-related diseases. Antioxidants are the protective agents which reduce the risk of these metabolic ailments by neutralizing the free radicals. Phytogenic extracts contain a variety of bioactive components which possess anti-oxidative properties and are beneficial in controlling the problem of oxidative degeneration in the body. Likewise, *Citrullus colocynthis* extracts have a vast range of biologically active substances such as flavonoids, isosaponarin, isovitexin, and isoorientin 3'-O-methyl ether, etc., having significant antioxidant properties [29].

Production of reactive oxygen species (ROS) is also responsible for the damaging of red blood cells and leads to anemia in diabetics. Studies have explored the protective action of *Citrullus colocynthis* extracts against oxidative damage of red blood cells through bio-efficacy trials [29]. Findings of many researchers have shown that oral administration of *Citrullus colocynthis* extracts is helpful in decreasing the problem of lipid peroxidation by reducing the activities of oxidation-promoting enzymes. Additionally, the extracts of *Citrullus colocynthis* are helpful in increasing the numbers of erythrocytes, packed cell volume (PCV) and blood hemoglobin level. BACs present in C.C. extracts are also beneficial in decreasing TBARS (Thiobarbituric acid reactive substances) values in blood and hence possess antioxidant potential to reduce oxidative stress in diabetic patients [7].

Studies have also proven that *Citrullus colocynthis* extracts can also be used for treatment or prevention of constipation, bacterial infections, edema, and cancer. Screening of methanolic extracts of *Citrullus colocynthis* has revealed that these extracts possess strong antioxidant potential due to free-radical scavenging activity, which reduces the problem of oxidative damage of cells in case of various metabolic ailments [33].

9.4.3 ANTIMICROBIAL POTENTIAL OF CITRULLUS COLOCYNTHIS

In addition to strong antioxidant activity, *Citrullus colocynthis* extracts are known for their potential in mitigating the growth of a broad spectrum of Gram-positive and Gram-negative bacteria such as *Escherichia coli*, *Enterococcus faecalis*, *Pseudomonas aeruginosa*, *Staphylococcus aureus*, etc. Additionally, *Citrullus colocynthis* extracts also possess momentous antifungal properties and have been reported to show their functionality against various Candida species such as *Candida glabrata*, *C. albicans*,

C. kreusei, and *C parapsilosis*. The reported minimum inhibitory concentration (MIC) of *Citrullus colocynthis* extracts against *C. albicans* and *C. glabrata* is 0.10 mg/mL whereas against *E. coli* and *P. aeruginosa* is 0.20 mg/mL [29]. Additionally, ethanolic extracts of fruits, roots, leaves, and stems of bitter apple are also effective in combating the growth of *Bacillus pumilus*, *Staphylococcus aureus,* and *Bacillus subtilis* [17].

According to a study, anti-fungal characteristics of *Citrullus colocynthis* were assessed against six fungal strains. In this experiment, GYPS (glucose peptone yeast and sucrose) medium was used for maintenance of stock culture. Methanolic extract of *Citrullus colocynthis* revealed significant antifungal activities against Aspergillus flavus, Aspergillus fumigatus, and Mucor sp. While, non-significant effect was noticed against *Penicillium* sp., *Candida albicans*, and *Rhizopus* sp. [12] Studies have also reported that presence of different phenolic compounds, tannins, and flavonoid substances in the aqueous extracts of bitter apple is responsible for controlling the growth of *Klebsiella pneumoniae* and *B. subtilis*. Additionally, *in vitro* trials to evaluate the antimicrobial potential of methanol extracts of bitter apple have shown that the extract is useful against different antibiotic-resistant strains of *E. coli*, *S. aureus*, *K. pneumoniae* and *B. subtillus* Moreover, phytochemical screening of methanolic extracts of bitter apple showed strong antibacterial action of experimental extracts against resistant strains of *Streptococcus pyogenes*, *Salmonella typhi, Streptococcus faecalis*, *Proteus mirabilis,* and *Vibro cholera* [23].

9.4.4 ANTICANCER PERSPECTIVES OF CITRULLUS COLOCYNTHIS

Cancer has become the prime cause of mortality and is the global problem involving about one third of the total population. Surveys have pointed out that cancer is responsible for more than 20% mortality globally [2]. A wide range of therapeutic strategies have been practiced treating the cancer which mainly include chemotherapy, radiotherapy, surgery, immune therapy, and hormonal therapy. Among these approaches, surgery has the highest efficiency for the treatment of cancer but is not suitable in many circumstances. Globally, utilization of chemotherapeutic drugs is the most extensively used method for cancer treatment but has various side effects which include loss of hair, suppression of bone marrow, and development of drug resistance in cancerous cells, cardiac toxicity neurological malfunctioning and gastric ulcer [25].

In the recent years, an elevated interest has seen regarding the use of several plant-based bioactive substances as an effective source of developing novel therapeutic drugs for cancer treatment. Active components present in the plant extracts have shown broad biological effects including considerable anti-carcinogenic activity. Several chemical substances produced by plants having no direct association with growth and development of plants are known as are called secondary metabolites. These secondary metabolites such as phenolic components, nitrogenous compounds and terpenes possess strong biological activities and have been widely used for the treatment of several disorders such as inflammation, cardiovascular complications, diabetes, neurological disorders, and various types of cancers. For cancer treatment, these bioactive substances have also been used along with radiotherapy. C.C. mainly has cucurbitacins A, B, C, and D and α-elaterin, which have been reported to possess strong anti-cancer potential [1].

Literature has unveiled those extracts obtained from various parts of bitter apple are very helpful for cancer treatment by reducing the cytotoxicity induced by cancerous cells. Additionally, trials on human cell lines also indicated that bioactive components found in C.C. extract reduce the viability of cancer-causing cells in a dose-dependent manner. Moreover, minute concentrations of C.C. extracts (5 μg/mL, 10 μg/mL, 20 μg/mL) show strong effects on human hepatoma cells (HepG-2) which explains their anti-cancer potential [24].

Citrullus colocynthis leaves contain cucurbitacin glucosides which have the potential to control or mitigate the growth of breast cancer cells in humans. The cucurbitacin glucosides are responsible for the inhibition of progression of human breast cancer cell lines, i.e., ER MDA-MB-231 and ER (+) MCF-7. These compounds also reduce the formation of protein complexes which are responsible to regulate mitosis and induce changes in the cellular morphology. Administration of cucurbitacin glucoside also shows pleiotropic effect leading to apoptotic conditions and cell-cycle arrest. These functionalities of cucurbitacin glucosides strongly support its potential to reduce the growth and development of breast cancer cells [35].

Presently, green development of metallic nanoparticles is an emerging research field having vast applications in the domain of producing nanomedicines. In this regard, development of silver nanoparticles (SNPs) by using extracts obtained from different parts of bitter apple is a quite safe and cost-effective approach. The developed SNPs can be detected by using a transmission electron microscope (TEM) having a size ranged from 7 nm to 19 nm. Additionally, *in vivo* studies have confirmed the anti-cancer potential of these SNPs against breast and colon cancers in human cell lines (Table 9.2) [31].

TABLE 9.2 Medicinal Perspective of *Citrullus colocynthis*

Bioactivity	Description	References
Anti-inflammatory	• Reduce the phrasing of TNF-α (tumor necrosis factor) in THP-1 monocytes/macrophages and inflammatory cytokine	[4]
Anti-oxidative	• Reduce the activities of oxidation-promoting enzymes • Increase the numbers of erythrocytes	[7, 29]
Anticancer	• Cell-cycle arrest • Inhibition of apoptosis	[35]
Hepatoprotective	• Minimize the problem of oxidation • Reduce pathological damages • Reduce CCl_4-induced liver toxicity	[8, 36]
Antidiabetic	• Control the production of TBARS • Improve glycogen contents and lipid profile	[5, 20]
Anti-hyperlipidemic	• Improve the lipid profile • Decrease blood cholesterol level	[7, 37]

9.4.5 CITRULLUS COLOCYNTHIS AS HEPATOPROTECTIVE AGENT

Liver is one of the most important organs in the body because it determines the nutritional level of an individual by its blood-producing mechanism based on the diet consumed. The human liver is linked with several functions which include maintenance and regulation of homeostasis, and mediation of different biological pathways such as nutrient supply, energy generation, growth, and defense. Liver also plays a vital role in the metabolism of macronutrients such as carbohydrates, proteins, and lipids. Liver is also responsible for the removal of unwanted metabolites from our bodies. Liver is also associated with the secretion of bile, having an important role in the process of digestion. Under the influence of various factors, liver metabolism can be disturbed leading to the onset of several hepatic disorders. Apart from treatment of liver disorders, daily care of liver can be a keystone for the good health of a person [6].

Phytogenic extracts have shown their potential in reducing the liver ailments through animal modeling. In this regard extract of *Citrullus colocynthis* also have shown its capability in combating various hepatic complexities. Findings of scientific investigations have explored that treatment of experimental animals (rats) with bitter apple extract (300 mg/kg) significantly decreases liver toxicity and shows hepatoprotective action by minimizing the problem of oxidation. Additionally, bitter apple extracts have also shown their effectiveness in protecting hepatic disorders occurred due to the consumption of polluted water. Studies based on various hepatic biochemical parameters such as Aspartate transaminase (AST), Alanine transaminase (ALT), alkaline phosphatase (ALP) and Total proteins, etc., have also confirmed that BACs found in *Citrullus colocynthis* extract are helpful in reducing the pathological damages and possess significant hepatoprotective potential [8]. Moreover, phenolic substances and flavonoids present in the extract of bitter apple are helpful in reducing the CCl_4-induced liver toxicity and prevent liver damage in a significant way [36]. These findings have confirmed that bitter apple extract has momentous hepatoprotective activity which is useful in protecting the liver against damage and injury.

9.4.6 ANTI-DIABETIC EFFECT OF *CITRULLUS COLOCYNTHIS*

Diabetes mellitus is a complex metabolic disorder which is associated with chronic hyperglycemia which disturbs the metabolism of carbohydrates leading to impaired insulin secretion and disturbed insulin action. Several classes of drugs have been used for the treatment of diabetes, which mainly include insulin sensitizers (biguanide metformin and thiazolidinediones) and agents responsible for delaying the absorption of carbohydrates from the bowel [10]. However, the use of anti-diabetic drugs is associated with several disadvantages, such as some drugs have severe side effects, some drugs are quite expensive and some are unavailable. Alternatively, use of medicinal plants for the treatment of diabetes is regarded as an effective and safe approach. These plants contain several hypoglycemic compounds which can be used as dietary supplements or for the development of plant-based drugs. Accordingly, *Citrullus colocynthis* has shown its potential as an attractive choice for the treatment of diabetes [5].

It is evident from the previous findings that extracts obtained from seeds of C.C. can be successfully used for the treatment of diabetes because bioactive components present in bitter apple seed extracts have a momentous anti-hyperglycemic effect. Additionally, extracts obtained from roots of bitter apple also possess strong anti-diabetic potential and these extracts can be successfully employed to reduce the blood sugar level [3]. Additionally, oral administration *of Citrullus colocynthis* fruit extract is also useful in controlling the level of blood glucose and control the production of TBARS, which is helpful in reducing the oxidative damage of tissues in diabetic patients. Extracts obtained from different parts of *Citrullus colocynthis* contain plentiful amounts of flavonoids, alkaloids, tannins, saponins, phenolic compounds, and glycosides, which alter the levels of biochemical parameters positively in diabetic patients. These compounds are also helpful in improving the glycogen contents and lipid profile by possessing strong antioxidant and anti-diabetic activities [20].

9.4.7 ANTI-HYPERLIPIDEMIC POTENTIAL OF CITRULLUS COLOCYNTHIS

Raised levels of low-density lipoprotein (LDL) cholesterol and triglycerides in plasma is associated with the reduced concentration of high-density lipoprotein (HDL) resulting in an elevated risk of cardiovascular complications. Cardiovascular disorders are considered as main reason of mortality in the world. American Diabetes Associations recommend immediate lipid-lowering treatment for the patients suffering from hyperlipidemia [32].

Hyperlipidemia is also considered as a critical risk factor for the onset of several other lifestyle-related metabolic ailments which include atherosclerosis, vascular diseases, and heart stroke. Plant-based substances are also claimed to have anti-hyperlipidemic properties which are helpful to curtail these problems. Likewise, *Citrullus colocynthis* is known to improve the lipid profile of patients suffering from hyperlipidemia [37]. Results of several animal trials have revealed that dietary intervention of C.C. oil (4%) is helpful in improving the blood lipid profile and body weight management. Additionally, essential oil obtained from Colocynth is helpful in decreasing the blood cholesterol level, triglyceride contents and hematocrit level which is useful in combating the problem of hyperlipidemia. Hence, extracts collected from various parts of C.C. are useful in the management of dyslipidemia and other associated abnormalities [7].

9.4.8 ANTI-OBESITY EFFECT OF *CITRULLUS COLOCYNTHIS*

For promotion and validation of herbal medicines as a potential ingredient for new drugs, it is quite necessary to investigate their phytochemical composition and medicinal perspectives. *Citrullus colocynthis* has been prescribed in folk medicines to deal with obesity related disorders. Accordingly, C.C. is proposed to have its anti-obesity effect mainly due to inhibitory action on inflammatory cytokines that are secreted in obese conditions. For this purpose, a study was conducted on mice comprising of four groups; group-I (basal diet), group-II (basal diet + C.C.), group-III (fat-rich diet) and Group-IV (fat-rich diet + C.C.). Group-II and IV were subjected to hydroalcoholic extract of C.C. (50 mg kg^{-1} B.W.) for a period of 42 days to assess its anti-obesity perspectives. Results revealed reduction in body weight, feed intake, and TNF-α and IL-6 expression as compared to groups that were not subjected to C.C. anti-obesity effect of *Citrullus colocynthis* may be attributed to down-regulation of proinflammatory cytokines that are linked to obesity [30].

Similarly, outcomes of another investigation performed on off-springs of Wister rats could help in authenticating the potential of *Citrullus colocynthis* against obesity. Details of three groups constituted in this study were as control group (comprising of male off-springs from mothers relying on same calorie-diet, i.e., sunflower oil: 4%), negative control group (comprising of male off-springs from mothers relying on calorie-rich diet, i.e., sunflower oil: 32%), and experimental group (comprising of male off-springs from mothers relying on sun-flower oil: 28% and *Citrullus colocynthis* oil: 4%). After a period of 56 days, off-springs were scarified for blood sample collection and analysis. Momentous reduction was observed in the body weight of the experimental group that were subjected to *Citrullus colocynthis* oil. Serum glucose levels remained within normal limits in Group-1 and 3, on the other hand its level elevated significantly in group-2. However, serum cholesterol, triglyceride, and LDL levels reduced significantly in group administrated with oil of

Citrullus colocynthis as compared to group-2. In a nutshell, the results of this study suggest the anti-obesity potential of *Citrullus colocynthis* due to reduction of lipid biomarkers [18].

9.5 SUMMARY

The above discussion concludes that bitter apple is a useful plant which contains several health-promoting BACs having the potential to control, prevent or treat several metabolic ailments. The findings discussed in this chapter clearly indicate that bitter apple extracts have been critically explored for their anti-inflammatory, anti-oxidant, anti-cancer, anti-diabetic, anti-microbial, and anti-hepatotoxic activities. However, mechanisms associated with these functionalities are still needed to explore in detail. Additionally, limited data is available regarding the use of bitter apple extracts in pharmaceutical and food industries, which should be the future research domain. Therefore, future research work should be focused on understanding the mechanisms of therapeutic activity of bitter apple extracts and their applications in food and pharmaceutical sectors for the development of novel drugs and functional food products.

KEYWORDS

- **anti-cancer potential**
- **bioactive compounds**
- **bitter apple**
- ***Citrullus colocynthis***
- **metabolic disorders**
- **silver nanoparticles**

REFERENCES

1. Adam, S., & Al-Qarawi, E. I., (2003). Effect of combination of capsicum frutescens and *Citrullus colocynthis* on growth, haematological and pathophysiological parameters of rats. *Phytotherapy Research, 17*(1), 92–95.

2. Agarwal, N., Majee, C., & Chakraborthy, G., (2012). Natural herbs as anti-cancer drugs. *International Journal of PharmTech Research, 4*(3), 1142–1153.
3. Agarwal, V., Sharma, A. K., Upadhyay, A., Singh, G., & Gupta, R., (2012). Hypoglycemic effects of *Citrullus colocynthis* roots. *Acta Poloniae Pharmaceutica, 69*(1), 75–79.
4. Akhzari, M., Mirghiasi, S., Vassaf, M., Bidgoli, M., & Tari, Z., (2015). The effect of *Citrullus colocynthis* on the reduction of inflammatory agents in osteoarthritis. *Molecular Biology, 4*(4), 15–16.
5. Bailey, C. J., & Day, C., (1989). Traditional plant medicines as treatments for diabetes. *Diabetes Care, 12*(8), 553–564.
6. Chaterjee, T. K., (2000). *Medicinal Plants with Hepatoprotective Properties* (pp. 135–137). Herbal Options. Books & Allied (P) Ltd., Calcutta.
7. Dallak, M., & Bin-Jaliah, I., (2010). Antioxidant activity of *Citrullus colocynthis* pulp extract in the rbcâ€™ s of alloxan-induced diabetic rats. *Pakistan Journal of Physiology, 6*(1), 1–5.
8. Dar, A. I., Saxena, R. C., & Bansal, S. K., (2012). Hepatoprotection: A hallmark of *Citrullus colocynthis* L. against paracetamol-induced hepatotoxicity in Swiss albino rats. *American Journal of Plant Science, 3*(1), 1022–1027.
9. Giwa, S., Abdullah, L. C., & Adam, N. M., (2010). Investigating "egusi" (*Citrullus colocynthis* L.) seed oil as potential biodiesel feedstock. *Energies, 3*, 607–618.
10. Guillausseau, P. J., & Laloi-Michelin, M., (2003). Pathogenesis of type 2 diabetes mellitus. *La Revue de Médecine Interne, 24*(11), 730–737.
11. Gupta, A. S., & Chakrabarty, M. M., (2003). The component fatty acids of *Citrullus colocynthis* seed fat. *Journal of the Science of Food and Agriculture, 15*(2), 74–77.
12. Gurudeeban, S., Rajamanickam, E., Ramanathan, T., & Satyavani, K., (2010). Antimicrobial activity of *Citrullus colocynthis* in Gulf of Mannar. *International Journal of Current Research, 2*(1), 78–81.
13. Gurudeeban, S., & Ramanathan, T., (2010). Antidiabetic effect of *Citrullus colocynthis* in alloxen- induced diabetic rats. *Inventi. Rapid Ethnopharmacology, 1*(1), 112–113.
14. Hussain, A. I., Rathore, H. A., Sattar, M. Z., Chatha, S. A., Ahmad, F., Ahmad, A., & Johns, E. J., (2013). Phenolic profile and antioxidant activity of various extracts from *Citrullus colocynthis* (L.) from the Pakistani flora. *Industrial Crops and Products, 45*(2), 416–422.
15. Hussain, A. I., Rathore, H. A., Sattar, M. Z., Chatha, S. A., Sarker, S. D., & Gilani, A. H., (2014). *Citrullus colocynthis* (L.) schrad (bitter apple fruit): A review of its phytochemistry, pharmacology, traditional uses and nutritional potential. *Journal of Ethnopharmacology, 155*(1), 54–66.
16. Jayaraman, R., & Christina, A. J., (2013). Evaluation of *Citrullus colocynthis* fruits on *in vitro* antioxidant activity and *in vivo* DEN/PB induced hepatotoxicity. *International Journal of Applied Research in Natural Products, 6*(1), 1–9.
17. Khatibi, R., & Teymorri, J., (2011). Anticandidal screening and antibacterial of *Citrullus colocynthis* in South East of Iran. *Journal of Horticulture and Forestry, 3*(13), 392–398.
18. Khemmar, L., Amamou, F., Yazit, M., Didi, A., & Chabane-Sari, D., (2012). Anti-obesity and anti-hyperlipidemic effect of *Citrullus colocynthis* oil in the offspring of obese rats. *Annals of Biological Research, 3*(5), 2486–2490.

19. Kumar, S., Kumar, D., Saroha, K., Singh, N., & Vashishta, B., (2009). Antioxidant and free radical scavenging potential of *Citrullus colocynthis* (L.) schrad. methanolic fruit extract. *Acta Pharmaceutica, 58*(2), 215–220.
20. Lakshmi, B., Sendrayaperumal, V., & Subramanian, S., (2013). Beneficial effects of *Citrullus colocynthis* seeds extract studied in alloxan-induced diabetic rats. *International Journal of Pharmaceutical Sciences Review and Research, 19*(1), 47–55.
21. Marzouk, B., Marzouk, Z., Haloui, E., Turki, M., Bouraoui, A., Aouni, M., & Fenina, N., (2011). Anti-inflammatory evaluation of immature fruit and seed aqueous extracts from several populations of Tunisian *Citrullus colocynthis* schrad. *African Journal of Biotechnology, 10*(20), 4217–4225.
22. Meena, M. C., & Patni, V., (2008). Isolation and identification of flavonoid" quercetin" from *Citrullus colocynthis* (Linn.) schrad. *Asian Journal of Experimental Sciences, 22*(1), 137–142.
23. Memon, U., Brohi, A. H., Ahmed, S. W., Azhar, I., & Bano, H., (2003). Antibacterial screening of *Citrullus colocynthis*. *Pakistan Journal of Pharmaceutical Sciences, 16*(1), 1–6.
24. Mukherjee, A., & Patil, S. D., (2012). Effects of alkaloid rich extract of *Citrullus colocynthis* fruits on artemia salina and human cancerous (MCF-7 AND HEPG-2) cells. *Journal of Pharmaceutical Science and Technology, 1*(2), 15–19.
25. Nussbaumer, S., Bonnabry, P., Veuthey, J. L., & Sandrine, F., (2011). Analysis of anticancer drugs: A review. *Talanta, 85*, 2265–2289.
26. Pashmforosh, M., Rajabi, V. H., Rajabi, V. H., Pashmforosh, M., & Khodayar, M., (2018). Topical anti-inflammatory and analgesic activities of *Citrullus colocynthis* extract cream in rats. *Medicine, 54*(4), 51–53.
27. QuratulAin, S., Khalil, A. A., Faiz-ul-Hassan, S., Khan, A. A., Hina, G., Khan, M. A., Ahood, K., et al., (2019). Proximate and mineral nutrient composition of various parts of *Citrullus colocynthis*-an underutilized plant. *Pakistan Journal of Food Sciences, 29*(2), 10–14.
28. Rajamanickam, E., Gurudeeban, S., Ramanathan, T., & Satyavani, K., (2010). Evaluation of anti-inflammatory activity of *Citrullus colocynthis*. *International Journal of Current Research, 2*(1), 67–69.
29. Ramanathan, T., Gurudeeban, S., & Satyavani, K., (2010). Antioxidant and radical scavenging activity of *Citrullus colocynthis* inventi. *Rapid Nutraceutical, 1*(2), 37.
30. Sanadgol, N., Najafi, S., Ghasemi, L. V., Motalleb, G., & Estakhr, J., (2011). A study of the inhibitory effects of *Citrullus colocynthis* (CCT) using hydro-alcoholic extract on the expression of cytokines: TNF-α and IL-6 in high fat diet-fed mice towards a cure for diabetes mellitus. *Journal of Pharmacognosy and Phytotherapy, 3*(6), 81–88.
31. Shawkey, A. M., Rabeh, M. A., Abdulall, A. K., & Abdellatif, A. O., (2013). Green nanotechnology: Anti-cancer activity of silver nanoparticles using *Citrullus colocynthis* aqueous extracts. *Advances in Life Science and Technology, 13*(2), 60–69.
32. Smith, Jr. S. C., Jackson, R., Pearson, T. A., Fuster, V., Yusuf, S., Faergeman, O., Wood, D. A., et al., (2004). Principles for national and regional guidelines on cardiovascular disease prevention: A scientific statement from the world heart and stroke forum. *Circulation, 109*(25), 3112–3121.

33. Sudhanshu, Rao, N., Mittal, S., & Menghani, E., (2012). Screening of antioxidant potential of *Citrullus colocynthis* methanolic extract. *Journal of Chemical and Pharmaceutical Research, 4*(5), 2507–2511.
34. Sultan, A., Khan, F. U., Iqbal, H., Khan, M. A., & Khan, I. U., (2010). Evaluation of chemical analysis profile of *Citrullus colocynthis* growing in southern areas of Khyber Pukhtunkhwa Pakistan. *World Applied Sciences Journal, 10*(4), 402–405.
35. Tannin-Spitz, T., Grossman, S., Dovrat, S., Gottlieb, H. E., & Bergman, M., (2007). Growth inhibitory activity of cucurbitacin glucosides isolated from *Citrullus colocynthis* on human breast cancer cells. *Biochemical Pharmacology, 73*(2), 56–67.
36. Yang, W. J., GU, Z. Y., Tursun, D., Chen, B., Hasimu, H., Liu, C., Jiang, H. E., et al., (2013). Protection of *Citrullus colocynthis* fruit extracts on carbon tetrachloride-induced and bacillus Calmette-Guerin plus lipopolysaccharide-induced hepatotoxicity in mice. *Chinese Herbal Medicines, 3*(5), 205–211.
37. Zamani, M., Rahimi, A. O., Mahdavi, R., Nikbakhsh, M., Jabbari, M. V., Rezazadeh, H., Delazar, A., et al., (2007). Assessment of anti-hyperlipidemic effect of *Citrullus colocynthis*. *Revista Brasileira de Farmacognosia, 17*(4), 492–496.

CHAPTER 10

HAREER (BLACK MYROBALAN): A THERAPEUTIC PLANT

ALI IKRAM, FARHAN SAEED, HAROON MUNIR, MUZZAMAL HUSSAIN, and HUMA BADER UL AIN

ABSTRACT

Hareer is a well-known therapeutic plant that is prominent for its pharmacological and medicinal usages. Hareer is the regional name of Black Myrobalan, whose scientific name is *Terminalia chebula* which goes to the family *Combretaceae*. It is moreover, known as "King of Medicines" in Tibet and mostly used in Ayurvedic medicine. Numerous phytochemical constituents are present such as gallic acid, tannic acid, amino acid, ellagic acid, chebulic acid, flavonoids, i.e., luteolin, rutin, and quercetin, etc. Previous research evaluated the pharmacological and medicinal utilization of hareer such as antioxidant activity, hypocholesterolemic activity, carcinogenic activity, anti-inflammatory activity, antifungal activity, antibacterial activity, and antispasmodic activity. The current review discusses the evidence-based knowledge about this plant and its pharmacological and phytochemical activities. It improves vision capacity, improves metabolism, improves life expectancy, anti-aging, good for skin, improves, and nourishes body weight. It is useful in COPD asthma, helps normalize bowel movements, relieves cough and cold, difficulty breathing, also useful in stacks, urinary tract disorders, and diabetes.

10.1 INTRODUCTION

Plants have a significant source of medicine. The world leads to medicinal products based on a plant that strengthens and repairs the body system (especially immune systems) and helps to kill destructive pathogens without side effects [1]. WHO noted that nearly 80% of world inhabitants in all rising nations believe in the traditional method of medicine for their prime healthcare requirements at major levels. Mostly, the revitalization of interest in usual medications started in the last era, due to the widespread credence that green medication is better than the man-made drug. Nowadays there is a various growth in therapeutic plant-based trades payable to the rise in interest in the usage of therapeutic plants all over the world which is rising at the rate of 7% to 15% annually.

The hareer plant was first used in Tibetan medicine and mentioned in Chinese medicine in 1061, where it is stated as the 'king of medicines' [2]. Physicians are using plants for many years but the major restriction of the medicines was their undesirable effects which then lead to Ayurveda. People in the world are looking for alternative modern therapy. The innocuous use of these medicines is an obligatory first thing in the founding of standards of quality, safety, and efficacy [3]. In Ayurvedic medicine, hareer is largely used as a plant medicine due to its large range of pharmacological behavior. The traditional therapeutic system has been using as an herbal medication globally, is a significant base for the new drugs invention [4].

Hareer, a prudent tree used in traditional medications, goes to family *Combretaceae*. It is generally termed as *Chebulic myrobalan* or *Black myrobalan*. Hareer is tremendously used in Asia, as well as Africa [5]. Additionally, the herb has been widely used for the cure of diseases, i.e., cardiovascular diseases (CVDs), leprosy, ulcers, paralysis, arthritis, cancer, gout. Further, the herb has been well stated to possess antidiabetic, antibacterial, antioxidant, antiviral, anticancer, antifungal, anti-mutagenic, antiulcer, and wound healing actions [6]. The ayurvedic preparations have been extensively arranged from the plant for the dealing of infectious diseases, i.e., leucorrhea, chronic ulcers, fungal toxicities of the skin and pyorrhea [7]. The herb has also a history to be used in instruction to impart longevity, prevent aging, and immunity [8]. Taxonomically Species *Terminalia chebula* may be grouped into Kingdom Plantae-Plants, having family *Combretaceae* (Indian almond family). Terminalia is its generic name commonly represented as Terminalia-L tropical almond [9].

Terminella chebula enjoy the major places amongst therapeutic plant in various countries of Africa and Asia. It is generally used in Unani, homeopathic, Siddha, and Ayurveda medications. It is the highest registered plant in Ayurveda medicine for the cure of asthma, sore throat, bleeding piles, and vomiting. It is also used in Thai customary treatment as an expectorant, astringent, and carminative. The "Triphala" a mixture of "three fruits" from herbal plants *Emblica Officinalis, Terminalia bellerica,* and *Terminalia chebula* is used as purgative in purifying agents of the colon, constipation, rejuvenator of the body and food gastric problems (assimilation and poor digestion). Certain researches have publicized that "Triphala" motivates appetite and is beneficial in handling detoxification and cancer. Triphala is also known as the multi-purpose of herbal preparations and also recommended as a candid and cardiotonic infection [10].

10.2 CHEMICAL CONSTITUENTS

The main phytoconstituents in hareer are hydrolysable tannins which may differ from 32–34%. Further phytoconstituents existing in hareer are amino acids, flavonoids steroids, resins, fructose, anthraquinone, fixed oils, carbohydrates, sorbitol, and glucose, etc. The plant is equally rich in hydrolysable tannins (32–34%). In numerous researches, scientists found that hareer comprises 14 constituents of hydrolysable tannins like chebulagic acid, chebulic acid, punicalagin, gallic acid, cerulenin, neochebulinic, ellagic acid, corilagin, and terchebulin [11]. Phyto-chemicals like sennoside, anthraquinone ethanoic acid, and terpinols have been testified. Tannin content differs with geographical variations.

Several other slight chemical elements were poly-phenols, i.e., corilagin, punicalagin, gallolyl glucose, maslimic acid, terflavin A, Fructose, succinic acid, amino acid, resin, the purgative and beta-sitosterol principle of anthraquinone are also existing [12]. Around 12 fatty acids were identified from hareer, in which linoleic acid, palmitic, and oleic acid are key elements. Triterpenoid glycosides for example, arjunenin glucoside, chebulosides has been isolated. The leaf of the plant was found to have polyphenols, for example, punicalagin, theaflavins B, C and D and punicalin [13]. The plant is also originated to comprise pyrogallol and phloroglucinol, accompanied by phenolic acids, for example, p-coumaric, ferulic, vanillic, and caffeic acids. Current studies show that hareer contains extra phenolics as compared to other plants [14].

10.3 MACROSCOPIC FEATURES

Average sized tree up to 25 m high with younger branch glabrescent. The branch is dark brown frequently longitudinally fractured with woody scales and typically short cylinder-shaped bole of 60–80 cm in diameter, 5–10 m in length and crown curved with spread out branches. The seed has a harsh taste, endosperm has a sweet taste and the stem has a bitter taste. Leaves are opposite or alternate, elliptic-obviate or thin curvaceous ovate, rounded at base, 7–12 cm × 4–6.5 cm, pubescent beneath. Flowers are yellowish-white and release a strong aggressive odor. The fruit is a yellowish-green ellipsoid or bovid, five to six ridged when dry and hard. Flowering happens from May to June, while fruiting arises in winter (November to March) [15].

A shining ellipsoidal, glabrous, broadly ovoid, bovid, faintly five-angled, or obscurely and shallowly furrowed, greenish to yellow drupe from 25 mm or less wide and about 25–50 mm long. The surface of the dry fruit is slightly wrinkled and shows five somewhat thick but definite longitudinal edges that are 2 mm thick and 2–3 mm wide. The surface color differs from light yellowish-brown to a closely uniform brown with yellowish patterns or coverings. In some fruits, the basal Portion is slightly tapering or elongated and narrower. One course is ellipsoid 0.2–0.7 cm by 1.0–2.0 cm and without edges. The seed is pale yellow in color and globes, generally 2–6 cm long [16].

10.4 CULTIVATION, PROPAGATION, AND COLLECTION

10.4.1 CULTIVATION

Hareer grows mostly in the subtropical and tropical area up to a height of 1,500 meters and extremely at 2,000 meters. The best temperature surroundings for the cultivation of Hareer is 22–35°C, however, it can bear 5–47°C and is destroyed by temperatures below–5°C. Desirable annual rainfall for development of hareer is 1,000–1,700 mm normally, it can produce in yearly rainfall of 750–3,300 mm. The plant demands nonstop sunlight overhead, however, newer plants can bear a certain extent of shade. In most areas, the trees grow well. Judiciously fertile well-drained soil changing from sandy to clayey is well appropriate for the growth of

Hareer. The pH 5.5–5.6 is appropriate for the growing of the plant. The growing rate of young trees and seedlings is reasonably slow. The revival of natural parts is poor. It regrows healthy after coppicing and also after burning to produce new sprouts 2–3 m long after 5 years. Annually 10 kg of fruits is collected from each tree.

10.4.2 PROPAGATION

The dropped fruits are collected and dried carefully after the hard-bitten surface detached. The fermentation of stones offers the good germination. Then this step is monitored by soaking it in the water for 36 hours. The seed is usually scattered in the nursery bed or container, as direct sowing rises the threat of predation and also seeds germinate sick. The germination speed of the seed is 50%. Early growth is moderately slow with seedlings 10 to 20 cm lofty at the lasts of the first season and 20 to 50 cm lofty at the end of next season.

10.4.3 COLLECTION

For the collection of the fruit, January to March is the finest period. In the first part of January, the fruit should be together from the ground once they have fallen. January is reflected in the best time for the gathering of the fruit for ideal tannin content. Dried and Freshly collected fruit contains optimum tannin content, has a yellowish color and a raise a better price. The fruits on the earth if left have a darker color and at times mold attacks can also happen. The collection of fruit is commonly done by shaking the trees and gathering from the ground. Then the fruit is dehydrated in the sun with measures for escaping impurity within 3–4 weeks. For the drying perseverance contractors mostly erect momentary sheds to stock hareer in the occurrence of rain, as rains terminate the valuable stuff of fruits. The fresh hareer is categorized under various trade names, the collections based upon their color, freedom from insect occurrence and solidness. Grading normally involves separating poorer fruits, which create a second grade, the leftovers being the first grade. The dehydrated hareer was categorized at the locations of the wholesale dealers into various grades based on the solidity of nuts, freedom from pest attack and color. In the trade, they were generally known by the sites of origin. These fruit of fair superiority from any part were marketed without categorizing fair average quality (FAQ) which involves 25% of hollow and 75% of solid nuts [17].

10.5 HAREER IN AYURVEDA

Hareer is generally used as the therapeutic herb in Ayurvedic treatment. Hareer is usually used in the conventional method of medication, just because of its wide range of medicinal behavior [4]. Hareer is held in elevated esteem in Ayurveda for its assets to cure the diseases. This plant is used outwardly in fungal infections, wound healing, and within as a rejuvenative, laxative, astringent, purgative, and stomachic. Hareer is useful in cough, piles, and asthma [18].

The nearness of a few phytochemicals controls the antimicrobial properties of different plants. They give plants its shading, flavor, scent, and are a piece of the safeguards framework. Phytochemicals are the bioactives, non-supplement plants mix in natural products, vegetables, grains, and different plant sustenance that has been connected to limit the danger of major breaking down ailment [18, 19]. Phytochemicals as plant determined synthetic concoctions, which are useful to human wellbeing and infection hindrance [20]. For cases, isoflavones, anthocyanins (ACNs), and flavonoids do work as phytoalexins, a substance that helps a plant to oppose pathogens. Ayurveda mentions herbs that are believed to cleanse the mind and body, restore health inhibit relapse and reverse or postpone the aging process [21].

According to a scientist, aging is inescapable and every living organism has a limited life span. Aging in individuals is affected to a greater extent by several factors such as genetic factors, diet, environmental, and social factors, the occurrence of age-related disorders, etc. [22]. It is described the causative factors of aging viz. biological (failure in the function of various organs), medicinal (due to different types of diseases especially after 60 years) and psychosocial. Biologically aging starts after the age of 21–22 years. After the achievement of adulthood, the process of decline in the function of various organs starts. According to Ref. [23], during biological reaction generally, there is the formation of free radicals like hydroxyl ion, superoxide, etc., leading to progressive injury to DNA, macromolecules, and protein, it is the primary cause of aging. For the occurrence of lots of degenerative and chronic diseases like cancer, ischemic heart disease, atherosclerosis, diabetes mellitus, aging neurodegenerative diseases (NDDs); oxidative stress is often produced by the reactive oxygen species (ROS) which plays a vital role [24].

The products of the tree have many medical advantages and have been utilized as conventional medications as household cures [25]. It is utilized broadly in a few Ayurvedic definitions endorsed for irresistible maladies, for example, leucorrhea, pyorrhea, perpetual ulcers, and parasitic diseases of the skin. It is utilized as an element of various plans which are utilized as a part of an everyday exercise for treating different disarranges. It is utilized as a rule in numerous Ayurvedic arrangements as cardiotonic and diuretic. It is utilized to forestall give life span and maturing and resistance [7]. For different therapeutic purposes like a diuretic, expectorant astringent, carminative, and tonic impacts, a product of this herb is widely utilized as a part of Thai tradition solution [7]. It is routinely utilized as a conventional drug by clans of Tamil Nadu to regard a few sicknesses, for example, hack, looseness of the bowels, fever, gastroenteritis, candidiasis, urinary tract contamination, skin ailments, and wound diseases [27].

10.6 TOXICOLOGICAL STUDIES ON HAREER

The intense harmfulness investigation of the half alcoholic concentrate subchronic lethality investigation of both water and powder separates [28] intense and constant lethality investigations of water remove given orally [29] from dried products of Hareer exhibited no harmful impacts in mice. Investigation of interminable examples of products of hareer demonstrated the subsequent level of tannin. The tannins in hareer have a place with the pyrogallol kind. They likewise altogether vary in their weakness to hydrolytic disappointment. The hydrolysable chebulinic acid ($C_{41}H_{32}O_{27}$), corilagin, chebulagic acid (C41 $H_{30}O_{27}$' $10H_2O$; m.p. >2,400) [30] and tannins, are the real tannin part introduce in hareer. These have a place with ellagi tannin class. They are joined by a few extents of the resulting results of their entire and inadequate hydrolysis: 3,6-digalloylglycose ($C_{20}H_{20}O_{14}$), chebulic acid ($C_{14}H_{12}O_{11}$) [30] BD-glucogallin, ellagic acid and gallic acid, the nearness of terchebin, 1,3,6-trigalloylglucose and 1,2,3,4,6-pentagalloylglucose has likewise been accounted for. A portion of these components is accounted for in the concentrate, yet not in the new develop natural products.

The starches display in hareer are sorbitol and glucose, around 1% every one of fructose, and sucrose, hints of xylose, rhamnose, maltose, and arabinose, and the minor measure of gentiobiose. Around 18 amino acids of plant all potentially taking place in the trees are likewise existing in extra little amounts of succinic, quinic, phosphoric, and dehydro shikimic acids and dihydro-shikimic. During the development of the organic products, the amount of tannin diminishes in the interim the acridity expands [31].

Hareer displayed antibacterial action against a few bacterial animal types. One gathering of scientists studied that it is productive in keeps the urease action of *Helicobactor pyroli* (*H. pyroli*), an ever-show bacterium worried in the improvement of stomach malignancies, ulcers, and gastritis [32]. Anti-bacterial activity of hareer against both Gram-negative and Gram-positive human pathogenic organisms has likewise been depicted [33]. Gallic acid and their ethyl ester disengaged from an ethanolic concentrate of hareer antimicrobial activity compared to methicillin-safe *Staphylococcus aureus* [34]. A diffusive hareer demonstrates an inhibitory result in contradiction of strain XC-100 of the bacterium *Xanthomonas Campestris* PV. Citri speaking to its handiness for the overseeing of citrus ulcer contamination. It has additionally advancement inhibitory act beside Salmonella typhi and the abdominal microorganisms [35].

Hareer natural products manage the cost of four invulnerable inadequacy infection write integrase inhibitors, 1 (HIV-1) gallic acid, and three galloy glucose. Their galloyl moiety assumes a key part for hold next to the 3'-handling of the HIV-1 integrase of the mixes [36]. Hareer has likewise retroviral turn around transcriptase inhibitor movement. It secures epithelial cells close to the flu. An infection is supportive its regular use for helping in retrieval from serious respiratory diseases. It additionally shows a critical inhibitory activity on the impacts of safe lack of infection 1-transcriptase [37]. Vermani and Garg [38] considered that *T. chebula* has exhibited remedial movement against herpes simplex virus (HSV) both in vivo tests and *in vitro*. These findings provoked a group of Japanese scientists to look at *T. chebula*'s impact on human cytomegalovirus (CMV). They found that *T. chebula* was genuine in restraining the duplication of human CMV in an AIDS and *in vitro* display with immune-suppressed mice and inferred that it might be painful for the inhibitory of CMV illnesses and immune-compromised patients. It is additionally steady in sexually spread maladies and AIDS.

10.7 PHARMACOLOGICAL ACTIONS

10.7.1 ANTI-OXIDANT AND FREE RADICAL SCAVENGING ACTIVITY

Hareer is a phenomenal hostile to the oxidant as shown in Table 10.1. Hostile to lipid peroxidation, against superoxide radical generation and free radical-searching exercises are performed by the *T. chebula* [39]. In an examination, 6 concentrates and 4 unadulterated mixes of *T. chebula* demonstrate hostile to lipid peroxidation, against superoxide radical improvement and free radical-searching exercises at various extents of intensity. The ethanolic concentrate on the products of *T. chebula* diminished the measure of lipid peroxidase in pale-skinned person rats [39]. Both treatment as well as pretreatment of the refined rodent essential hepatocytes with *T. chebula* fluid natural product extricate (500 or 1,000 mg/kg body-weight for 5 days) fundamentally turned around the t-BHP-instigated cell cytotoxicity as well as lactate dehydrogenase spillage.

Hareer (methanolic remove) nearness represses lipid peroxide development and to search hydroxyl as well as superoxide radicals *in vitro*. Further, a hepato-defensive compound, sperate from the ethanolic concentrate of the products of hareer, was perceived as a blend of chebulic acid as well as its slight isomer, neochebulic acid that likewise defeats the tert-butyl hydroperoxide (t-BHP)-actuated cell cyto-danger in segregated rodent hepatocyte explore [40]. Tri-ethyl chebulate is a solid hostile to oxidant and free radical-forager which is appeared by the *in vitro* assessment of *T. chebula*, which may partake in the counter oxidative capacity. Besides, triethylchebulate exhibited strong DPPH free-radical-rummaging limit and decently stifled azide-incited mitochondria ROS making. The outcomes demonstrate that the triethyl chebulate was solid against oxidant and free radical-scrounger, which may take an interest in the counter oxidative capacity of *T. chebula* [41].

The fluid concentrate of *T. chebula* is by all accounts ready to shield cell organelles from radiotherapy-induced harms. The fluid result of *T. chebula* shielded the counter oxidant proteins from responsive oxygen species (ROS) created by gamma radiation in the rodent liver microsomes and mitochondria. *T. chebula* (Aqueous concentrate) showed xanthine/xanthine oxidase restraint, 2,2-diphenyl-1-picrylhydrazyl (DPPH) radicals-rummaging working because of the essence of phenolics bark, leaves, and product of *T. chebula* showed great hostile to oxidant property [42]. Free radical-prompted nitric oxide (NO), hemolysis secretes from lipopolysaccharide empowered murine macrophages are additionally restrained *T. chebula* (fluid concentrate) [43].

T. chebula (chebulagic acid, gallic acid separated from organic product remove) blocked cyto-poisonous T lymphocyte (CTL)-interceded cyto-harmfulness. Granule exocytosis because of against CD3 incitement was likewise hindered by the above phytochemicals at equal fixations. *T. chebula* (Acetone separate) has more grounded hostile to oxidant action than alpha. *T. chebula* (ethanol extricate) repressed oxidative pressure and the age-subordinate shortening of the telomeric DNA length. To assess *in vitro* hostile to maturing action *T. chebula* irk were again tried for hostile to oxidative and tyrosinase restraint works and in addition to proliferative and MMP-2 hindrance activity on early maturing human skin fibroblasts. The chilly water concentrate of *T. chebula* bothers to demonstrate the most astounding incitement file (SI) on typical human fibroblast multiplication. The concentrate likewise shows MMP-2 restraint on fibroblasts 1.37 times additional intense as compare to ascorbic acid. The examination affirmed the customary utilization of hareer bother in numerous therapeutic plant formulas for life span [44].

TABLE 10.1 Antioxidant Potential of *Terminalia chebula*

Source	Observed Antioxidant Potential
Leaf	Methanol extract (METC), 95% ethanol extract (EETC), water extract (WETC), at 15°C dried powder use for fermented product and at 25°C water extract for fermented product demonstrated antioxidant action created on the pyrogallol-luminol assay
Fruit	The highest antioxidant function was found in the ethanolic extract
	Checked nitric oxide release and free radical-induced hemolysis through lipopolysaccharide-stimulated murine macrophages while using polyherbal formulation (NR-A2/Aller-7)
	Acetone extract showed stronger antioxidant activity
	Showed various magnitudes of potency as antioxidant activity
	Lipid peroxidation inhibition in microsomes of rat liver in different doses through radiation
	The methanolic extract inhibited the formation of scavenge superoxide hydroxyl and lipid peroxide and radicals *in vitro*

TABLE 10.1 *(Continued)*

Source	Observed Antioxidant Potential
Leaf extract	Showed inhibition of xanthine/xanthine oxidase activity and also as DPPH radicals' excellent scavenger
	Antioxidant latent of leaf assessed *in vitro* DPPH-radical scavenging action and reported it to be a harmless cause of functional food as a natural antioxidant resource

10.7.2 IMMUNO-MODULATORY ACTIVITY

T. chebula separate delivered an ascent in humoral neutralizer titer and increment the sort extreme touchiness in mice [7]. It is powerful against the movement of cutting-edge glycation finished results-initiated endothelial cell brokenness. Unrefined concentrate of *T. chebula* empower cell intervened safe reaction in the trial amoebic liver boil in brilliant hamsters [45]. In hepatic amoebiasis, the generation indicated the most astounding fix rate of 73% at 800 mg per kg body weight. In insusceptible regulation investigations, humoral insusceptibility was adjusted where T-cell include stayed unaffected the creatures, yet cell-interceded safe reaction was empowered *T. chebula* beside with many other therapeutic plants supports to attack in contradiction of a number of stressors in many ways [46].

10.7.3 ANTI-MUTAGENIC AND ANTI-CARCINOGENIC ACTIVITIES

Unrefined concentrate of *T. chebula* natural product has tannic acid, chebulinic acid as well as ellagic acid, these are observed to be the most development inhibitory phenolics of *T. chebula* as shown in Table 10.2. The impact of 70% methanolic natural product concentrate of *T. chebula* was checked for development of some threatening cell lines in which a human (MCF-7) and mouse (S115) bosom disease cell line, a human osteosarcoma cell line (HOS-1), a human prostate malignancy cell line (PC-3) in addition a non-tumorigenic, deified human prostate cell line (PNT1A) utilizing tests for expansion (3H-thymidine consolidation and colter tallying), cell suitability (ATP assurance) and cell end arrange implies passing (stream cytometry and Hoechst DNA recoloring). In all cell lines uncovered that the concentrate diminished cell suitability, restrained cell expansion, and prompted the cell passing in a dosage subordinate way [47].

$(CH_3)_2CO$ concentrate of hareer has been accounted for having phytochemicals with promising anti-mutagenic and against cancer-causing properties. The chebulagic acid, one of the fractionated mixes from ethanolic organic product concentrate of hareer, indicated strong double hindrance against COX and 5-LOX. It likewise gives the antiproliferative action against COLO-205, HCT-15, MDA-MB-231, K562, DU-145 cell lines [48]. As indicated by the ongoing investigation the capacity of Triphala to restrain cytochrome P450 [49] *T. chebula* (watery concentrate and hydrolysable tannins) indicate anti-mutagenic action in *Salmonella typhimurium*.

TABLE 10.2 Observed Anticarcinogenic Activity

Source	Anticarcinogenic Potential
Fruit	Inhibited growth of cancer cells
	Induced death of the cell, inhibited proliferation of cell in ethanolic extract of many cell lines being malignant in prostate cancer. Cell (PC-3), human osteosarcoma cell line (HOS-1), mouse (S115) breast cancer cell line including human (MCF-7)
	Increase of lifespan of the mice through the restoration of hematological parameters @200 mg/kg when given orally and found to be cytotoxic in the *in vitro* showing its significant anticancer potential.
	Fruit powder at 500 µg per ml of every fraction was used against cells of the A549 lung cancer cell line using the MTT assay, at 48 h. Fraction 3 exhibited the lowermost rate of development inhibition while fraction 4 demonstrated the peak rate
Bark and fruit powder	Anticarcinogenic activity was also indicated by the acetone extract

10.7.4 RADIO-PROTECTIVE, CHEMO-PREVENTIVE, AND CHEMO-MODULATORY ACTIVITY

In an analysis, a fluid concentrate of the product of *T. chebula* (50 µg) demonstrates that it can kill 1,1-diphenyl-2-picrylhydrazyl, a steady free radical by 92.9% and shielded the plasmid DNA pBR322 from experiencing the radiation-incited strand breaks [50]. In the mice, human lymphocytes light impacts lessen by *T. chebula* while experiencing the gamma radiation-*in vitro* [51]. *T. chebula* separate in a dosage of 80 mg per kg bodyweight past to entire body illumination of mice brought about lessening of peroxidation of layer lipids in the liver as well as decrease in radiation-initiated harm to DNA. Treatment of mice with the fluid concentrate of Triphala in relatively unique dosages successively for five days before to the light postponed the beginning of mortality and diminished the indications of radiation disorder contrasted with controls. *T. chebula* likewise demonstrated the chemo-preventive impact on nickel chloride-prompted renal oxidative pressure, danger, and cell expansion reaction in male Wistar rats. *T. chebula* concentrate could be utilized as a restorative operator for disease avoidance as it blocked or smothered the occasions related to compound carcinogenesis [52].

10.7.5 CARDIO-PROTECTIVE ACTIVITY

Cardio-defensive impact of ethanolic concentrate of *T. chebula* organic products (500 mg per kg body weight) were exhibited in isoproterenol-instigated myocardial harm in rats. Hareer was accounted for that pre-treatment with hareer extricate had a cardio-defensive impact due to the lysosomal layer adjustment counteracting myocardial rot and hindrance of modifications in the heart mitochondrial ultrastructure and work in the exploratory rats [53]. The pericarp of hareer has additionally been indicated cardio-defensive movement in confined frog heart display.

10.7.6 HEPATOPROTECTIVE ACTIVITY

Defensive impacts of a watery concentrate of *T. chebula* organic product on the tert-butyl hydroperoxide-instigated oxidative damage was seen in refined rodent essential hepatocytes and rodent liver *T. chebula* natural products demonstrated solid hepato-defensive movement through hostile to oxidant impact in disconnected rodent hepatocytes [35]. The comparative property of its 95% ethanolic remove was seen against hostile to tuberculosis drugs, i.e., Isoniazid Rifampicin and Pyrazinamide (blend)-prompted lethality in the sub-unending model (12 weeks) because of its noticeable hostile to oxidative and film balancing out exercises [54].

10.7.7 ANTI-BACTERIAL ACTIVITY

T. chebula demonstrated hostile to the bacterial movement against diverse Gram-negative, Gram-positive microscopic organisms, for example, *Bacillus subtilis*, *Staphylococcus epidermidis*, *Salmonella typhi*, *Pseudomonas aeruginosa*, and *Staphylococcus aureus* proposing its wide range hostile to microbial action [55]. In this way, different concentrates of *T. chebula* demonstrate hostile to bacterial action against various bacterial species. Another examination uncovered those gram-positive creatures hindered on a bigger degree as complexity to gram-negative life forms. Against *Helicobacter pylori* the ether, alcoholic, and fluid concentrates of *T. chebula* tried, a however watery subsidiary of the plant at a convergence of 1 to 2.5 mg/ml, hindered urease working of *H. Pylori* as shown in Table 10.3 [35].

TABLE 10.3 Antibacterial Nature of *Terminalia chebula*

Source	Antibacterial Activity
Fruit extract	Inhibited the urease activity of *Helicobacter pylori* effectively.
	Methanolic extract of leaf proved potent equally effective antibiotics such as kanamycin, gentamycin, ciprofloxacin, ofloxacin, and cephalexin in comparison to the aqueous extract.
	Inhibited the growth of salivary bacteria *Streptococcus* mutants through the aqueous extract
	Strong antibacterial against many bacteria which is human pathogenic for both gram-positive and gram-negative
	Strong inhibitory activity on *Salmonella typhi*, *Klebsiella*, and intestinal bacteria
	Strong antibacterial activity was observed alongside multidrug-resistant uropathogenic *Escherichia coli* in ethanolic fruit extract.
Ripe seeds	Exhibited strong antibacterial potential against *S. aureus*.
Leaves	Acetone extract showed higher inhibitions for *S. aureus*, *E. faecalis*, *B. subtilis*, and *C. diphtheria*. Particularly *K. pneumoniae*, *E. faecalis*, *B. subtilis*, and *K. pneumoniae* exhibited higher activity as compare to the standard.

10.7.8 ANTI-FUNGAL AND ANTI-VIRAL ACTIVITY

The anti-fungal function of an aqueous extract of hareer was tested beside many yeasts and dermatophytes as shown in Table 10.4. It is active in contrast to the pathogenic yeast dermatophytes and *Candida albicans, Floccosum*, *Trichophyton rubrum, Microsporum gypsum,* and *Epidermophyton* [56]. Deep-rooted idle contamination of tactile neurons is because of Herpes simplex infection 1 (HSV-1). Heated water concentrate of *T. chebula* comes about hostile to herpes simplex infection (HSV) action in vivo and anti-CMV movement both in vivo and *in vitro* in a writing. Lederman-96 and every one of its 23 singular parts were tried on an epithelial tissue culture cell line for their defensive capacity against cyto-dangerous impacts caused by flu. Of the 23 parts tried, just a single segment result demonstrated a critical defensive impact when used to the epithelial cells independently. *T. chebula* can likewise be utilized as a part of sexually transmitted illnesses and AIDS [38]. The concentrate of products of *T. chebula* indicated inhibitory impacts on human immunodeficiency infection 1 invert transcriptase [37]. As of late, $(CH_3)2CO$ concentrate of *T. chebula* has turned out as another treat pandemic swine flu A disease because of ease, it is anything but difficult to readiness and potential impact [57].

TABLE 10.4 Potential of *Terminalia chebula* Against Fungi

Source	Antifungal Potential
Fruit	Exhibited antifungal activity against dermatophytes and *Candida albicans*. *Microsporum gypsum*, *Floccosum*, *Epidermophyton* in aqueous extract
	Strongly antidermatophytic on three yeasts (*Candida* spp.) and (*Trichophyton* spp.)
	Strongly anticandidal contrary to clotrimazole resistant *Candida albicans* in methanolic extract
Seed extract	Trichophyton glabrata showed antifungal activity in aqueous seed extract

10.7.9 ANTI-INFLAMMATORY ACTIVITY

Watery concentrate of the dried product of *T. chebula* demonstrated mitigating movement by repressing inducible no blend [58]. Chebulagic acid separated from the delicate product of *T. chebula* altogether restrains the beginning and movement of collagen-incited joint inflammation in mice. *T. chebula* in a polyherbal detailing (Aller-7) showed a mitigating impact against joint inflammation in rats [59].

10.7.10 ANTI-ALLERGIC ACTIVITY

T. chebula, the element of a polyherbal plan (Aller-7), demonstrated strong *in vitro* hostile to unfavorably susceptible action [60]. Hostile to histamine and against convulsive in guinea pig ileum was appeared by the Hydro-ethanol concentrate of *T. chebula*. Oral organization of a watery concentrate of natural product altogether stifled histamine emit from rodent peritoneal pole cells and furthermore essentially expanded the generation of tumor necrosis factor (TNF) by hostile to dinitrophenyl IgE [26].

10.8 SUMMARY

Living a long healthy childlike lifecycle is a cherished wish of every individual. Our natural health, happiness, and internal sensibility of comfort are covered by the accumulation of impurities due to a sedentary lifestyle, tense mental conditions, excess use of pesticides, chemicals, additives, and augmented use immune-suppressive drugs. These impurities or poisons lead to the decline of normal body functioning and which affects the quality of life as well as a lifecycle. Ayurveda stands as an answer to solve this issue and provide healthful durability including mental growth and resistance against diseases. Hareer is generally used as the therapeutic plant in Ayurvedic medicine. Hareer is the local name of a black maroblan whose scientific name is *Terminalia chebula* which belongs to the family *Combretaceae*. It is also known as "King of Medicines" in Tibet. *T. chebula* possesses antidiabetic, antifungal, antimutagenic, antiviral, antibacterial, antioxidant, wound healing, and antiulcer properties. It also prevents kidney diseases and used for healing cardiac damage. It is effective, safe and mild, in traditional medicine. It is the highest registered plant in Ayurveda material medical for the cure of asthma, gout, sore throat, bleeding piles, and vomiting. Hareer and its phytochemical constituents have a therapeutic effect with no side effects. Hareer is recognized to possess an extensive range of phytochemical ingredients. *T. chebula* has a number of phytoconstituents such as tannins, amino acids, flavonoids, sterols, fixed oils, fructose, resin, etc., and it is fairly filled with approximately 32% tannin content. The chebulinic acid, chebulic acid, corilagin, gallic acid, chebulagic acid, and ellagic acid are the chief components.

KEYWORDS

- **Ayurveda**
- **cytomegalovirus**
- **fair average quality**
- **hareer**
- **herpes simplex infection**
- **responsive oxygen species**

REFERENCES

1. Rathinamoorthy, R., & Thilagavathi, G., (2014). *Terminalia chebula*-review on pharmacological and biochemical studies. *International Journal of PharmTech Research, 6*(1), 97–116.
2. Das, N. D., Jung, K. H., Park, J. H., Mondol, M. A. M., Shin, H. J., Lee, H. S., & Kim, M. S., (2011). *Terminalia chebula* extract acts as a potential NF-κB Inhibitor in human lymphoblastic T cells. *Phytotherapy Research, 25*(6), 927–934.
3. Kumar, A., Kumar, S., Rai, A., & Ram, B., (2017). Pharmacognostical and phytochemical evaluation of haritaki (*Terminalia chebula* Retz.) fruit pulp. *International Journal of Pharmaceutical, Chemical & Biological Sciences, 7*(4).
4. Ratha, K. K., & Joshi, G. C., (2013). Haritaki (Chebulic myrobalan) and its varieties. *Ayu, 34*(3), 331.
5. Baliga, M. S., Meera, S., Mathai, B., Rai, M. P., Pawar, V., & Palatty, P. L., (2012). Scientific validation of the ethnomedicinal properties of the ayurvedic drug triphala: A review. *Chinese Journal of Integrative Medicine, 18*(12), 946–954.
6. Kannan, P., Ramadevi, S., & Hopper, W., (2009). Antibacterial activity of *Terminalia chebula* fruit extract. *African Journal of Microbiology Research, 3*(4), 180–184.
7. Aher, V. D., & Wahi, A. K., (2011). Melatonin: A novel biomaterial for immunotherapy. *International Journal of Pharmaceutical Sciences and Research, 2*(4), 772.
8. Bag, A., Bhattacharyya, S. K., & Chattopadhyay, R. R., (2013). The development of *Terminalia chebula* Retz.(*Combretaceae*) in clinical research. *Asian Pacific Journal of Tropical Biomedicine, 3*(3), 244–252.
9. Kumar, G. P. S., Arulselvan, P., Kumar, D. S., & Subramanian, S. P., (2006). Anti-diabetic activity of fruits of *Terminalia chebula* on streptozotocin-induced diabetic rats. *Journal of Health Science, 52*(3), 283–291.
10. Kadian, R., Parle, M., & Yadav, M., (2014). Therapeutic potential and phytopharmacology of *Terminalia bellerica*. *World Journal of Pharmacy and Pharmaceutical Sciences, 3*(10), 804–819.
11. Juang, L. J., Sheu, S. J., & Lin, T. C., (2004). Determination of hydrolyzable tannins in the fruit of *Terminalia chebula* Retz. by high-performance liquid chromatography and capillary electrophoresis. *Journal of Separation Science, 27*(9), 718–724.

12. Han, Q., Song, J., Qiao, C., Wong, L., & Xu, H., (2006). Preparative isolation of hydrolysable tannins chebulagic acid and chebulinic acid from *TERMINALIA chebula* by high-speed counter-current chromatography. *Journal of Separation Science, 29*(11), 1653–1657.
13. Patel, M. G., Patel, V. R., & Patel, R. K., (2010). Development and validation of improved RP-HPLC method for identification and estimation of ellagic and gallic acid in triphala churna. *International Journal of ChemTech Research, 2*(3), 1486–1493.
14. Mahajan, R., (2010). Critical incident reporting and learning. *British Journal of Anaesthesia, 105*(1), 69–75.
15. De Boer, W., Driessen, W., Jansz, A., & Tytgat, G., (1995). Effect of acid suppression on efficacy of treatment for helicobacter pylori infection. *The Lancet, 345*(8953), 817–820.
16. Chattopadhyay, R., & Bhattacharyya, S., (2007). Plant review *Terminalia chebula*: An update. *Pharmacognosy Reviews, 1*(1), 151–156.
17. Singh, P., & Malhotra, H., (2017). *Terminalia chebula*: A review pharmacognostic and phytochemical studies. *International Journal of Recent Scientific Research, 8*, 21496–21507.
18. Shastry, A. H., Thota, B., Arimappamagan, A., & Santosh, V., (2015). P53 stratification reveals the prognostic utility of matrix metalloproteinase-9 protein expression in glioblastoma. *Neurology India, 63*(3), 399.
19. Liu, R. H., (2004). Potential synergy of phytochemicals in cancer prevention: Mechanism of action. *The Journal of Nutrition, 134*(12), 3479S–3485S.
20. Anderson, G. D., (2004). Pharmacogenetics and enzyme induction/inhibition properties of antiepileptic drugs. *Neurology, 63*(10 suppl 4), S3–S8.
21. Surya, D., Prakash, N., Sree, S., & Meena, V., (2012). Extraction of chebulinic acid from *Terminalia chebula* sp by Soxhlet extractor An experimental & modelling studies. *Asian Journal* of *Biochemical and Pharmaceutical Research, 2*(3), 170–176.
22. Kanungo, R., (1960). Vocational choice and occupational values among adolescent students. *Journal of Vocational and Educational Guidance (Baroda, India), 7*, 11–19.
23. Pacifici, R. E., & Davies, K. J., (1991). Protein, lipid and DNA repair systems in oxidative stress: The free-radical theory of aging revisited. *Gerontology, 37*(1–3), 166–180.
24. Azizova, O., (2002). Role of free radical processes in the development of atherosclerosis. Biological membranes. *Journal of Membrane and Cell Biology, 19*(6), 451–471.
25. Varier, A., (2002). *Dictionary of Indian Raw Materials & Industrial Products* (Vol. 387). Publication and information directorate council of scientific & industrial research, New Delhi.
26. Shin, T., Jeong, H., Kim, D., Kim, S., Lee, J., Chae, B., & Lee, K., (2001). Inhibitory action of water-soluble fraction of *Terminalia chebula* on systemic and local anaphylaxis. *Journal of Ethnopharmacology, 74*(2), 133–140.
27. Bhagavathy, S., Newsam, S., & Manjunath, B. S., (2002). Modeling object classes in aerial images using texture motifs. In: *Object Recognition Supported by User Interaction for Service Robots* (Vol. 2, pp. 981–984).

28. Babu, B. H., Larkin, A., & Kumar, H., (2019). To evaluate the effect of auxin concentrations (IBA and IAA) on survival percentage of stem cuttings of species *Terminalia chebula* (Retz.). *Indian Forester, 145*(4), 333–338.
29. Panunto, W., Jaijoy, K., Lerdvuthisopon, N., Lertprasertsuke, N., Jiruntanat, N., Soonthornchareonnon, N., & Sireeratawong, S., (2010). Acute and chronic toxicity studies of the water extract from dried fruits of *Terminalia chebula* Rezt. in rats. *International Journal of Applied Research in Natural Products, 3*(4), 36–43.
30. Sharma, S., Singh, B., & Kumar, H., (2019). A critical review of pharmacological actions of haritaki (*Terminalia chebula* Retz) in classical texts. *Journal of Ayurveda and Integrated Medical Sciences, 4*(4), 258–269.
31. Krishnan, K., (1998). The wealth of India. *Raw Material, 10*, 171.
32. Malekzadeh, F., Ehsanifar, H., Shahamat, M., Levin, M., & Colwell, R., (2001). Antibacterial activity of black myrobalan (*Terminalia chebula* Retz) against helicobacter pylori. *International Journal of Antimicrobial Agents, 18*(1), 85–88.
33. Phadke, S., & Kulkarni, S., (1989). Screening of *in vitro* antibacterial activity of *Terminalia chebula, Eclapta* alba and *Ocimum sanctum*. *Indian Journal of Medical Sciences, 43*(5), 113–117.
34. Sato, Y., Oketani, H., Singyouchi, K., Ohtsubo, T., Kihara, M., Shibata, H., & Higuti, T., (1997). Extraction and purification of effective antimicrobial constituents of *Terminalia chebula* RETS. against methicillin-resistant *Staphylococcus aureus*. *Biological and Pharmaceutical Bulletin, 20*(4), 401–404.
35. Kim, H., Cho, J., Jeong, E., Lim, J., Lee, S., & Lee, H., (2006). Growth-inhibiting activity of active component isolated from *Terminalia chebula* fruits against intestinal bacteria. *Journal of Food Protection, 69*(9), 2205–2209.
36. Ahn, M. J., Kim, C. Y., Lee, J. S., Kim, T. G., Kim, S. H., Lee, C. K., & Kim, J., (2002). Inhibition of HIV-1 integrase by galloyl glucoses from *Terminalia chebula* and flavonol glycoside gallates from *Euphorbia pekinensis*. *Planta Medica, 68*(05), 457–459.
37. El-Mekkawy, S., Meselhy, M. R., Kusumoto, I. T., Kadota, S., Hattori, M., & Namba, T., (1995). Inhibitory effects of Egyptian folk medicines on human immunodeficiency virus (HIV) reverse transcriptase. *Chemical and Pharmaceutical Bulletin, 43*(4), 641–648.
38. Vermani, K., & Garg, S., (2002). Herbal medicines for sexually transmitted diseases and AIDS. *Journal of Ethnopharmacology, 80*(1), 49–66.
39. Hazra, B., Sarkar, R., Biswas, S., & Mandal, N., (2010). Comparative study of the antioxidant and reactive oxygen species scavenging properties in the extracts of the fruits of *Terminalia chebula, Terminalia belerica* and *Emblica officinalis*. *BMC Complementary and Alternative Medicine, 10*(1), 20.
40. Lee, H. S., Koo, Y. C., Suh, H. J., Kim, K. Y., & Lee, K. W., (2010). Preventive effects of chebulic acid isolated from *Terminalia chebula* on advanced glycation end product-induced endothelial cell dysfunction. *Journal of Ethnopharmacology, 131*(3), 567–574.
41. Chen, X., Sun, F., Ma, L., Wang, J., Qin, H., & Du, G., (2011). *In vitro* evaluation on the antioxidant capacity of triethyl chebulate, an aglycone from *Terminalia chebula* Retz fruit. *Indian Journal of Pharmacology, 43*(3), 320.

42. Chang, C., Lin, C., Lai, G., Chen, Y., Tuan, W., & Hsu, C., (2010). Influence of *Terminalia chebula* extracts on the effect of PC12 cell growth. *Journal of Traditional and Complementary Medicine, 21*(1), 23–30.
43. Mahesh, R., Bhuvana, S., & Hazeena, V. M., (2009). Effect of *Terminalia chebula* aqueous extract on oxidative stress and antioxidant status in the liver and kidney of young and aged rats. *Cell Biochemistry and Function: Cellular Biochemistry and its Modulation by Active Agents of Disease, 27*(6), 358–363.
44. Manosroi, A., Jantrawut, P., Akihisa, T., Manosroi, W., & Manosroi, J., (2010). *In vitro* anti-aging activities of *Terminalia chebula* gall extract. *Pharmaceutical Biology, 48*(4), 469–481.
45. Dwivedi, S., (2008). Anthelmintic activity of alcoholic and aqueous extract of fruits of *Terminalia chebula* Retz. *Ethnobotanical Leaflets, 2008*(1), 101.
46. Rege, N. N., Thatte, U. M., & Dahanukar, S. A., (1999). Adaptogenic properties of six rasayana herbs used in ayurvedic medicine. *Phytotherapy Research: An International Journal Devoted to Pharmacological and Toxicological Evaluation of Natural Product Derivatives, 13*(4), 275–291.
47. Saleem, A., Husheem, M., Härkönen, P., & Pihlaja, K., (2002). Inhibition of cancer cell growth by crude extract and the phenolics of *Terminalia Chebula* Retz. fruit. *Journal of Ethnopharmacology, 81*(3), 327–336.
48. Reddy, D. B., Reddy, T., Jyotsna, G., Sharan, S., Priya, N., Lakshmipathi, V., & Reddanna, P., (2009). Chebulagic acid, a COX–LOX dual inhibitor isolated from the fruits of *Terminalia chebula* Retz., induces apoptosis in COLO-205 cell line. *Journal of Ethnopharmacology, 124*(3), 506–512.
49. Ponnusankar, S., Pandit, S., Babu, R., Bandyopadhyay, A., & Mukherjee, P. K., (2011). Cytochrome P450 inhibitory potential of triphala—A rasayana from Ayurveda. *Journal of Ethnopharmacology, 133*(1), 120–125.
50. Naik, G., Priyadarsini, K., Naik, D., Gangabhagirathi, R., & Mohan, H., (2004). Studies on the aqueous extract of *Terminalia chebula* as a potent antioxidant and a probable radioprotector. *Phytomedicine, 11*(6), 530–538.
51. Deepti, D., Amit, K., Harsha, L., & Bhalla, D. B., (2012). Protective effects of *Terminalia chebula* in modulating oxidative damages against gamma irradiation-induced lethality in rats. *International Journal of Biological & Pharmaceutical Research, 3*, 734–742.
52. Prasad, L., Khan, T. H., Jahangir, T., & Sultana, S., (2006). Chemo modulatory effects of *Terminalia chebula* against nickel chloride-induced oxidative stress and tumor promotion response in male Wistar rats. *Journal of Trace Elements in Medicine and Biology, 20*(4), 233–239.
53. Nair, V., Singh, S., & Gupta, Y. K., (2010). Anti-arthritic and disease-modifying activity of *Terminalia chebula* Retz. in experimental models. *Journal of Pharmacy and Pharmacology, 62*(12), 1801–1806.
54. Tasduq, S., Singh, K., Satti, N., Gupta, D., Suri, K., & Johri, R., (2006). *Terminalia chebula* (fruit) prevents liver toxicity caused by sub-chronic administration of rifampicin, isoniazid and pyrazinamide in combination. *Human & Experimental Toxicology, 25*(3), 111–118.
55. Savitt, J., Singh, D., Zhang, C., Chen, L. C., Folmer, J., Shokat, K. M., & Wright, W. W., (2012). The *in vivo* response of stem and other undifferentiated spermatogonia to

the reversible inhibition of glial cell line-derived neurotrophic factor signaling in the adult. *Stem Cells, 30*(4), 732–740.

56. Barazani, V., Sathiyomoorthy, P., Shalev, R., Vardy, D., & Golan, G., (2003). Screening of South-Indian medicinal plants for antifungal activity. *Phytotherapy Research, 17*(9), 1123–1125.
57. Ma, H., Diao, Y., Zhao, D., Li, K., & Kang, T., (2010). A new alternative to treat swine influenza A virus infection: Extracts from *Terminalia chebula* Retz. *African Journal of Microbiology Research, 4*(6), 497–499.
58. Moeslinger, T., Friedl, R., Volf, I., Brunner, M., Koller, E., & Spieckermann, P. G., (2000). Inhibition of inducible nitric oxide synthesis by the herbal preparation Padma 28 in macrophage cell line. *Canadian Journal of Physiology and Pharmacology, 78*(11), 861–866.
59. Pratibha, N., Saxena, V., Amit, A., D'souza, P., Bagchi, M., & Bagchi, D., (2004). Anti-inflammatory activities of aller-7, a novel polyherbal formulation for allergic rhinitis. *International Journal of Tissue Reactions, 26*(1–2), 43–51.
60. Mokkhasmit, M., Ngarmwathana, W., Sawasdimongkol, K., & Permphiphat, U., (1971). Pharmacological evaluation of Thai medicinal plants. *Journal of the Medical Association of Thailand= Chotmaihet Thangphaet, 54*(7), 490–503.

CHAPTER 11

CLOVE ESSENTIAL OIL: PHYTOCHEMICAL AND PHARMACOLOGICAL ASPECTS

UBAID UR RAHMAN, ANUM ISHAQ, ANEES AHMED KHALIL, AYESHA ASLAM, and AHOOD KHALID

ABSTRACT

Medicinal plants are unremittingly prompting the interest of researchers due to their versatile health-promoting functionalities, cost-effectiveness, availability, and minimal adverse effects related to the drugs derived from these plants. In this context, clove (*Syzygium aromaticum*) has gained prime importance among various herbs that have been investigated in developing a variety of traditional medicines since ancient times. The phytochemicals present in the extracts obtained from various parts of clove have a broad spectrum of biological activity. These phytoconstituents comprise of different classes of chemical compounds such as phenolics, flavonoids, sesquiterpenes, hydrocarbon compounds, and monoterpenes. The major bioactive substances present in clove-based essential oils include eugenol, eugenyl acetate and β-caryophyllene. These biologically active components possess strong beneficial aspects including anti-microbial, anti-oxidant, anti-diabetic, and anti-carcinogenic properties. This chapter presents a summary of various phytochemical and pharmacological investigations conducted on clove.

11.1 INTRODUCTION

Plants possess several health and physiological benefits due to the presence of various bioactive compounds (BACs). Due to the industrial progression, these functional ingredients have been widely used for their health promoting activities. These plant-based components can be derived from different parts of the plant including leaves, seeds, bark, flowers, fruits, and roots, etc. The pharmacological aspects of these biogenic compounds are mainly due to the production of various primary and secondary metabolites in plants during different stages of growth and development. Furthermore, functional, and pharmaceutical effects of these phytocomponents are based on the type and nature of plants due to their diversified taxonomy.

Plants have been the backbone of several types of conventional herbal medicines, particularly amongst rural inhabitants since ancient times. It is also projected that about 40% of the global population is directly dependent on phytomedicines. Globally, researchers have been aimed to identify and validate the plants-derived active components for their functional aspects to improve the health status by treating various lifestyle-related complications. Additionally, approximately 25% of the recent medications are being directly or indirectly formulated from plants. In this regard, the clove plant, technically known as *Syzygium aromaticum*, is known as a good source of various pharmacologically active components having vast applications in combating several ailments. Therefore, the current chapter is aimed to summarize the established facts regarding phytochemistry and pharmacological properties of clove [42].

This chapter focuses on pharmacological potential and phytochemicals found in clove (*Syzygium aromaticum)* essential oil.

11.2 PHYTOCHEMISTRY

Previous investigations revealed that clove buds/powder are a good source of several valuable nutrients. Regarding the proximate composition, clove buds/powder has moisture contents ranged from 10–23%, carbohydrates from 31–51.5%, crude fat from 12–19%, crude protein from 1.2–7%, crude fiber from 10.65–20% and ash from 5.2–9.1%. Additionally, clove bud powder is also rich in different minerals such as magnesium (avg. 1,259 mg/kg), calcium (avg. 782.54 mg/kg), iron (avg. 710 mg/kg), potassium (avg. 2.69 mg/kg), sodium (avg. 2.56 mg/kg), manganese (avg. 209 mg/kg), phosphorus (avg. 16 mg/kg), zinc (Zn) (avg. 14 mg/kg) and copper (avg. 4 mg/kg) [5, 14, 54].

The oil contents of clove buds (*Syzygium aromaticum*) are ranged from 15–20%. Clove essential oil is mainly considered to be an abundant source of eugenol, eugenyl acetate, and β-caryophyllene and their content varies from 70 to 85%, 15%, 5 to 12%, respectively. Other bioactive constituents of clove essential oil include crategolic acid, tannins, methyl salicylate, kaempferol, eugenin, gallotannic acid, eugenitin, vanillin, rhamnetin, and oleanolic acid [34, 61]. Moreover, clove essential oil also possesses significant quantities of α and β-humulene, methyl salicylate, methyl amyl ketone, benzaldehyde, chavicol, and β-ylangene. Among these compounds, methyl salicylate and methyl amyl ketone is primarily considered to be responsible for giving distinctive aroma to cloves [21].

Studies have also affirmed that clove bud oil has plentiful quantities of terpenoids, tannins, glycosides, saponins, steroids, ketones, gallic acid, phenylacetic acid, quercetin, and p-hydroxybenzoic acid, etc., having significant functional properties [31, 38]. These phytoconstituents possess several health-promoting functionalities and can be used to combat various lifestyle related ailments. A generalized presentation of various bioactive components isolated from clove buds along with their structures and biological activity has been depicted in (Table 11.1).

TABLE 11.1 Pharmacological Properties of Various Polyphenols Present in Clove Bud Essential Oil

Phytochemical	Structure	Functional Activity
Eugenol	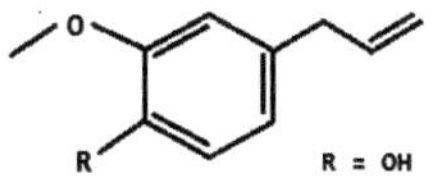 R = OH	Anti-oxidant, anti-inflammatory, anti-microbial, anti-cancer, anti-thrombotic, anti-pyretic
Eugenetin	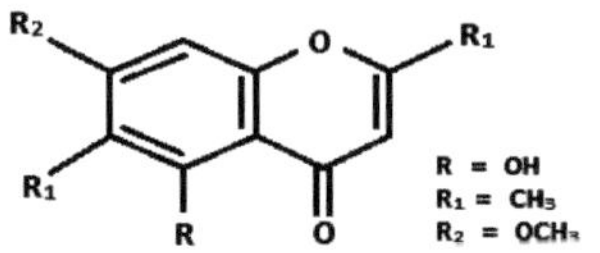 R = OH; R_1 = CH_3; R_2 = OCH_3	Anti-fungal
β-caryophyllene	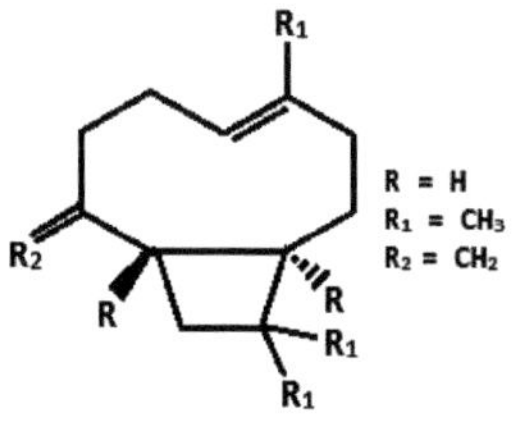 R = H; R_1 = CH_3; R_2 = CH_2	Anti-oxidant, anti-tumor, anti-inflammatory, anesthetic

TABLE 11.1 *(Continued)*

Phytochemical	Structure	Functional Activity
Vanillin	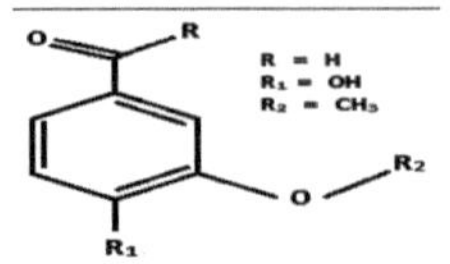 R = H; R_1 = OH; R_2 = CH_3	Anti-oxidative, antimicrobial, anti-depressant
Rhamnetin	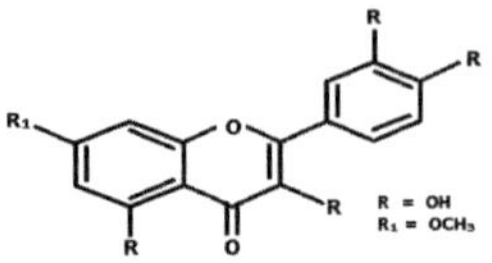 R = OH; R_1 = OCH_3	Anti-inflammatory, antifungal, anti-oxidative, cardioprotective
Crategolic acid	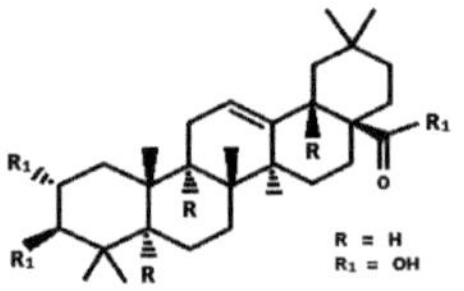 R = H; R_1 = OH	Anti-tumor
Gallic acid	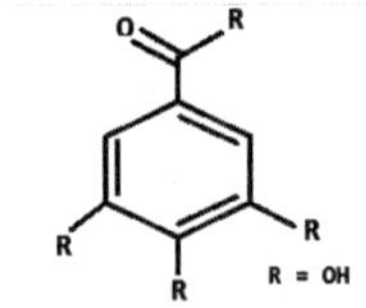 R = OH	Anti-microbial, anti-inflammatory, anti-oxidant
Kaempferol	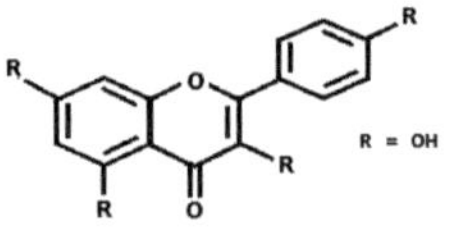 R = OH	Anti-cancer, anti-oxidant, anti-microbial, anti-inflammatory
Biflorin	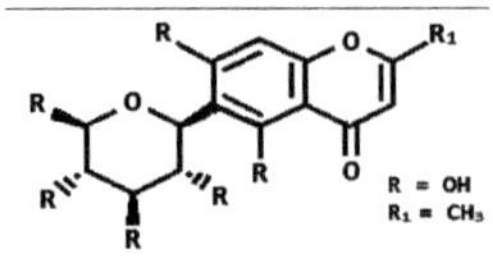 R = OH; R_1 = CH_3	Anti-cancer, anti-oxidant, anti-bacterial
Oleanolic acid	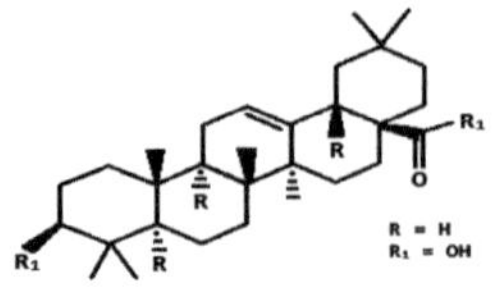 R = H; R_1 = OH	Anti-diabetic, anti-cancerous, anti-microbial
Myricetin	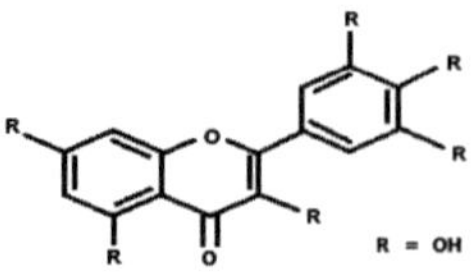 R = OH	Anti-cancerous, anti-microbial, Anti-inflammatory, anti-oxidative

11.3 FUNCTIONAL PERSPECTIVES OF CLOVE ESSENTIAL OIL

Several studies have demonstrated that clove essential oil has potent therapeutic activity due to its chemical constituents particularly eugenol. The pharmacological properties of clove bud essential oil are attributed to its prevention of reactive oxygen species (ROS), effective free radical scavenging activity, and denaturation properties of bacterial proteins and DNA. Studies have also indicated that clove essential oil can be effectually used to combat different lifestyle-related disorders such as gastrointestinal (GI) complications, reproductive ailments, irregularity of blood cholesterol levels, diabetes, hypertension, carcinogenicity, physiological inflammations, and microbial infections. Hence, the present section of the chapter highlights the importance of clove bud essential oil as an effective therapeutic agent and their proposed mode of action (Table 11.2).

11.3.1 ANTI-OXIDANT PROPERTIES OF CLOVE BUD OIL

Clove-based essential oils are known for their strong antioxidant potential and are considered as one of the best food additives. For that purpose, clove oil is usually added in various longevity formulae. Clove bud essential oil have shown significant anti-oxidative properties when compared with synthetic antioxidants such as BHA (butylated hydroxy-anisole) and pyrogallol [12]. It is also reported that clove oil @ 15 μg/mL has the ability to reduce the lipid peroxidation in emulsions up to 97.3%. Additionally, clove essential oil is also reported to have strong DPPH radical scavenging potential at lower concentration of eugenol as compared to the synthetic antioxidants [22]. It has also been revealed that clove essential oil has strong free-radical scavenging and antioxidant effectiveness and these essential oils can be used to delay the process of lipid oxidation in many food products [28]. The antioxidant capability of BACs present in essential oil derived from clove buds is mainly due to the formation of complexes with metallic substances. Furthermore, lipid oxidation is delayed by clove essential oil due to the formation of iron-oxygen chelate complex and elimination of free radicals [29, 34].

11.3.2 ANTIMICROBIAL PERSPECTIVES

Previous investigations have demonstrated the potential of clove clove-based essential oils to inhibit the functionality of a broad spectrum of microorganisms, particularly bacteria [23]. This inhibitory action of clove essential oils is primarily due to the existence of different biologically active components in clove bud oil such as eugenol, oleanolic acid, 2-heptanone, eugenyl acetate, β-caryophyllene [8], gallic acid, α-humulene, acetyl-eugenol, methyl salicylate, methyl-eugenol, trans-confireryl, iso-eugenol [60], dehydrodieugenol, myricetin, aldehyde, phenyl propanoides, kaempferol, rhamnetin, ellagic acid, and biflorin [7]. These active compounds control the bacterial growth and activities by denaturing the cellular proteins of bacteria, reacting with the phospholipids present in the bacterial membranes and modifying their permeability.

Researchers have explored that clove essential oil is effective in mitigating the growth of *E. coli* O157:H7 [6]. Likewise, a recent investigation has reported that clove essential oil has the potential to inhibit the progression of pathogenic strains of *Listeria monocytogenes* (gram-positive), *Salmonella Typhimurium* (gram-negative), and *E. coli* (O157:H7) present on food surface [28]. Similarly, clove essential oil is also regarded as useful in reducing the growth of different Gram-positive (*S. aureus*, *E. faecalis*, and *B. cereus*) and Gram-negative (*S. choleraesuis*, *P. aeruginosa, E. coli*, and *Y. enterocolitica*) pathogenic bacteria [23]. Studies have also affirmed that ethanolic and aqueous extracts (500–1,000 mg/mL) derived from clove buds could retard the growth of methicillin-resistant bacterial isolates in a considerable manner [10]. In another investigation, eugenol (2 μg/mL) was found effective in controlling the growth of *Helicobacter pylori* (31 different strains) after incubational period of 9 hours without developing resistance in bacteria as compared to amoxicillin [2].

Clove essential oil is also useful in decreasing the growth of different food-borne and pathogenic species of yeasts and filamentous fungi [16, 18]. A study concluded that essential oil derived from clove buds might be helpful in minimizing the growth of *Mucor*, *Aspergillus*, *Fusarium moniliform* NCIM 1100, *Fusarium oxysporum, Trichophytum rubrum, and Microsporum gypseum* [48]. Amongst various phenolic and flavonoid components of clove derived essential oil, eugenol is primarily considered to be responsible for controlling the fungal growth by destroying the micelles and fungal spores. Additionally, eugenol disrupts the cellular membranes and deforms the macromolecules which results in cell lysis [11]. The other proposed mechanisms of antifungal activity of clove bud essential oil include partition of lipid bilayer in fungal membranes and changing the membrane permeability which results in the leakage of cellular constituents [6]. These studies have shown the efficacy of clove essential oil as a potent antimicrobial agent.

11.3.3 ANTI-INFLAMMATORY ACTIVITY OF CLOVE EXTRACT

Essential oil derived from clove buds is also found to be helpful in treating several respiratory disorders such as cough, colds, asthma, eye sties, sinus conditions, and bronchitis. It is stated that clove essential oil possesses considerable anti-inflammatory activity at doses of 0.05 (90.15%) to 0.200 mL/kg (82.78%) [46]. Additionally, clove essential oil is helpful in reducing musculoskeletal pain and nasal obstruction which infers its anti-inflammatory potential [35]. The flavonoids present in clove bud oil (kaempferol, β-caryophyllene, rhamnetin) are majorly responsible for its anti-inflammatory properties [19]. In addition, eugenol (200 to 400 mg kg^{-1}) has been reported to decrease the pleural exudates volume without altering the numbers of leukocytes [9].

The BACs found in clove oil suppress the expression of cyclo-oxygenase (COX)-2, TNF-signaling (tumor necrosis factor), and inflammatory cytokine activity [37]. In addition, eugenol also inhibits the synthesis of prostaglandin which delays the process of inflammation [35]. It has also been investigated those bioactive components of clove bud oil are effective in reducing the activity of TNF-a and protecting the dysfunctionality of macrophages [41]. However, more research work is required to further explore the mechanisms associated with the anti-inflammatory action of bioactive components present in clove bud oil.

11.3.4 ANTICANCER ASPECTS

Clove bud essential oil possesses cancer-preventive and anti-mutagenic potential due to strong free-radical scavenging action [44, 62]. Various studies have confirmed the anti-cancer properties of clove essential oil, specifically against skin, lung, and GI cancers [50]. It has been investigated that ethyl acetate extract of clove buds significantly controls the progression of tumor cells and accelerates apoptosis and cell-cycle arrest. The anti-tumor or anti-cancer action of clove extract is mainly owing to the abundance of bioactive components. These biologically active constituents of clove essential oil induce apoptotic conditions and promoted the seizure of G0 (Gap 0)/G1 (Gap 1)-cell cycle in a concentration dependent manner [40]. Another principal component of clove essential oil, eugenol, interferes with cell-signaling pathways and shows anti-cancer potential. Moreover, eugenol also suppresses the development of malignant melanoma by inhibiting anchorage-dependent and anchorage-independent growth, resulting in size reduction of tumor cells [20]. Additionally, research has also inferred that eugenol, in combination with gemcitabine, can significantly control the growth of cancerous cells [26]. It has also been proven through clinical trials that bioactive components of clove essential oil are effective in reducing the risks of colon cancer, gastric tumors, leukemia, prostate cancer, breast cancer and skin cancer [30].

Since last few years, clove essential oil has been used as a potent chemotherapeutic compound for prevention of cancer. Clove oil, particularly eugenol, can be synergistically used with several chemo-preventive drugs to minimize the toxic effects of drugs on normal cells [27]. Studies have also explored that eugenol has a more beneficial effect on human cancer cells when combined with chemotherapeutic agents as compared to the other phenylpropanoids [24]. Similarly, eugenol, and iso-eugenol have anti-proliferative properties and could impart significant influence in mitigating the risk of melanoma cancer. These BACs are reported to reduce the size of tumor cells up to 40% [36].

11.3.5 ANTI-THROMBOTIC PROPERTIES OF CLOVE ESSENTIAL OIL

The BACs present in clove oil are useful in preventing the aggregation of human platelet cells which is mainly induced by platelet-activating factor (PAF), arachidonic acid (AA) or collagen. The inhibitory effect of clove extract against platelet aggregation is sounder against PAF (IC50 = 6 μM) and AA (IC50 = 6 μM) aggregation as compared to the collagen (IC50 = 132 μM). It has also been revealed from previous findings that clove essential oil administrated at varied concentrations guarded (100% and 70%) against PAF (11 mg kg^{-1}) and AA (2.0 mg kg^{-1}) induced thrombosis in rabbits [49]. Additionally, eugenol and acetyl eugenol is reported to be more useful as compared to aspirin in controlling the human platelets aggregation induced by adrenaline, collagen, and arachidonate [53].

11.3.6 ANTI-DIABETIC EFFECT

Nowadays, researchers have been working on inventing diet-based strategies to control metabolic disorders. Likewise, use of natural plant-based ingredients to treat or prevent high blood glucose level is being widely explored by researchers. Numerous epidemiological studies have demonstrated the use of phenolic components in preventing diabetes. Investigations through animal modeling have illustrated those bioactive components of clove bud oil can reduce the risk of diabetes by modifying the activities of enzymes linked with glucose metabolism [52]. Likewise, it is also examined that eugenol is quite operative in alleviating blood glucose level by controlling the activity of α-amylase [55].

In another investigation, it has confirmed through bio-efficacy trial that clove (20–40 g/kg) is quite beneficial in ameliorating the problem of hyperglycemia under type-II diabetic condition [1]. Recently, it has also been investigated that iso-eugenol, present in clove extract, positively influences the activities of alpha-amylase and alpha-glucosidase resulting in anti-hyperglycemic effect [57]. The mechanistic approach behind the anti-diabetic potential of clove essential oil is inhibition of α-glucosidase activity followed by averting the production of AGE (advanced glycation end) products [51].

TABLE 11.2 Mode of Action of Clove Essential Oil in Various Bioactivities

Bioactivity	Mode of Action(s)	References
Anti-oxidant	• Delays the process of lipid oxidation • Suppress the formation of reactive oxidative species	[21, 28]
Antimicrobial	• Hinder the growth of gram-positive and negative bacterial strains by increasing cell permeability • Decrease the growth of foodborne fungi and yeast	[16, 23]
Anti-inflammatory	• Suppress the expression and signaling of TNF and COX • Disables activity of TNFα	[35, 41]

TABLE 11.2 *(Continued)*

Bioactivity	Mode of Action(s)	References
Anti-cancer	• Activates cell apoptosis • Suppresses expression of COX-2 gene • Inhibits the oxidation of DNA	[20, 40]
Anti-thrombotic	• Prevents aggregation of human platelets	[49, 53]
Anti-diabetic	• Downregulates the action of α-amylase and a-glucosidases • Hinder formation of advanced glycation end products • Increase the level of anti-oxidative enzymes	[51, 55]
Anti-hyperlipidemic	• Reduces the concentration of MDA • Elevates the levels of superoxide dismutase • Retard the activity of HMG-CoA reductase	[43, 59]
Neuro-protective	• Inhibit the activity of 5-lipoxygenase enzyme • Regulates the levels of plasma corticosterone	[17, 32]
Other pharmacological	• Reduces toothache • Prevents bone mineralization	[25, 33]

11.3.7 ANTI-HYPERLIPIDEMIC POTENTIAL

Dyslipidemia of hyperlipidemia is mainly caused due to oxidative degradation in the body, resulting in malfunctioning of the anti-oxidative defense system [13]. Dietary interventions containing significant quantities of antioxidants are a good choice to control the problem of dyslipidemia. Studies have affirmed that clove extract is a beneficial remedy to control hypercholesterolemia. A study based on rat-modeling have confirmed that eugenol can significantly reduce total cholesterol level, (55.88%), LDL-cholesterol level (79.48%) and triglyceride (64.30%) level. It has also been investigated that anti-hyperlipidemic activity of eugenol is more effective as compared to lovastatin (lipid-lowering drug). Additionally, eugenol can momentously reduce the activities of several enzymes associated with lipid metabolism such as ALP (alkaline phosphatase), LDH (lactic acid dehydrogenase) and ALT (alanine aminotransferase) enzymes as compared to positive control [59]. Furthermore, clove-leaf extract is useful in reducing the levels of MDA (malondialdehyde) and SOD (superoxide dismutase) up to normal ranges in hyperlipidemic patients [43]. However, more clinical investigations based on human cell lines are needed to suggest clove essential oil to the patients suffering from dyslipidemia.

11.3.8 NEURO-PROTECTIVE PERSPECTIVE OF CLOVE BUD OIL

Evidently, bioactive components of clove essential oil, particularly eugenol, are helpful in reducing the risk of neuro-degenerative problems by improving neuronal functionality [3]. Studies have confirmed the neuroprotective potential of eugenol due to its inhibitory action against the activity of 5-lipoxygenase enzyme and protection of neuronal cells from oxidative damage [32].

Clinical trials based on rat modeling have also confirmed the neuro-protective and anti-stress potential of functional ingredients present in clove oil. For example, in an investigation, male albino rats were administrated by acrylamide to induce neuropathy followed by consumption of eugenol and iso-eugenol. The findings concluded that both BACs were helpful in significant reduction of oxidative degeneration augmenting the concentration of glutathione in brain and sciatic nerve regions. Hence, dietary interventions based on BACs of clove essential oil can be an effective approach in curtailing neuropathy [47].

Additionally, it has also been investigated that clove-based biologically functional constituents have the capability to modulate the hypothalamic-pituitary adrenal cortex axis as a preventive strategy to combat the problem of restraint stress irritable bowel syndrome (IBS). Moreover, eugenol is helpful in regulating the serotonergic system which can extenuate the levels of plasma corticosterone. In addition, eugenol is also beneficial in enhancing the antioxidant immune system in brain region [17].

11.4 OTHER PHARMACOLOGICAL ACTIVITIES OF CLOVE ESSENTIAL OIL

In addition to the above-mentioned pharmacological functions, some other investigated functional aspects of clove bud oil have been discussed in this section. For instance, eugenol, the major bioactive component of clove oil, is being extensively used in the dental clinics as an analgesic agent due to its potential to reduce the toothache [25]. The analgesic effect of eugenol is primarily ascribed to its potential to inhibit prostaglandins and leukotriene and suppressive action on N-methyl-D-aspartate (NMDA) receptors [4]. Additionally, clove bud oil also possesses anesthetic effects due to the presence of eugenol, the active ingredient [58].

Furthermore, the antispasmodic effect of eugenol is also documented in the literature due to blockage of signals to calcium channels which reduces the sensitivity of contractile proteins [39]. Moreover, clove bud oil also possesses antipyretic effectiveness and is able to reduce the severity of fever [15]. Eugenol also has anti-depressant effect due to its ability to inhibit the activity of monoamine oxidase enzyme [56]. Likewise, clove bud oil is useful in preventing bone demineralization, hence, it can be a preventive option to alleviate the risk of osteoporosis [33]. Additionally, eugenol can also act as a repellent to the beetles and can prevent the growth of insect eggs and larval stages of insects inside the grains [45].

11.5 SUMMARY

This chapter discusses the phytochemical and pharmacological activities of essential oil extracted from clove based on the presence of various bioactive components. The primary bioactive constituents of clove essential oil are eugenol, iso-eugenol, thymol, carvacrol, and cinnamaldehyde. The major functional capabilities of clove essential oil include antimicrobial activity, anti-inflammatory properties, antioxidant potential, anti-diabetic efficacy, anti-hyperlipidemic potential and anticancer functionality. These major activities of bioactive components found in clove essential oil are comprehensively discussed in this chapter.

KEYWORDS

- **arachidonic acid**
- **clove**
- **essential oil**
- **eugenol**
- **lactic acid dehydrogenase**
- **pharmacological properties**
- **phytochemistry**

REFERENCES

1. Adefegha, S. A., Oboh, G., Adefegha, O. M., Boligon, A. A., & Athayde, M. L., (2014). Antihyperglycemic, hypolipidemic, hepatoprotective and antioxidative effects of dietary clove (*Syzygium aromaticum*) bud powder in a high-fat diet/streptozotocin-induced diabetes rat model. *Journal of Science and Food Agriculture, 94*(13), 2726–2737.
2. Ali, S. M., Khan, A. A., Ahmed, I., Musaddiq, M., Ahmed, K. S., Polasa, H., Rao, L. V., et al., (2005). Antimicrobial activities of eugenol and cinnamaldehyde against the human gastric pathogen *Helicobacter pylori. Annals of Clinical Microbiology and Antimicrobials, 4*, 20.
3. Anuj, G., & Sanjay, S., (2010). Eugenol: A potential phytochemical with multifaceted therapeutic activities. *Pharmacologyonline, 2*, 108–120.
4. Aoshima, H., & Hamamoto, K., (1999). Potentiation of GABA A receptors expressed in Xenopus oocytes by perfume and phytoncide. *Bioscience of Biotechnology and Biochemistry, 63*,743–748.
5. Bello, M. O., & Jimoh, A. A., (2012). Nutrients composition of seed, chemical characterization and fatty acid composition of oil of *Syzygium aromaticum. Applied Chemistry, 42*, 6065–6068.
6. Burst, S. A., & Reinders, R. D., (2003). Antibacterial activity of selected plant essential oils against *Escherichia coli* O157:H7. *Letters of. Applied Microbiology, 36*(3), 162–167.
7. Cai, L., & Wu, C. D., (1996). Compounds from *Syzygium aromaticum* possessing growth inhibitory activity against oral pathogens. *Journal of Natural Products, 59*(10), 987–990.
8. Chaieb, K., Zmantar, T., Ksouri, R., Hajlaoui, H., Mahdouani, K., Abdelly, C., & Bakhrouf, A., (2007). Antioxidant properties of essential oil of *Eugenia caryophyllata* and its antifungal activity against a large number of clinical Candida species. *Journal of Mycoses, 50*(5), 403–406.
9. Daniel, A. N., Sartoretto, S. M., Schmidt, G., Caparroz-Assef, S. M., Bersani-Amado, C. A., & Cuman, R. K., (2009). Anti-inflammatory and antinociceptive activities of eugenol essential oil in experimental animal models. *Revista Brasileira de Farmacognosia, 19*, 212–217.
10. Demirpek, U., Olgun, A., Kısa, O., & Guvenç, A., (2009). *In vitro* antibacterial activity of cloves (*Syzygium aromaticum*) against MRSA. *Planta Medicine, 75*(9), 59.
11. Devi, K. P., Nisha, S. A., Sakthivel, R., & Pandian, S. K., (2010). Eugenol (an essential oil of clove) acts as an antibacterial agent against *Salmonella* typhi by disrupting the cellular membrane. *Journal of Ethnopharmacology, 130*(1), 107–115.
12. Dorman, H. D., Figueiredo, A. C., Barroso, J. G., & Deans, S. G., (2000). *In vitro* evaluation of antioxidant activity of essential oils and their components. *Flavor and Fragrance Journal, 15*(1), 12–16.
13. Duarte, M. M., Rocha, J. B., Moresco, R. N., Duarte, T., Da Cruz, I. B., Loro, V. L., & Schetinger, M. R., (2009). Association between ischemia-modified albumin,

lipids and inflammation biomarkers in patients with hypercholesterolemia. *Clinical Biochemistry, 42*, 666–671.

14. Ereifej, K. I., Feng, H., Rababah, T. M., Tashtoush, S. H., & Al-U'datt, M. H., (2015). Microbiological status and nutritional composition of spices used in food preparation. *Food Nutrition Science, 6*, 1134–1140.
15. Feng, J., & Lipton, J. M., (1987). Eugenol: Antipyretic activity in rabbits. *Neuropharmacology, 26*, 1775–1778.
16. Fujisawa, S., & Murakami, Y., (2016). Eugenol and its role in chronic diseases. In: *Drug Discovery from Mother Nature* (pp. 45–66). Springer, Cham.
17. Garabadu, D., Shah, A., Ahmad, A., Joshi, V. B., Saxena, B., Palit, G., & Krishnamurthy, S., (2011). Eugenol as an anti-stress agent: Modulation of hypothalamic-pituitary-adrenal axis and brain monoaminergic systems in a rat model of stress. *Stress, 14*, 145–155.
18. Gayoso, C. W., Lima, E. O., Oliveira, V. T., Pereira, F. O., Souza, E. L., & Lima, I. O., (2005). Sensitivity of fungi isolated from onychomycosis to *Eugenia caryophyllata* essential oil and eugenol. *Fitoterapia., 76*, 247–249.
19. Ghelardini, C., Galeotti, N., Mannelli, L. C., Mazzanti, G., & Bartolini, A., (2001). Local anaesthetic activity of beta-caryophyllene. *Farmacology, 56*(5), 387–389.
20. Ghosh, R., Nadiminty, N., Fitzpatrick, J. E., Alworth, W. L., Slaga, T. J., & Kumar, A. P., (2005). Eugenol causes melanoma growth suppression through inhibition of E2F1 transcriptional activity. *The Journal of Biological Chemistry, 280*(7), 5812–5819.
21. Gopalakrishnan, M., Narayanan, C. S., & Mathew, A. G., (1984). Sesquiterpene hydrocarbons from clove oil. *Lebensmittel-Wissenschaft und -Technologie B., 17*, 42–43.
22. Gulcin, I., Elmastas, M., & Aboul-Enein, H. Y., (2012). Antioxidant activity of clove oil - A powerful antioxidant source. *Arabian Journal of Chemistry, 5*, 489–499.
23. Gupta, N., Parashar, P., Mittal, M., Mehra, V., & Khatri, M., (2014). Antibacterial potential of *Elletaria cardamomum*, *Syzygium aromaticum* and *Piper nigrum*, their synergistic effects and phytochemical determination. *Journal of Research in Pharmacy, 8*(8), 1091–1097.
24. Hemaiswarya, S., & Doble, M., (2013). Combination of phenylpropanoids with 5-fluorouracil as anti-cancer agents against human cervical cancer (HeLa) cell line. *Phytomedicine, 20*, 151–158.
25. Hosseini, M., Kamkar, A. M., & Rakhshandeh, H., (2011). Analgesic effect of clove essential oil in mice. *Avicenna Journal of Phytomedine, 1*(1), 1–6.
26. Hussain, A., Brahmbhatt, K., Priyani, A., Ahmed, M., Rizvi, T. A., & Sharma, C., (2011). Eugenol enhances the chemotherapeutic potential of gemcitabine and induce anticarcinogenic and anti-inflammatory activity in human cervical cancer cells. *Cancer Biotherapy Radiopharmaceuticals, 26*(5), 519–527.
27. Hussain, A., Priyani, A., Sadrieh, L., Brahmbhatt, K., Ahmed, M., & Sharma, C., (2012). Concurrent sulforaphane and eugenol induces differential effects on human cervical cancer cells. *Integrative Cancer Therapies, 11*(2), 154–165.
28. Ishaq, A., Syed, Q. A., Khan, M. I., & Zia, M. A., (2019). Characterizing and optimizing antioxidant and antimicrobial properties of clove extracts against food-borne pathogenic bacteria. *International Food Research Journal, 26*, 1165–1172.

29. Ito, M., Murakami, K., & Yoshino, M., (2005). Antioxidant action of eugenol compounds: Role of metal ion in the inhibition of lipid peroxidation. *Food and Chemistry Toxicology,* 43,461–466.
30. Jaganathan, S. K., & Supriyanto, E., (2012). Antiproliferative and molecular mechanism of eugenol-induced apoptosis in cancer cells. *Molecules, 17*, 6290–6304.
31. Jimoh, S. O., Arowolo, L. A., & Alabi, K. A., (2017). Phytochemical screening and antimicrobial evaluation of *Syzygium aromaticum* extract and essential oil. *International Journal Current. Microbiology and Applied Sciences, 6*, 4557–4567.
32. Kabuto, H., Tada, M., & Kohno, M., (2007). Eugenol [2-methoxy-4-(2-propenyl) phenol] prevents 6-hydroxydopamine induced dopamine depression and lipid peroxidation inductivity in mouse striatum. *Biological and Pharmaceutical Bulletin, 30*, 423–427.
33. Kamatou, G. P., Vermaak, I., & Vilijoen, A. M., (2012). Eugenol-from the remote Maluku islands to the international market place: A review of a remarkable and versatile molecule. *Molecules, 17*(6), 6953–6981.
34. Khalil, A. A., Ur Rahman, U., Khan, M. R., Sahar, A., Mahmood, T., & Khan, M., (2017). Essential oil eugenol: Sources, extraction techniques and nutraceutical perspectives. *RSC Advances, 7*, 32669–32681.
35. Kim, E. H., Kim, H. K., & Ahn, Y. J., (2003). Acaricidal activity of clove bud oil compounds against *Dermatophagoides farinae* and *Dermatophagoides pteronyssinus* (Acari: Pyroglyphidae). *Journal of Agriculture and Food Chemistry, 51*(4), 885–889.
36. Kim, G. C., Choi, D. S., Lim, J. S., Jeong, H. C., Kim, I. R., Lee, M. H., & Park, B. S., (2006). Caspases-dependent apoptosis in human melanoma cell by eugenol. *Journal of Analytical Sciences and Technology, 39*, 245–253.
37. Kim, S. S., Oh, O. J., Min, H. Y., Park, E. J., Kim, Y., Park, H. J., Han, Y. N., & Lee, S. K., (2003). Eugenol suppresses cyclooxygenase-2 expression in lipopolysaccharide-stimulated mouse macrophage RAW264.7 cells. *Life Sciences, 73*, 337–348.
38. Kumar, U., Kumar, B., Bhandari, A., & Kumar, Y., (2010). Phytochemical investigation and comparison of antimicrobial screening of clove and cardamom. *International Journal of Pharmaceutical Sciences and Research, 1*,138–147.
39. Lima, F. C., Peixoto-Neves, D., Gomes, M. D., Coelho-de-Souza, A. N., Lima, C. C., Zin, W. A., Magalhães, P. C., et al., (2011). Antispasmodic effects of eugenol on rat airway smooth muscle. *Fund Clinical Pharmacology, 25*(6), 690–699.
40. Liu, H., Schmitz, J. C., Wei, J., Cao, S., Beumer, J. H., Strychor, S., Cheng, L., et al., (2014). Clove extract inhibits tumor growth and promotes cell cycle arrest and apoptosis. *Journal of Oncology Research, 21*(5), 247–259.
41. Mahapatra, S. K., Bhattacharjee, S., Chakraborty, S. P., Majumdar, S., & Roy, S., (2011). Alteration of immune functions and Th1/Th2 cytokine balance in nicotine-induced murine macrophages: Immunomodulatory role of eugenol and N-acetylcysteine. *International Immunopharmacology, 11*, 485–495.
42. Mittal, M., Gupta, N., Parashar, P., Mehra, V., & Khatri, M., (2014). Phytochemical evaluation and pharmacological activity of *Syzygium aromaticum*: A comprehensive review. *International Journal of Pharmacy and Pharmaceutical Sciences, 6*, 67–72.
43. Munisa, A., Manalu, W., Wresdiyati, T., & Kusumorini, N., (2015). The effect of clove leaf methanol extract on the profiles of superoxide dismutase and malondialdehyde in

the liver of rabbits under hypercholesterolemia condition. *Translational. Biomedicine, 6*, 1–5. doi: 10.21767/2172-0479.100012.

44. Miyazawa, M., & Hisama, M., (2001). Suppression of chemical mutagen induced SOS response by alkylphenols from clove (*Syzygium aromaticum*) in *Salmonella Typhimurium* TA153/pSK1002 umu test. *Journal of Agriculture and Food Chemistry, 49*, 4019–4025.
45. Obeng -Ofori, D., & Reichmuth, C. H., (1997). Bioactivity of eugenol, a major component of essential oil of *Ocimum suave* (Wild.) against four species of stored-product Coleoptera. *International Journal of Pest Management, 43*, 89–94.
46. Ozturk, A., & Ozbek, H., (2005). The anti-Inflammatory activity of *Eugenia caryophyllata* essential oil, An animal model of anti-inflammatory activity," *European Journal of General Medicine, 2*(4), 159–163.
47. Prasad, S. N., & Muralidhara, M., (2013). Neuroprotective efficacy of eugenol and isoeugenol in acrylamide-induced neuropathy in rats: Behavioral and biochemical evidence. *Neurochemistry Research, 38*, 330–345.
48. Rana, I. S., Rana, A. S., & Rajak, R. C., (2011). Evaluation of antifungal activity in essential oil of the *Syzygium aromaticum* (L.) by extraction, purification and analysis of its main component eugenol. *Brazil Journal of Microbiology, 42*(4), 1269–1277.
49. Saeed, S. A., & Gilani, A. H., (1994). Antithrombotic activity of clove oil. *Journal of Pakistan Medical Associations., 44*(5), 112–115.
50. Schonfelder, I., & Schonfelder, P., (2004). *Syzygium aromaticum* (L.) Merr & Perr (*Eugenia caryophyllata* THUNB *Jambosa caryophyllus* (SPRENG.) (NIEDENZU). *Das neue Handbuch der Heilpflanzen, Kosmos-Verlag, Stuttgart, Germany,* pp. 431, 432.
51. Singh, P., Jayaramaiah, R. H., Agawane, S. B., Vannuruswamy, G., Korwar, A. M., Anand, A., Dhaygude, V. S., et al., (2016). Potential dual role of eugenol in inhibiting advanced glycation end products in diabetes: Proteomic and mechanistic insights. *Scientific Reports, 6*, 1–13.
52. Srinivasan, S., Sathish, G., Jayanthi, M., Muthukumaran, J., Muruganathan, U., & Ramachandran, V., (2014). Ameliorating effect of eugenol on hyperglycemia by attenuating the key enzymes of glucose metabolism in streptozotocin-induced diabetic rats. *Molecular Cell Biochemistry, 385*(1, 2), 159–168.
53. Srivastava, K. C., (1993). Antiplatelet principles from a food spice clove (*Syzygium aromaticum* L). *Prostaglandins, Leukotrienes & Essential Fatty Acids, 48*, 363–372.
54. Sulieman, A. M., El Boshra, I. M., & El Khalifa, E. A., (2007). Nutritive value of clove (*Syzygium aromaticum*) and detection of antimicrobial effects of its bud oil. *Research Journal of Microbiology, 2*, 266–271.
55. Tahir, H. U., Sarfraz, R. A., Ashraf, A., & Adil, S., (2016). Chemical composition and anti-diabetic activity of essential oils obtained from two spices (*Syzygium aromaticum* and *Cuminum cyminum*). *International Journal of Food Properties, 19*, 2156–2164.
56. Tao, G., Irie, Y., Li, D. J., & Keung, W. M., (2005). Eugenol and its structural analogs inhibit monoamine oxidase A and exhibit antidepressant-like activity. *Bioorganic & Medical Chemistry, 13*, 4777–4788.
57. Topal, F., (2019). Anticholinergic and antidiabetic effects of isoeugenol from clove (*Eugenia caryophylata*) oil. *International Journal of Food Properties, 22*(1), 583–592.

58. Taylor, P. W., & Roberts, S. D., (1999). Clove oil: An alternative anesthetic for aquaculture. *North. American. Journal of Aquaculture, 61*, 150–155.
59. Venkadeswaran, K., Muralidharan, A. R., Annadurai, T., Ruban, V. V., Sundararajan, M., Anandhi, R., Thomas, P. A., & Geraldine, P., (2014). Antihypercholesterolemic and antioxidative potential of an extract of the plant, Piper betle, and its active constituent, eugenol, in triton WR-1339-induced hypercholesterolemia in experimental rats. *Evidence. Based Complimentary and Alternative Medicine,* 1–11. doi: 10.(1155)/(2014)/(4789)73.
60. Yang, Y. C., Lee, S. H., Lee, W. J., Choi, D. H., & Ahn, Y. J., (2003). Ovicidal and adulticidal effects of *Eugenia cryophyllata* bud and leaf oil compounds on *Pediculus capitis*. *Journal of Agriculture and Food Chemistry, 51*(17), 4884–4888.
61. Zachariah, T. J., Krishnamoorthy, B., Rema, J., & Mathew, P. A., (2005). Oil constituents in bud and pedicel of clove (*Syzygium aromaticum*). *Journal of Indian Perfumer, 49*, 313–316.
62. Zheng, G. Q., Kenny, P. M., & Lam, L. K., (1992). Sesquiterpenes from clove (*Eugenia carrophyllata*) as potential anticarcinogenic agents. *Journal of Natural Products, 55*, 999–1003.

CHAPTER 12

EFFECT OF WHOLE FLAXSEED ON BONE TURNOVER MARKERS AND GUT MICROBIOTA IN MENOPAUSAL-RELATED BONE LOSS: REVIEW

HALAH HAFIZ, DANIEL COMMANE, GEMMA WALTON, and KIM JACKSON

ABSTRACT

Osteoporosis is a disease characterized by bone loss and architectural changes in bone tissue affecting 200 million women across the globe. Aging, hormonal state, and poor nutritional status are considered important risk factors. Dietary strategies including functional foods are potential lifestyle modifications that could help to prevent or delay the onset of osteoporosis and related diseases. Dietary flaxseed is becoming increasingly recognized as a functional food and has gained attention in the area of osteo-science as the seeds are a rich source of phytoestrogenic lignans. In particular, they contain the lignan derivatives secoisolariciresinol diglucoside, matairesinol (MAT), lariciresinol, and pinoresinol (Pin) which may be metabolized by the gut microbiota to form the mammalian lignans. To date, studies of dietary supplementation with flaxseed on markers of bone health have been inconsistent in demonstrating benefit. Inter-individual differences in gut bacteria composition may influence the metabolism of lignans and this might partly explain discrepancies between studies. Accumulative evidence from animal studies suggests a potential beneficial effect of flaxseed metabolites on skeletal health, mediated through a manipulation of the bone remodeling processes. This chapter narrates the potential of flaxseed as a protective and therapeutic functional food in osteoporosis with an emphasis on the mitigating role of the gut microbial community.

12.1 INTRODUCTION

Osteoporosis affects 200 million women across the globe [71] and remains one of the primary causes of disability, morbidity, and mortality in postmenopausal women [1]. In the UK, the annual healthcare cost of osteoporosis and related bone diseases is estimated at £3.5 million [71] and these costs are increasing. Thus, diet or lifestyle strategies that reduce the burden of these diseases in the population are needed. Evidence has suggested foods that contain natural components with estrogen like structures (also referred to as phytoestrogens) may protect against osteoporosis. It has been estimated that the daily human diet contains approximately 1.5 grams of phytochemicals [2], which includes phenolic compounds such as flavonoids and lignans.

In the Western diet, flaxseeds are the most abundant lignan source which represent a diverse class of phenylpropanoid oligomers and dimers. These compounds are predominately metabolized to estrogenic enterolignans; enterodiol (END) and enterolactone (ENL) by the gut microbiota. Although, the effect of soy isoflavones on bone health has been studied in several randomized controlled trials (RCTs) [3, 4], very little is known about the effects of flaxseeds on bone health. In this manuscript, we explore the literature on the role of flaxseeds on markers of bone health in postmenopausal women. Before the presentation of the methodology and results, we provide a general overview of bone metabolism and biochemical bone turnover markers (BTM) and describe the individual impact of the menopause and phytoestrogens on bone health markers.

12.1.1 BONE (FORMATION AND RESORPTION) OR TURNOVER PROCESS

During the first three decades of life, the process of bone formation is favored over bone resorption in order to reach a peak bone mass [5]. However, after the age of 30 years, the process of bone resorption initiates and propagates more rapidly than the process of bone formation, hence causing a gradual decline in bone mineral content (BMC) in both men and women [6]. Bone remodeling is ideally a balanced lifelong process, in which osteoclast (cells that break down bone) activity is equal to osteoblasts (cells responsible for the modeling and mineralization of bone during both initial bone formation) activity. Once bone resorption becomes favored over bone formation, there is a progressive reduction in total bone mass. Aging, menopause, and low body weight are all associated with rapid bone loss.

Bone turnover involves four processes: (i) bone surface activation; (ii) resorption; (iii) repercussion; and (iv) formation [7] (Figure 12.1). These processes are controlled by cytokines and hormones [8], and take place at the surface of both cortical and trabecular bone. Osteoclasts start the cycle degrading hydroxyapatite locally in the osteon which is the main unit of compact bone. Afterwards these osteoclasts release minerals leading towards the formation of a hole in bone tissue. Bone then enters into a brief adjustment (resorption) phase, in which bone cells try to repair the bone by covering the hole in bone tissue to build a cementum line which consists of 50% non-organic material hydroxyapatite and 50% of collagen and proteoglycans [5, 8]. This leads to the start of the formation phase when osteoblasts fill the osteon with collagenous matrix and finally the stabilization phase, when the bone matrix matures through mineralization [5].

Activation of the bone turnover cycle induces an upregulation in expression of cytokines including, tumor necrosis factor-alpha (TNFα), receptor activator of nuclear factor kappa B ligand (RANKL) and macrophage colony-stimulating element (M-CSF), and suppression of the expression of the cytokine osteoprotegerin (OPG) [9]. TNFα is a pro-inflammatory cytokine, it is known to suppress the differentiation of osteoblasts, hence suppressing bone formation [10], RANKL is a member of the TNF family and is the principle stimulatory factor for the formation of mature osteoclasts, and is fundamental for their survival. RANKL activates its receptor RANK found on osteoclasts and dendritic cells. RANKL is primarily secreted by osteocytes, mature osteoblasts that grow to be embedded within the mineralized bone matrix [11]. The antagonist of RANKL is OPG or osteoclastogenesis inhibitory factor (OCIF), a glycoprotein of the TNF receptor superfamily 11b (TNFRSF11B). OPG is created by a wide range of tissues and cells including osteoblasts [11] and is a negative regulator of bone resorption by binding to RANKL and preventing activation of its receptor RANK, thus suppressing its function in osteoclastogenesis (Figure 12.1).

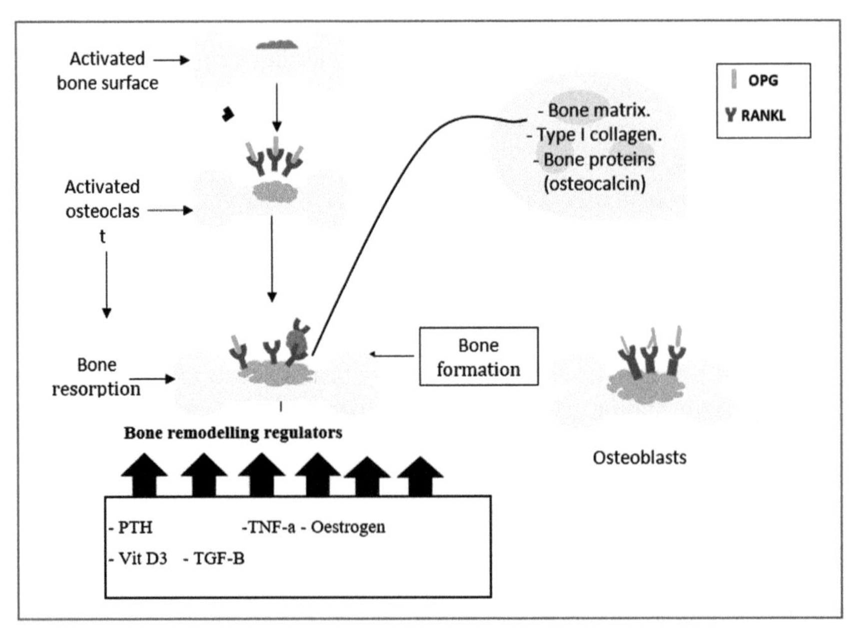

FIGURE 12.1 Overview of the four phases of the bone metabolism cycle; bone surface activation, resorption, repercussion, and formation. [Abbreviations: OPG: Osteoprotegerin; PTH: parathyroid hormone; TGF-β: transforming growth factor-beta-β; TNFα: tumor necrosis factor-alpha; RANKL: receptor activator of nuclear factor kappa B ligand].

12.1.2 BIOCHEMICAL BONE METABOLISM AND TURNOVER MARKERS

Practical biomarkers for use in intervention studies aimed at the prevention of bone loss and osteoporosis include the biochemical BTM which are sensitive to the phases of bone metabolism. During bone resorption, osteoclasts attach to the bone surface removing bone type I collagen, releasing bone minerals such as phosphorous and calcium and the small proteins within the bone matrix such as cytokines and growth factors. Biochemical BTM contains collagen degradation indicators measured in urine or serum and include cross-linked C-telopeptide of type I collagen (CTX-I), cross-linked amino-terminal collagen (NTX-I), pyridinoline, deoxypyridinoline, and excessive calcium in urine. In contrast, the bone formation phase might be detected by the organic component of bone products which produce osteocalcin (OC), bone-specific alkaline phosphatase (BAP), osteopontin, and procollagen I carboxyterminal propeptide (PICP). Figure 12.2 highlights the bone metabolism markers and their relationship to resorption and formation phases.

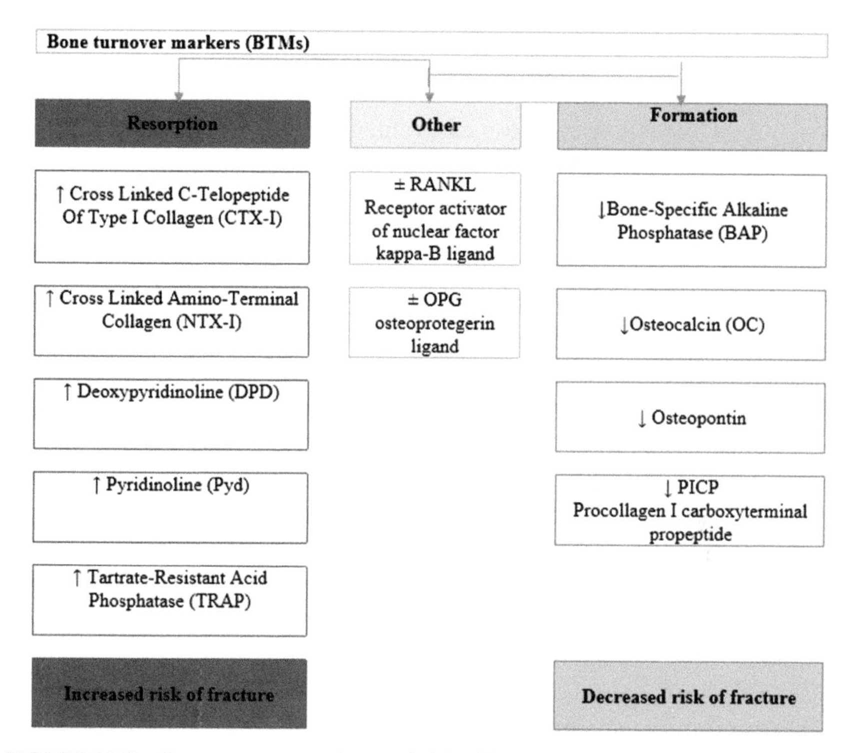

FIGURE 12.2 Bone turnover markers and risk of fracture.

During the third decade of life in healthy men, bone metabolism is usually in a balanced and stable phase in which the levels of BTM remain constant until the seventh or eighth decade, when a gradual increase in BTM can be observed in blood and urine [12]. However, during the early postmenopausal period (≤ 5 years following onset of the menopause) a significant acceleration in bone turnover occurs, which is measured by an increase of 50–100% in circulating levels of bone resorption and formation markers [13, 14].

A prospective cohort study has compared one-year changes in BTM in premenopausal, perimenopause, and early postmenopausal healthy women [15]. This study highlighted that alteration in BTM starts to happen during the late perimenopause characterized by a reduction in bone formation markers followed by an increase in bone resorption markers [15]. Studies on postmenopausal osteoporosis have pinpointed that following the menopause, bone loss markedly changes mainly because of an increase in bone resorption leading to an imbalance in bone turnover cycle [16]. It has been demonstrated in prospective clinical studies that postmenopausal women diagnosed with low bone mass might benefit from the measurement of bone resorption markers as an early predictor of osteoporotic fracture risk [17]. In line with these findings, a prospective study has highlighted that measurement of bone metabolism markers in addition to BMD and clinical data provides a better estimation of future fracture risk [18].

Interestingly, women who had both an increased serum CTX level and a low BMD had a 4.8% higher risk of hip fracture compared with those who had only low BMD or higher levels of CTX. In addition, a positive relationship was found between risk of hip fractures and serum BAP (marker of bone formation) when it was decreased by the intake of the anti-resorptive medicine bisphosphonates [19]. Moreover, a prospective study involving 671 postmenopausal women with a median follow-up of 9.1 years, elevated bone metabolism markers were independently correlated with a higher risk of fracture [20]. Only women who had higher levels of serum BAP were diagnosed with non-vertebral fractures [20]. Hence, biochemical BTMs may play a key role in the early diagnosis of an increased fracture risk that a BMD measurement alone could not adequately explain. By revealing the proteins, enzymes, and collagen in serum or urine during the bone turnover cycle, the active phase (resorption or formation) can be identified by specific biochemical BTMs. Meta-analysis and control clinical trials have addressed the alteration in BTM in response to treatments such as hormone replacement therapy (HRT), estrogen therapy and anti-resorptive drugs as a primary or secondary endpoint [21]. However, little is known about alteration in BTM in response to dietary strategies such as flaxseed.

12.1.3 FLAXSEED LIGNANS AND BONE HEALTH

Phytoestrogens have been associated with a variety of health outcomes including, a decreased risk of postmenopausal osteoporosis, breast, ovarian cancer, and CVD. A meta-analysis of observational epidemiological studies revealed that plant phytoestrogen consumption is much higher in those following an Asian diet than a Western-style diet [22]. Furthermore, researchers also found that Asian women have a lower hip fracture risk compared to western women [23]. Isoflavones are well studied with respect to their potential *in vitro* and *in vivo* role in reducing postmenopausal symptoms and modulation of bone remodeling [24]. However, little is known about the potential effect of lignans on postmenopausal bone loss.

The main food sources of lignans in the Western-style diet are oily seeds such as flaxseed [25]. To date, the *in vitro* and *in vivo* abilities of flaxseed lignans to impact postmenopausal bone loss have been proposed to occur via their effects on sex hormones and RANKL hormone metabolism (Figure 12.3). Lignans have also been reported to affect bone remodeling by activating calcium signaling pathways, and through inhibition of RANKL binding to the RANK ligand receptor on the bone cell surface preventing osteoclast formation (osteoclastogenesis) [26]. Ground flaxseeds have also been shown to have an effect on the levels and metabolism of sex hormones demonstrated by the observed increase in urinary 2-OH estrogen excretion [27]. The metabolism of estradiol and in particular 16α-hydroxylation (16α-OHE1) is of particular interest in relation to bone health since 16α-OHE1 is formed from estrogen by two hydroxylation pathways [27]. A positive relationship has been observed between urinary 16 α-OHEI and BMD amongst postmenopausal women [28]. Adlercreutz, Höckerstedt [29] have also shown that lignans have the ability to positively affect postmenopausal hormone imbalance by directly increasing the levels of sex hormone-binding globulin (SHBG) [30], and decreasing luteinizing hormone (LH) and follicle-stimulating hormone (FSH) [31]. As presented above, the decrease in these hormones could have a direct impact on bone health.

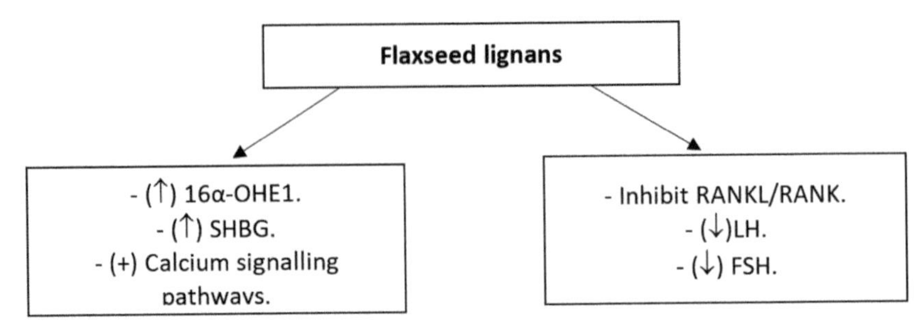

FIGURE 12.3 Influence of flaxseed lignans on bone remodeling. [Abbreviations: 16α-OHE1: 16α-hydroxylation; FSH: follicle-stimulating hormone; LH: luteinsin hormone; RANK: receptor activator of nuclear factor κB; RANKL: RANK-ligand; SHBG: sex hormone-binding globulin].

12.2 COMPREHENSIVE REVIEW OF VARIOUS STUDIES

12.2.1 EPIDEMIOLOGICAL STUDIES

To date, no study has assessed the influence of long-term flaxseed lignan consumption on BTMs or BMC. However, the association of flaxseed lignans with many health outcomes (e.g., breast and prostate cancer and postmenopausal symptoms) have been commonly examined [32–34]. The two epidemiological studies which have determined the relationship between lignans, and BMD are summarized in Table 12.1. In a cross-sectional study performed in 75 Korean postmenopausal women aged 52 to 65 years, study participants were divided into three groups; osteoporotic, osteopenic, and healthy BMD. The osteoporotic group was found to have significantly lower urinary ENL than the participants in the healthy group. Moreover, higher urinary excretion of ED and ENL was correlated to a higher BMD in study participants [35]. These findings suggest that lignan metabolites might be associated with a healthier bone mass in postmenopausal women. However, another epidemiological study conducted in healthy postmenopausal women in the Netherlands found higher cortical bone loss was positively correlated with the urinary excretion of ENL and equol [36]. Differences in the findings might reflect the BMD criteria used in these studies but also other factors such as ethnicity and differences in habitual dietary intakes between the Asian and European groups which might have additional direct effects on bone loss rate.

12.2.2 HUMAN DIETARY INTERVENTION STUDIES

To date, five dietary intervention studies have examined the effect of whole flaxseed consumption (without any other interventions) on BMD, BMC or BTM in postmenopausal women (Table 12.1). Despite the high doses used in these studies (25–45 g/d) and the relatively long intervention durations (six weeks to one year), their findings were inconsistent with respect to BMT and bone health. Of the studies reporting beneficial effects, Arjmandi et al. [46] demonstrated that daily intake of 38 g/d whole flaxseeds for 6 weeks was associated with altered BTMs in postmenopausal women aged 65 years and older. Although bone formation markers alkaline phosphatase (BAP), IGF-1 and IGF-binding protein-3 were unchanged, a significant decrease in the bone resorption marker tartrate-resistant acid phosphatase (TRAP) was observed. Furthermore, flaxseeds were shown to have an estrogenic effect in 46 postmenopausal women after consumption of 25 g/day of flaxseed flour baked into muffins for four months [37]. This was demonstrated by a significant increase in urinary concentration of 2-OHE in women consuming the flaxseed versus the wheat placebo muffins ($p=0.05$).

Interestingly, women in the flaxseed group were shown to excrete mainly END and ENL whilst women in the soy group excreted mainly genistein, daidzein, and equol. This study also highlighted a negative relationship between serum ALP and urinary excretion of lignan metabolites in the flaxseed group only, suggesting the effect of whole flaxseeds on bone health was largely derived from its lignan content. The longest intervention period conducted on flaxseeds and bone health was a 12-month randomized, double-blinded study in 174 postmenopausal women. Dodin et al. [45] reported no effect of flaxseed consumption on BMD loss at the lumbar vertebrae or femur neck with consumption of 40 g/day of flaxseeds in the form of bread or as ground flaxseeds added to foods. However, Dodin et al. [45] did highlight that the normal range of BMD in all the study participants could have masked the effects of the flaxseed intervention. In addition, guidelines have recommended that BMD should only be assessed after 18–24 months of starting a therapeutic intervention, so measurement of BTMs may have provided a useful indicator of changes in bone health over the intervention period [38]. Similar to these findings, no differences in BMD were observed in postmenopausal women, with a baseline BMI and BMD within the normal range, after consuming either

40 g/day of flaxseeds (n=29) or wheat (control, n=29) for three months [39]. Although changes in BTMs were also not observed, Lucas et al. [39] did report significant reductions in both serum cholesterol (6%) and triglyceride (12.8%) concentrations following the flaxseed intervention. Dalais et al. [40] also observed no differences in BMD in a double-blind, randomized controlled crossover study in 52 women aged between 45 and 65 years given bread containing 45 g/day of flaxseed or wheat over 12 weeks [40] (Table 12.1).

12.2.3 ANIMAL STUDIES

Four relevant animal studies determined the role of flaxseed lignans on markers of bone health. A significant increase in femur bone strength and tendency for an increase in BMC of ovariectomized (OVX) (breast cancer) mice was observed after feeding a basal diet supplemented with 10% flaxseeds for 25 weeks compared to the control group fed 20% corn oil [41]. The flaxseed-based diet had an effect on the biomechanical strength of the femur and an estrogenic adverse effect on the uterus. This was evidenced by the significantly heavier wet weight of uterus in the isoflavone-rich soy protein isolate and flaxseed group compared to flaxseed only group, suggesting that flaxseed lignans may have a tissue-selective effect [41]. In accordance with these findings, S. Babu [42] found that flaxseed lignans only showed a tendency for an effect on the bone formation marker BAP in growing female rats. This study compared supplementation with flaxseeds (5% or 10%) and defatted flaxseed meal (6.2%), and the authors speculated that the lower ALP level in the treatment groups than control group could be due to the flaxseed lignan content rather than oil content (e.g., ALA).

Nevertheless, a study conducted in growing pigs (under the age of one year) supplementing the diet with 2.36–2.75% ALA by using flaxseed or flaxseed oil reported a significant rise in plasma levels of n-3 PUFA but no effect on the level of plasma NTX-1 [43]. In order to determine whether the mammalian lignan precursors of flaxseeds were effective in bone development, Ward, and co-workers conducted a randomized control trial in which 20 female rats were fed 50 or 100 g/kg of purified SDG or a control for 6 months [44]. The total urinary lignan excretion was significantly higher in the order: 100 g/kg SDS group > 50 g/kg

SDS group > control group (P <0.001). Although BMD, femur weight, length, and width were unchanged, rats treated with 100 g/kg SDS had significantly higher quality of bone (bone strength) compared to control group. The authors concluded that flaxseed SDG may be mediating the beneficial effects observed on bone composition [44] (Table 12.1). Taken together, both human and animal studies provide little or limited evidence to support a beneficial effect of flaxseed consumption on bone health in peri- and postmenopausal women. Further research is therefore needed to confirm the effect of lignans on bone health which include BTM as well as measures of BMD and BMC.

12.2.4 ROLE OF THE GUT MICROBIOTA IN PLANT LIGNAN METABOLISM AND BONE HEALTH

Mammalian lignans are intestinal microbiota derived metabolites of plant lignans which include secoisolairiciresinol-diglycoside (SECO), matairesinol (MAT), lariciresinol (Lari) and pinoresinol (Pin). In plasma and urine, the metabolites are recovered as END and ENL. As an intervention food in both animal and human studies, flaxseed consumption has been seen to induce a significant increase in the urinary excretion of the lignans END and ENL [47]. In 31 healthy premenopausal females, the consumption of between 5 and 10 g of flaxseed daily for a period of 6 weeks, increased the excretion of lignans between 3- and 285-fold. It was confirmed by Adlercreutz [51] that after consuming dietary lignans, metabolism of the enterolignans occurred mostly in the colon. Subsequently, Heinonen, Nurmi [48] examined the microbial metabolism of plant lignans in human fecal samples and determined that ENL was the primary metabolite compound of MAT, whilst SECO metabolites comprised of END and ENL (Figure 12.4), and Pin was metabolized to ENL, END, and enterofuran (ENF) as a minor metabolite.

Many studies have investigated the human consumption of lignans, their gut metabolites, and types of gut bacteria that can metabolize SDG and SECO to END and ENL. It was not until the early 2000s, when studies identified specific gut microbial inhabitants involved in converting plant lignans to enterolignans. Conversion of lignans to END and ENL involves a range of reactions, hence a consortium of bacteria has been identified as being involved. Yoder and co-workers [49] highlighted that *Bacteroides*

TABLE 12.1 Summary of Epidemiological, Human, and Animal Studies Determining the Role of Flaxseeds on Postmenopausal Bone Health

Study	Subjects	Duration and Study Design	Intervention	Bone Turnover Markers/ Measurements Used	Outcomes
			Epidemiologic Studies		
Kim et al. [35]	n = 75 Age 52–65 y	Sep 1998–Jan 1999 Cross-sectional.	Three groups of participants: • Osteoporosis (n = 21). • Osteopenia (n = 29). • Normal BMD (n = 25).	• 24 h urinary phyto-estrogens. • BMD at lumbar spine, femoral neck, and Ward's triangle.	• ↓ urinary ENL in osteoporotic group ($P < 0·05$). • ENL positively correlated with lumbar spine ($r = 0·388$), femoral neck ($r = 0·271$) and Ward's triangle ($r = 0·322$) BMD ($P < 0·05$). • Total urinary lignans positively correlated with lumber spine BMD.
Kardinaal [36]	n = 67 Early postmenopausal women.	10-year follow-up study.	• 1st group with rate of cortical bone loss at the radius ≤ 0.5%/ year. • 2nd group with rate of cortical bone loss ≥ 2.5%/ year.	• Urinary lignan enterolactone. • Cortical bone density by single-photon absorptiometry.	• ↑ urinary ENL excretion in the high-rate bone loss group (P 0.05).
			Human Studies		
Lucas et al. [39]	n = 58 age < 65 years	Three months, Double-blind controlled Parallel.	40 g/d of ground FS 40 g/d of wheat-based regimen (placebo).	Serum IGF-1; serum IGFBP-3; and serum Ca+	NS

TABLE 12.1 *(Continued)*

Study	Subjects	Duration and Study Design	Intervention	Bone Turnover Markers/ Measurements Used	Outcomes
Dodin et al. [45]	n = 179 Age 45 to 65 years	12 months Randomized, Double-blind controlled Parallel	40 g/day of FS 40 g/d wheat germ placebo.	BMD at the lumber vertebrae or femur neck	NS
Brooks et al. [37]	n = 46 Mean age 52 years ± 1	16 weeks Randomized, Double-blind Parallel.	• Wheat-based placebo muffins • FS muffins (25 g ground flaxseed). • Soy muffins (25 g soy flour).	Urinary E2 metabolites, 2-OHE1, 16α-OH-E1), and serum hormones (E2). BAP	↑ urinary 2-OHE1 in FS group ($P = 0.05$).
Arjmandi et al. [46]	n= 38 Age >65 years	6 weeks Double-blind Crossover.	38 g/day of FS or sunflower seeds provided in bread and muffins.	Bone resorption marker (TRAP); Bone formation markers used: IGF-1, IGFBP-3, and ALP.	Suppression in bone resorption marker TRAP with FS ($P = 0.005$).
Dalais et al. [40]	n = 67 Age >55 years	3 months randomized Double-blind controlled Crossover.	45 g/day of FS or wheat placebo.	BMD and BMC	NS
			Animals Studies		
Babu et al. [42]	Growing female rats n = 47	8 weeks	BD (placebo), FS 0%, 5%, 10% or defatted FSM (6.2%).	ALP	NS

TABLE 12.1 *(Continued)*

Study	Subjects	Duration and Study Design	Intervention	Bone Turnover Markers/ Measurements Used	Outcomes
Ward et al. [44]	Female rats (n = 20)	6 months	BD (control), 50 or 100 g/kg SDG.	Bone strength at femur and BMC by DXA. Femur weight, length, and width were by digital scale. Bone biomechanical strength by utilized Texture analyzer. Urinary total lignans by GCMS.	↑ femur bone strength in 100 g/kg SDG vs BD group (P = 0.05).
Power et al. [41]	OVX mice with MCF-7 breast tumor n = 34	25 weeks	BD (20% corn oil), 10% FS, 20% SPI or 10% FS +20% SPI.	BMD at femur.	↑ femur bone strength (stiffness and peak load) in 10% FS vs control group (P = 0.05).
Farmer et al. [43]	Growing gilts n = 57	7 months	0% FS (control), 6.5% FSM, 10% FS or 3.5% FSO.	Plasma NTX-I, PRL, and E2.	NS

Abbreviations: 16α-OH-E1: 16α-Hydroxyestrone; 2-OHE1: 2-Hydroxyestrone; ALA: α-Linolenic acid; ALP: total alkaline phosphatase (bone formation marker); BAP: bone-specific alkaline phosphatase; BD: Basel diet; BGP: serum bone gamma-carboxyglutamic acid-containing protein; BMC: bone mineral contents; BMD: bone mineral density; BW: body weight; CTX: cross-linked C-telopeptide of type I collagen; Dpd: deoxypyridinoline; E2: estradiol; ENL: enterolactone; FF: flaxseed flour; FSM: flaxseeds meal; FN: femoral neck; FOS: fructooligosaccharides; Fs: flaxseeds; FSO: flaxseeds oil; FT: femoral trochanter; GCMS: gas chromatography-mass spectrometry; ICTP: pyridinoline cross-linked carboxyterminal telopeptide of type I collagen; IF: isoflavone; IGF-1: insulin-like growth factor-1; IGFBP3: insulin-like growth factor-binding protein 3; IL-6: Interleukin-6; MP: milk protein; MPI: milk protein isoflavone; N/A: not applicable; NTX-I: urinary cross-linked N-telopeptide of type 1 collagen; OC: osteocalcin; OPG: osteoprotegerin; OVX: ovariectomized rat; PRL: prolactin; PTH: parathyroid hormone; Pyr: urinary deoxypyridinoline; SDG: secoisolariciresinol diglycoside; SPI: isoflavone-rich soy protein isolate; TAG: triacylglycerol; TNFα: tumor necrosis; TRAP: tartrate resistant acid phosphatase.

(*fragilis, ovatus, and distasonis*); *Clostridium (cocleatum, saccharogumia, and ramosum)* and *Bifidobacterium* (*bifidum, breve, catenulatum, longum, and pseudocatenulatum*) were involved in the initial deglycosylation of the reactions process. Furthermore *Ruminococcus* spp. and *Clostridiaceae bacterium, Eggerthella* spp. and *Eubacterium* spp. were found to convert the lignans to the intermediate dihydroxyenterodiol (DHEND), or second intermediate dihydroxyenterolactone (DHENL). After this *Butyribacterium methylotrophicum, Eubacterium* (*callanderi* and *limosum*), *Blautia, Clostridiaceae bacterium, Eggerthella lenta, Clostridium scindens, Lactonifactor longoviformis, Enterococcus faecalis,* and *Eggerthella lenta* were involved in the final conversion of the plant lignans to END and ENL.

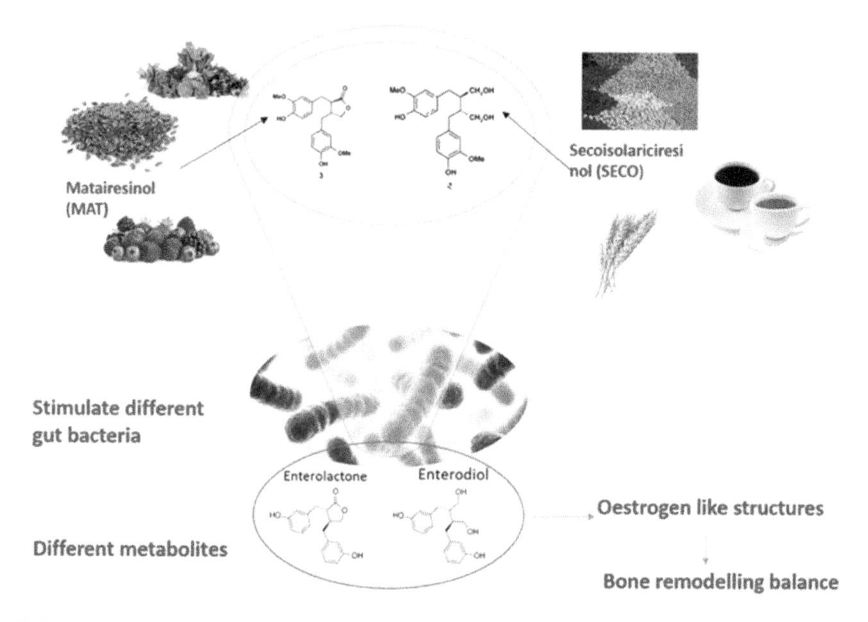

FIGURE 12.4 Main flaxseed lignans secoisolariciresinol (SECO) and matairesinol (MAT) and their mammalian lignans metabolites END and ENL.

12.2.5 INTER-INDIVIDUAL VARIATION IN METABOLISM OF LIGNANS BY GUT MICROBIOTA

Elevated levels of mammalian lignans have been measured in plasma, serum, and urine samples from animals and humans following moderate flaxseed consumption. An intake of 13.5 g of flaxseeds on a daily basis for 6 weeks resulted in 10-fold variation in circulating concentrations of plasma ENL and END ranging from 1,000 to 10,000 micromolar which are notably higher than circulating estradiol in women [50]. Thus, inherent differences in intestinal microbial bacteria composition and activity may influence enterolignan concentrations circulating in the blood [51, 52].

At present there are no human studies that have determined the impact of flaxseed intake on both changes in the gut microbiota and BTMs in post-menopausal women. Therefore, dietary patterns associated with enhanced ENL production such as those higher in fiber and lower in fat may provide insights into the bacterial species related to markers of bone health. Dietary patterns which have also been associated with enhanced ENL production include higher fiber diets with a lower fat intake [51]. Prebiotic fibers are dietary ingredients that target specific members of the gut microbiota to improve host health. Clinical studies have highlighted the importance of prebiotics in enhancing the intestinal absorption of various minerals and calcium [53, 54]. The levels of calcium, magnesium, and phosphorus have been noted to increase with consumption of prebiotics combined with increasing *Bifidobacterium* [55]. Furthermore, the fermentation of prebiotics within the colon is frequently associated with the production of the organic acid, lactate, and short-chain fatty acids (SCFAs). The gut microbiota may regulate BMD indirectly through SCFAs, by influencing OPG and suppressing RANKL signaling pathways which mediate bone remodeling cycle [56]. These microbial metabolites impact on the colonic pH, which is an essential factor for mineral (e.g., calcium) and nutrient (e.g., vitamin B and K) absorption [56], as well as the maintenance of the bile acids which control calcium absorption. In addition, the gut microbiota plays an important role in the synthesis of vitamins B and K, which are required for optimal bone health.

In addition to prebiotics, probiotics are live microorganisms which when consumed in an appropriate dose, confer a benefit to the host [54]. These microorganisms could be further stimulated by the consumption of prebiotics [57]. Studies have reported that consuming prebiotics alongside bioactive probiotics such as *Bifidobacterium longum* and *Lactobacillus (Lactobacillus rhamnosus* GG (LGG) and *Lactobacillus reuteri)* to improve the absorption of calcium, magnesium, and phosphate [53, 54, 58]. These improvements are likely to be brought about via metabolite production indicated above. Furthermore, murine experiments have provided evidence that some of the benefits to bone health of probiotic consumption could be through reduced inflammation, observed through reductions in TNF-a [59].

A variety of human intervention studies have assessed the impact of prebiotics on BMD and BTMs in postmenopausal women (Table 12.2). A study intervening with dietary soluble corn fiber found this potential prebiotic to enhance calcium retention in this subject group [60]. Intakes of 10 and 20 g per day brought about a 5% and 7% improvement in skeletal calcium retention, respectively, and the authors concluded that the improved calcium absorption was likely to be related to a change in the gut bacteria fermenting this soluble corn fiber [60]. Although the gut microbiota was not an outcome of this study, the authors suggested the findings were probably associated with a greater increase in bacteria of the phylum Bacteroidetes. In agreement with this study, Holloway [63] reported that inulin intake significantly improved calcium and magnesium absorption, resulting in an upward trend in BTMs DPD and OC in postmenopausal women. The author speculated that inulin could have an impact on calcium absorption indirectly by expanding the intestinal mucosa, increasing the surface area for diffusion. This was also confirmed by Whisner, Martin [61] studying calcium absorption in adolescence but whether the same microbes responsible for aiding calcium absorption were the same in these two distinct age groups is currently unknown.

Age is also an important factor influencing gut microbial composition. In a study that used 10 g/day of short-chain fructooligosaccharides (FOS), there was a tendency for calcium absorption to be improved in women during later, rather than earlier, stages of the postmenopausal period [62]. At the same time consumption of 10 g/day short- and long-chain inulin-type fructans for 6 weeks, proved highly beneficial in improving absorption of calcium and magnesium in postmenopausal women with a mean age of 73 years [63]. During adolescence and early adulthood (15–24 years), Bacteroidetes phylum bacteria is abundant in the gut, but numbers decrease on aging whereas, Firmicutes phylum microbiota increase in older age [53]. Additionally, a large abundance of *Clostridium perfringens* and a lower prevalence of *Lactobacillus* and *Bifidobacterium* were found in residential elderly individuals [64]. In general, the gut microbiota profiles of aged individuals (≥65 years)

TABLE 12.2 Human Intervention Studies on the Potential Effects of Prebiotic on Bone Turnover Markers in Postmenopausal Women

Study	Subjects	Duration and Study Design	Intervention	Bone Turnover Markers/ Measurements Used	Outcomes
Kruger, Chan [65]	n = 257 Pre-post-menopausal Chinese women. Aged 41 to 59 years	• 12 weeks. • Parallel	Two groups: • Milk, Ca 1,000–1,200 mg and vitamin D3 15 ug/day. • Milk and prebiotic (FOS) 4 g/day.	• Bone resorption markers (CTx-1 and PINP). • Serum parathyroid hormone • Vitamin D status.	↓ CTX in prebiotic/ milk postmenopausal group.
Jackman et al. [60]	n = 12 Postmenopausal women. Aged 40–78 years.	• 50 days. • Randomized, crossover, double-blinded	Diet contains 0, 10, 20 g/d of soluble corn fiber.	• Bone turnover markers (NTX, OC, BSAP). • Urinary Ca.	• ↑ bone Ca retention in 10 and 20 g fiber groups by 4.8% and 7% respectively. • ↑ BSAP (p= 0.035) in 20 g/d group.
Slevin et al. [66]	n = 300. Postmenopausal women. Aged 45–75 years.	• 2 y. • Randomized, crossover, double-blinded.	• 0 or 3.6 g/day fiber (FOS) + 800 Ca mg/ day.	• BMD by DXA. • CTX and OC.	• ↓ CTX after 12 mo and greater ↓ Oc after 24 mo in fiber/ Ca group.
Adolphi et al. [67].	n = 85 Postmenopausal women. Aged 48–67 years.	• 2 w. • Controlled, parallel, double-blind.	• Fermented milk supplied with Ca. • Fermented milk supplied with Ca and prebiotic fiber (inulin-type fructans and casein phosphopeptides). • Fermented control milk.	• Bone turnover markers (DPD, BSAP). • Urinary excretion of calcium and phosphorus	NS

TABLE 12.2 *(Continued)*

Study	Subjects	Duration and Study Design	Intervention	Bone Turnover Markers/ Measurements Used	Outcomes
Holloway et al. [63]	n = 15 Postmenopausal women. Mean age 72 y.	• 6 w. • Randomized, double-blind crossover.	• 10 g/day fiber (long-chain inulin)/ or control (maltodextrin).	• Bone turnover markers (DPD, OC). • Fractional Ca and Mg absorption.	• Sig in ↑ fractional Ca and Mg absorption in treatment group. • Sig impact of fiber on bone turnover markers.
Kim et al. [68]	n = 26 Postmenopausal women. Mean age 60 y.	• 3 mo. • Randomized, double-blind parallel	• 8 g/day of prebiotic (FOS) / or control (maltodextrin).	• BMD by DXA. • BSAP. • Mineral absorption (Ca, P, Zn).	• ↓ in BSAP. • ↑Ca and P absorption.
Tahiri et al. [62]	n = 12 Postmenopausal women Aged 50–70 years.	• 5 w. • Randomized, double-blind crossover.	• 10 g/day prebiotic (FOS)/ or control (sucrose).	• Calcium-status indexes. • Bone turnover markers (OC, DPD). • Vitamin D status. • Parathyroid hormone.	NS

Abbreviations: BMC: bone mineral content; BMD: bone mineral density; BMI: body mass index; Ca: calcium; CTx: C-telopeptide of type I collagen; DNA: deoxyribose nucleic acid; DXA: dual-energy X-ray absorptiometry; Fe: iron; FOS: fructooligosaccharides; GOS: galactooligosaccharides; IF: isoflavones; ITF: inulin-type fructans; Mg: magnesium; OVX: ovariectomized; P: phosphorus; PCR-DGGE: polymerase chain reaction-density gradient gel electrophoresis; PDX: polydextrose; rRNA: ribosomal ribonucleic acid; RS: resistant starch; rt-qPCR: real-time quantitative polymerase chain reaction; sc-FOS: short-chain fructooligosaccharides; SCFA(s): short-chain fatty acid(s); SI: single-isotope method, Vitamin D; 1,25(OH)2D: 1,25 dihydroxy vitamin D; Zn: zinc.

may not be ideal and the reason for the shift in gut microbiota composition on aging is likely to be due to changes in sensory characteristics, resulting in dietary changes [64]. Moreover, there are still considerable non-dietary factors influencing the inter-individual variation in lignan production, for instance, gut transit duration, body weight, and antibiotic use; as END and ENL production have been reported to be negatively affected by antibiotic intake, thus confirming microbial involvement in production [56].

Taken together, the shifts in microbial community could result in direct and indirect effects on bone health. However, more research is needed to understand potential mechanisms, and the roles that important determinants such as age, BMI, and dietary intakes have on the gut microbiota, and how they impact on efficacy of dietary intervention that target bone loss in individuals at high risk.

12.3 SUMMARY

To conclude, few human studies have reported a link between bone health and phytoestrogen intake among Asian and European postmenopausal women. Most of these studies have focused on the positive effects of soy consumption on BMD and bone remodeling in relation to the ALA content of flaxseeds only and not the phytoestrogen levels. To date, limited studies have determined the role of whole flaxseeds as a dietary lignan source on BTMs. The hypothesis of this review was that flaxseed phytoestrogens would have a positive effect on bone health. Although findings are inconsistent, human, and animal studies using flaxseed have highlighted a minimal beneficial effect on bone loss in postmenopausal women and OVX animals only. With flaxseeds an abundant source of phytoestrogen in the Western diet, further well designed RCTs are needed to determine the specific effects of these lignans on bone health and to determine the role of gut microbiota in mediating these effects. Furthermore, studies need to include BTMs as a primary outcome measure to evaluate the efficacy of lignans from dietary sources, as a strategy to slow down the progression of bone loss and development of osteoporosis in postmenopausal women.

KEYWORDS

- **bone mineral content**
- **dietary strategies**
- **flaxseed**
- **functional foods**
- **osteocalcin**
- **osteoporosis**
- **phytoestrogenic lignans**

REFERENCES

1. Lorentzon, M., & Cummings, S. R., (2015). Osteoporosis: The evolution of a diagnosis. *Journal of Internal Medicine, 277*(6), 650–661.
2. Liu, R. H., (2013). Health-promoting components of fruits and vegetables in the diet. *Advances in Nutrition, 4*(3), 384S–392S.
3. Ricci, E., et al., (2010). Soy isoflavones and bone mineral density in perimenopausal and postmenopausal Western women: A systematic review and meta-analysis of randomized controlled trials. *Journal of Women's Health, 19*(9), 1609–1617.
4. Taku, K., et al., (2010). Effect of soy isoflavone extract supplements on bone mineral density in menopausal women: Meta-analysis of randomized controlled trials. *Asia Pacific Journal of Clinical Nutrition, 19*(1), 33–42.
5. Morgan, E. F., Barnes, G. L., & Einhorn, T. A., (2013). The bone organ system: Form and function. In: *Osteoporosis* (4th edn., pp. 3–20). Elsevier.
6. Sanderson, S., Anderson, P. S., & Benton, M. J., (2016). Difference in bone mineral density between young versus midlife women. *American Journal of Health Education, 47*(3), 149–154.
7. Morgan, K. T., (2008). Nutritional determinants of bone health. *Journal of Nutrition for the Elderly, 27*(1, 2), 3–27.
8. Lerner, U., (2006). Bone remodeling in postmenopausal osteoporosis. *Journal of Dental Research, 85*(7), 584–595.
9. Raggatt, L. J., & Partridge, N. C., (2010). Cellular and molecular mechanisms of bone remodeling. *Journal of Biological Chemistry*, P. jbc. R109. 041087.
10. Kotake, S., & Nanke, Y., (2014). Effect of TNFα on osteoblastogenesis from mesenchymal stem cells. *Biochimica et Biophysica Acta (BBA)-General Subjects, 1840*(3), 1209–1213.
11. Kearns, A. E., Khosla, S., & Kostenuik, P. J., (2007). Receptor activator of nuclear factor κB ligand and osteoprotegerin regulation of bone remodeling in health and disease. *Endocrine Reviews, 29*(2), 155–192.

12. Meier, C., et al., (2006). Sex steroids and skeletal health in men. In: *Osteoporosis and the Osteoporosis of Rheumatic Diseases* (p. 145–156). Elsevier.
13. Midtby, M., Magnus, J., & Joakimsen, R., (2001). The tromsø study: A population-based study on the variation in bone formation markers with age, gender, anthropometry and season in both men and women. *Osteoporosis International, 12*(10), 835–843.
14. Uebelhart, D., et al., (1991). Effect of menopause and hormone replacement therapy on the urinary excretion of pyridinium cross-links. *The Journal of Clinical Endocrinology & Metabolism, 72*(2), 367–373.
15. Seifert-Klauss, V., et al., (2002). Bone metabolism during the perimenopausal transition: A prospective study. *Maturitas, 41*(1), 23–33.
16. Seibel, M. J., (2005). Biochemical markers of bone turnover part I: Biochemistry and variability. The Clinical biochemist. *Reviews/Australian Association of Clinical Biochemists, 26*(4), 97.
17. Garnero, P., et al., (1996). Markers of bone resorption predict hip fracture in elderly women: The EPIDOS Prospective Study. *Journal of Bone and Mineral Research, 11*(10), 1531–1538.
18. Ross, P., et al., (2000). Serum bone alkaline phosphatase and calcaneus bone density predict fractures: A prospective study. *Osteoporosis International, 11*(1), 76–82.
19. Bauer, D. C., et al., (2004). Change in bone turnover and hip, non-spine, and vertebral fracture in alendronate-treated women: The fracture intervention trial. *Journal of Bone and Mineral Research, 19*(8), 1250–1258.
20. Sornay-Rendu, E., et al., (2005). Identification of osteopenic women at high risk of fracture: The OFELY study. *Journal of Bone and Mineral Research, 20*(10), 1813–1819.
21. Camacho, P. M., & Lopez, N. A., (2008). Use of biochemical markers of bone turnover in the management of postmenopausal osteoporosis. *Clinical Chemistry and Laboratory Medicine, 46*(10), 1345–1357.
22. Qin, L. Q., et al., (2006). Soy food intake in the prevention of breast cancer risk in women: A meta-analysis of observational epidemiological studies. *Journal of Nutritional Science and Vitaminology, 52*(6), 428–436.
23. Kanis, J. A., et al., (2012). A systematic review of hip fracture incidence and probability of fracture worldwide. *Osteoporosis International, 23*(9), 2239–2256.
24. Sharan, K., et al., (2010). A novel flavonoid, 6-C-β-d-glucopyranosyl-(2S, 3S)-(+)-3′, 4′, 5, 7-tetrahydroxyflavanone, isolated from Ulmus Wallichiana Planchon mitigates ovariectomy-induced osteoporosis in rats. *Menopause, 17*(3), 577–586.
25. Weaver, C. M., & Cheong, J. M., (2005). Soy isoflavones and bone health: The relationship is still unclear. *The Journal of Nutrition, 135*(5), 1243–1247.
26. Zhang, R., et al., (2014). Effects of total lignans from *Eucommia ulmoides* barks prevent bone loss *in vivo* and *in vitro*. *Journal of Ethnopharmacology, 155*(1), 104–112.
27. Haggans, C. J., et al., (1999). Effect of flaxseed consumption on urinary estrogen metabolites in postmenopausal women. *Nutrition and Cancer, 33*(2), 188–195.
28. Hutchins, A. M., et al., (2000). Flaxseed influences urinary lignan excretion in a dose-dependent manner in postmenopausal women. *Cancer Epidemiology and Prevention Biomarkers, 9*(10), 1113–1118.

29. Adlercreutz, H., et al., (1987). Effect of dietary components, including lignans and phytoestrogens, on enterohepatic circulation and liver metabolism of estrogens and on sex hormone-binding globulin (SHBG). *Journal of Steroid Biochemistry, 27*(4–6), 1135–1144.
30. Chang, V. C., et al., (2018). Effect of dietary flaxseed intake on circulating sex hormone levels among postmenopausal women: A randomized controlled intervention trial. *Nutrition and Cancer,* 1–14.
31. Mirmasoumi, G., et al., (2018). The Effects of flaxseed oil omega-3 fatty acids supplementation on metabolic status of patients with polycystic ovary syndrome: A randomized, double-blind, placebo-controlled trial. *Experimental and Clinical Endocrinology & Diabetes, 126*(04), 222–228.
32. Calado, A., et al., (2018). The effect of flaxseed in breast cancer: A literature review. *Frontiers in Nutrition, 5*, 4.
33. Mason, J. K., & Thompson, L. U., (2013). Flaxseed and its lignan and oil components: Can they play a role in reducing the risk of and improving the treatment of breast cancer? *Applied Physiology, Nutrition, and Metabolism, 39*(6), 663–678.
34. Goyal, A., et al., (2014). Flax and flaxseed oil: An ancient medicine & modern functional food. *Journal of Food Science and Technology, 51*(9), 1633–1653.
35. Kim, M., et al., (2002). Relationships of urinary Phyto-oestrogen excretion to BMD in postmenopausal women. *Clinical Endocrinology, 56*(3), 321–328.
36. Kardinaal, A., et al., (1998). Phyto-oestrogen excretion and rate of bone loss in postmenopausal women. *European Journal of Clinical Nutrition, 52*(11), 850.
37. Brooks, J. D., et al., (2004). Supplementation with flaxseed alters estrogen metabolism in postmenopausal women to a greater extent than does supplementation with an equal amount of soy. *The American Journal of Clinical Nutrition, 79*(2), 318–325.
38. Vasikaran, S., et al., (2011). Markers of bone turnover for the prediction of fracture risk and monitoring of osteoporosis treatment: A need for international reference standards. *Osteoporosis International, 22*(2), 391–420.
39. Lucas, E. A., et al., (2002). Flaxseed improves lipid profile without altering biomarkers of bone metabolism in postmenopausal women. *The Journal of Clinical Endocrinology & Metabolism, 87*(4), 1527–1532.
40. Dalais, F., et al., (1998). Effects of dietary phytoestrogens in postmenopausal women. *Climacteric, 1*(2), 124–129.
41. Power, K. A., et al., (2007). Flaxseed and soy protein isolate, alone and in combination, differ in their effect on bone mass, biomechanical strength, and uterus in ovariectomized nude mice with MCF-7 human breast tumor xenografts. *Journal of Toxicology and Environmental Health, Part A,* 70(22), 1888–1896.
42. Babu, U. S., Mitchell, G. V., Wiesenfeld, P., & Jenkins, M. Y., & Gowda, H. Nutritional and hematological impact of dietary flaxseed and defatted flaxseed meal in rats. *Int J Food Sci Nutr. 2000* Mar; *51*(2), 109–117. doi: 10.1080/096374800100804. PMID: 10953754.
43. Farmer, C., et al., (2007). Effects of dietary supplementation with flax during prepuberty on fatty acid profile, mammogenesis, and bone resorption in gilts. *Journal of Animal Science, 85*(7), 1675–1686.

44. Ward, W. E., et al., (2001). Exposure to purified lignan from flaxseed (*Linum usitatissimum*) alters bone development in female rats. *British Journal of Nutrition, 86*(4), 499–505.
45. Dodin, S., et al., (2005). The effects of flaxseed dietary supplement on lipid profile, bone mineral density, and symptoms in menopausal women: A randomized, double-blind, wheat germ placebo-controlled clinical trial. *The Journal of Clinical Endocrinology & Metabolism, 90*(3), 1390–1397.
46. Arjmandi, B. H., et al., (1998). Whole flaxseed consumption lowers serum LDL-cholesterol and lipoprotein (A) concentrations in postmenopausal women. *Nutrition Research, 18*(7), 1203–1214.
47. Westcott, N., & Muir, A., (1997). Medicinal lignan from flaxseed. In: *Prairie Medicinal and Aromatic Plants Conference.* Brandon Manitoba.
48. Heinonen, S., et al., (2001). *In vitro* metabolism of plant lignans: New precursors of mammalian lignans enterolactone and enterodiol. *Journal of Agricultural and Food Chemistry, 49*(7), 3178–3186.
49. Yoder, S. C., et al., (2015). Gut microbial metabolism of plant lignans: Influence on human health. In *Diet-Microbe Interactions in the Gut.* (pp. 103–117). Elsevier.
50. Landete, J., (2012). Plant and mammalian lignans: A review of source, intake, metabolism, intestinal bacteria and health. *Food Research International, 46*(1), 410–424.
51. Rowland, I. R., et al., (2000). Interindividual variation in metabolism of soy isoflavones and lignans: Influence of habitual diet on equol production by the gut microflora. *Nutrition and Cancer, 36*(1), 27–32.
52. Hullar, M. A., et al., (2015). Enterolignan-producing phenotypes are associated with increased gut microbial diversity and altered composition in premenopausal women in the United States. *Cancer Epidemiology and Prevention Biomarkers, 24*(3), 546–554.
53. Conlon, M., & Bird, A., (2015). The impact of diet and lifestyle on gut microbiota and human health. *Nutrients, 7*(1), 17–44.
54. Weaver, C. M., (2015). Diet, gut microbiome, and bone health. *Current Osteoporosis Reports, 13*(2), 125–130.
55. Pérez-Conesa, D., et al., (2006). Bioavailability of calcium, magnesium and phosphorus in rats fed probiotic, prebiotic and synbiotic powder follow-up infant formulas and their effect on physiological and nutritional parameters. *Journal of the Science of Food and Agriculture, 86*(14), 2327–2336.
56. Chen, Y. C., et al., (2017). Association between gut microbiota and bone health: Potential mechanisms and prospective. *The Journal of Clinical Endocrinology & Metabolism, 102*(10), 3635–3646.
57. Scholz-Ahrens, K. E., et al., (2007). Prebiotics, probiotics, and synbiotics affect mineral absorption, bone mineral content, and bone structure. *The Journal of Nutrition, 137*(3), 838S–846S.
58. Li, J. Y., et al., (2016). Sex steroid deficiency-associated bone loss is microbiota dependent and prevented by probiotics. *The Journal of Clinical Investigation, 126*(6), 2049–2063.

59. Britton, R. A., et al., (2014). Probiotic *L. reuteri* treatment prevents bone loss in a menopausal ovariectomized mouse model. *Journal of Cellular Physiology, 229*(11), 1822–1830.
60. Jakeman, S. A., et al., (2016). Soluble corn fiber increases bone calcium retention in postmenopausal women in a dose-dependent manner: A randomized crossover trial. *The American Journal of Clinical Nutrition, 104*(3), 837–843.
61. Whisner, C. M., et al., (2016). Soluble corn fiber increases calcium absorption associated with shifts in the gut microbiome: A randomized dose-response trial in free-living pubertal females–3. *The Journal of Nutrition, 146*(7), 1298–1306.
62. Tahiri, M., et al., (2003). Effect of short-chain fructooligosaccharides on intestinal calcium absorption and calcium status in postmenopausal women: A stable-isotope study. *The American Journal of Clinical Nutrition, 77*(2), 449–457.
63. Holloway, L., et al., (2007). Effects of oligofructose-enriched inulin on intestinal absorption of calcium and magnesium and bone turnover markers in postmenopausal women. *British Journal of Nutrition, 97*(2), 365–372.
64. Lakshminarayanan, B., et al., (2013). Prevalence and characterization of clostridium perfringens from the faecal microbiota of elderly Irish subjects. *Journal of Medical Microbiology, 62*(3), 457–466.
65. Kruger, M. C., et al., (2016). Differential effects of calcium-and vitamin D-fortified milk with FOS-inulin compared to regular milk, on bone biomarkers in Chinese pre-and postmenopausal women. *European Journal of Nutrition, 55*(5), 1911–1921.
66. Slevin, M., Wallace, J. M., & Mc Sorley, E. M., (2014). Supplementation with calcium and short-chain fructo-oligosaccharides affects markers of bone turnover but not bone mineral density in postmenopausal women. *Journal of Nutrition, 144*, 297–304.
67. Adolphi, B., et al., (2009). Short-term effect of bedtime consumption of fermented milk supplemented with calcium, inulin-type fructans and casein phosphopeptides on bone metabolism in healthy, postmenopausal women. *European Journal of Nutrition, 48*(1), 45–53.
68. Kim, Y. Y., et al., (2004). The effect of chicory fructan fiber on calcium absorption and bone metabolism in Korean postmenopausal women. *Nutritional Sciences, 7*(3), 151–157.
69. YE, Y. B., & SU, Y. X., (2004). The attenuating effect of soy isoflavones extract on bone loss in postmenopausal women [J]. *Acta Nutrimenta Sinica, 1*, 018.
70. Cheong, J., et al., (2007). Soy isoflavones do not affect bone resorption in postmenopausal women: A dose-response study using a novel approach with 41Ca. *The Journal of Clinical Endocrinology & Metabolism, 92*(2), 577–582.
71. Mitchell, P., Harvey, N., & Dennison, E. (2017). IOF Compendium of Osteoporosis First Edition, October 2017.

CHAPTER 13

INTERVENTIONAL APPROACHES IN PERIPHERAL NERVE INJURIES: PATHOGENESIS AND CURRENT STATUS

GHULAM HUSSAIN, AZHAR RASUL, HASEEB ANWAR, SHOAIB AHMAD MALIK, JAVERIA MAQBOOL, ALI IMRAN, and HAFIZ ANSAR RASUL SULERIA

ABSTRACT

The nervous system is a body's complex system that works primarily to provide coordination between all other functional systems of the body. Numerous factors such as injury, trauma, or neuronal disorders can cause damage and affect the functioning of this system. The peripheral nervous system's (PNS) nerves have the ability to rejuvenate spontaneously by themselves after injury, while the central nervous system (CNS) fails to rejuvenate their injured nerves. Recent research has demonstrated some therapeutic approaches to recover the peripheral as well as CNS injuries. This chapter focuses on the basic causes and pathophysiological changes that triggered after injury. This chapter also highlighted the natural and pharmacological treatment approaches for central as well as PNS. However, more detailed research is necessary to explore the therapeutic markers to treat nerve injuries.

13.1 INTRODUCTION

A complex network of nerves makes the nervous system; that control and coordinates all the activities of the body with the help of signals that are transmitted throughout the body. It responds to the changes taking place in the surroundings of our body [1]. It is categorized into two portions: CNS (central nervous system) and PNS (peripheral nervous system). Collectively brain and the spinal cord make CNS, while the PNS comprises of the bundles of elongated fibers or axons that connect the body's peripheral parts to the CNS (Figure 13.1). The PNS comprises of cranial and spinal nerve fibers [2]. Sensory nerve fibers of PNS convey sensory message of the body to the CNS whereas; the motor nerve fibers of PNS possess the responsibility to transmit the response to the effector site [3]. Thus, the smooth continuous communication without any interruption is fundamental for the physiological regulation of the entire living system [4].

As neurons are delicate so they can be injured even by slight pressure, cutting, or stretching. A neuronal injury can halt signal transmission to and from the brain, thereby affecting associated muscles to not function appropriately and precisely. This chapter clarify the basic mechanism of nerve injuries and provides potential natural and surgical treatment approaches for both CNS and PNS injuries.

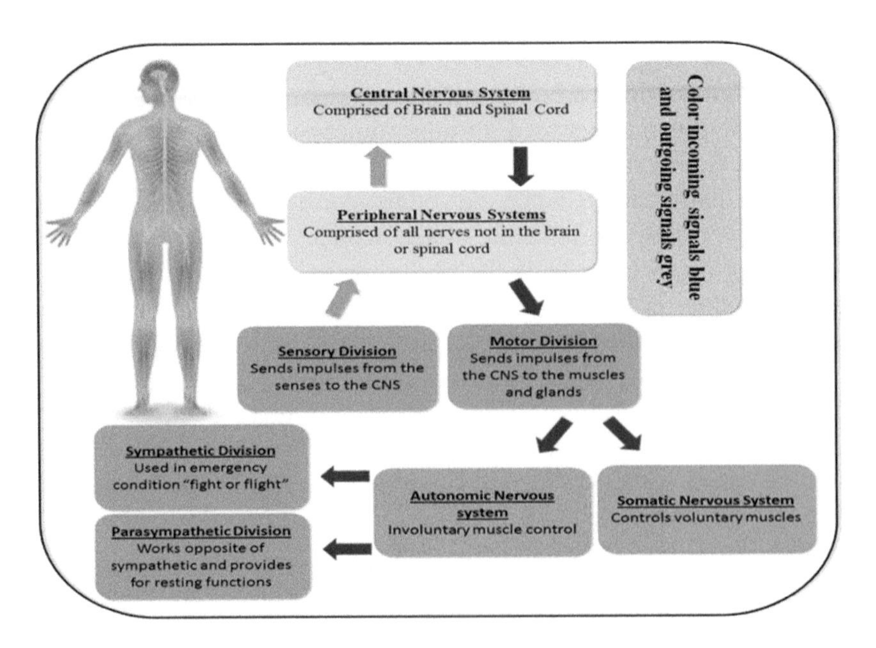

FIGURE 13.1 Organization of nervous system.

13.2 INJURIES OF THE NERVOUS SYSTEM

13.2.1 CNS INJURIES

Brain damage or brain injuries can take place due to a number of illnesses, injuries, and other conditions. They can happen in a brain that is starving of oxygen for an extended period of time. Injury to a brain can led to long-term difficulties in several types of functions, and the majority of such people who experience such types of injuries require help or support to do their routine physical activities. The brain injuries are usually classified into following categories:

1. **ABI (Acquired Brain Injuries):** The brain injuries happening after birth, are categorized as acquired brain injuries. The possible reasons for ABI may include tumors, choking, heart attacks, strangulation, stroke, poisoning, infection, certain neurological ailments, and drug misuse.
2. **TBI (Traumatic Brain Injuries):** The type of injury happens by any external force, like a blow to the head leading to brain damage. The reasons for TBI may include sports injuries, car accidents, falls, and physical violence. A wide variety of cognitive, behavioral, perceptive, emotional, and physical symptoms can develop depending upon the severity and type of brain injury [5].

13.2.2 PNS INJURIES

The Peripheral nerves are fragile that is why they can be easily damaged. Damaged nerves can affect the ability of the brain to communicate with organs and muscles, and it can cause loss of sensory as well as motor tasks or sometimes both tasks. Any damage to peripheral nerves is considered as peripheral nerve injury (PNI). PNIs can happen as a consequence of ailments such as diabetic neuropathy, infections like Guillain-Barré syndrome, nerve compression problems like carpal tunnel syndrome, and trauma such as brachial plexus injury. Some peripheral nerve damages are congenital as well.

The signs and symptoms of PNIs onset are mostly negligible and then get progressively severe with time, and they include: Numbness, Pain, Tingling or burning, Hypersensitivity upon touch at arms, legs, shoulders, or hands, Muscle weakness. After a peripheral nerve injury, the damaged nerve can be recovered but the immediate and quick medical care is necessary for rejuvenation of injured nerve [6]. The symptoms may vary from mild to severe. Some of the severe injuries and diseases that cause the nervous system problems are stated as vascular disorders leading to stroke, severe traumatic injuries of head and spine, congenital problems, the mental fitness complications such as anxiety disorders, psychosis, depression, exposure to the toxins including lead, arsenic or carbon monoxide, neurodegenerative diseases (NDDs) such as multiple sclerosis (MS), amyotrophic lateral sclerosis (ALS), Parkinson's disease (PD), Alzheimer's disease (AD), Huntington's disease (HD), and Peripheral neuropathies [7].

13.2.3 TRAUMATIC INJURIES TO NERVOUS SYSTEM

Such types of injuries can occur due to a brain trauma by any external stimuli and are also known as intracranial injuries. They can be categorized into diverse classes depending on the mechanism, severity, and other features. TBI is the main reason of death and disability all around the world, particularly in young adults and children. Its main causes include falls, violence, and particularly vehicle accidents that are common now a days. Its precautionary measures include the use of seat belts and helmets to avoid automobile accidents. Moreover, safety education programs and strict traffic laws enforcement can also reduce the rate of automobile accidents [8]. A cascade of inflammatory reactions starts after traumatic injury to CNS or PNS. However, the role done by inflammation in the degenerative, as well as regenerative progression of neuronal tissue, is not fully clarified yet. In peripheral nerves, the inflammatory reaction is vital for usual advancement of Wallerian degeneration and then rejuvenation whereas CNS traumatic inflammation leads to poor recovery of damaged nerves [9].

13.2.4 NEURODEGENERATIVE DISEASES (NDDS)

Neurodegenerative diseases (NDs) denote a wide range of ailments characterized by neuronal damage with various pathological as well as clinical features disturbing specific subsets of neurons involved in the specific functional systems. NDs develop persistently having unknown reasons. Huntington disease, PD, ALS, and AD are the most common NDs in current scenario [10].

13.3 CLASSIFICATION OF NERVE INJURIES

Injuries to the PNS can be categorized into diverse grades. This grading represents how severe the nerve to be injured. These injuries were categorized by Sunderland and Herbert Seddon (Table 13.1). Seddon categorized the PNIs into three grades depending upon the occurrence of demyelination, the degree of axonal damage and its surrounding connective tissue. Sunderland then explained additional subdivision based on the cutoff of several layers of connective tissues in the peripheral nerves [11].

13.3.1 NEUROPRAXIA

The mild form of nerve injury is named as Neuropraxia, categorized by focal demyelination of axons with little damage to the axons or the connective tissues. Neuropraxia characteristically happens from mild nerve compression or traction. It results in a decrease neuronal conduction velocity which temporarily cause functional loss. It tends to vanish within 4–6 weeks.

13.3.2 AXONOTMESIS

An injury that causes anatomical disruption of the axon and myelin sheath as well. However, the framework of connective tissue remains stable and intact. It requires a significant time lapse for regrowth of axon to the target muscle. Moreover, scar formation also hinders and delays the regrowth of axons. The patients with axonotmesis require an appropriate surgical treatment depending on the number of injured axons and extent of scar formation at the nerve injury site. When an axon is ruptured, there is a delay of 2–4-weeks before the axon starts regenerating. In mammalian adult, injured axons rejuvenate themselves at the rate of 1 inch per month, suggesting a great time-lapse for axons to grow down the muscles in arms. In infants, the rate of axonal regeneration is more rapid because the distance to be covered is less. The nerve receptors vanished over a time of 12–18 months after a muscle loses its innervation.

13.3.3 NEUROTMESIS

It is the most complicated type of injury in which complete anatomical disturbance of both axon and all the adjacent connective tissue may take place (rupture of the nerve). Birth brachial plexus injury is associated with neurotmesis. It is the most devastated category of nerve injury and it has very little or no chance of natural recovery. It requires an early surgical cure [12].

TABLE 13.1 Classification of Nerve Injuries based on Seddon and Sunderland Observation

Seddon	Sunderland	Pathological Findings
1. Neuropraxia	First degree	• Possible segmented demyelination • Moderately spared sensory paralysis • Very little muscle atrophy • Complete motor paralysis
2. Axonotmesis	Second degree	• Disturbance of Axonal continuity (with myelin sheath) • Endoneurium, epineurium, perineurium intact • Complete autonomic, motor, and sensory paralysis • Wallerian degeneration • Progression of muscular atrophy with time
3. Neurotmesis	Third-degree	• Disturbance of endoneurium and axonal continuity • Complete autonomic, motor, and sensory paralysis • Progression of muscular atrophy with time
4. Neurotmesis	Fourth degree	• Wallerian degeneration • Disrupted endoneurium, perineurium, and axonal continuity • Complete autonomic, motor, and sensory paralysis • Progression of muscular atrophy with time • Complete disruption of nerve

TABLE 13.1 *(Continued)*

Seddon	Sunderland	Pathological Findings
5. Neurotmesis	Fifth degree	• Progression of muscular atrophy with time • Complete autonomic, motor, and sensory paralysis • Wallerian degeneration

13.4 PATHOPHYSIOLOGY OF NERVE INJURIES

13.4.1 PATHOPHYSIOLOGY OF CNS INJURY

13.4.1.1 PRIMARY AND SECONDARY CNS INSULT

Primary insults (due to automobile accidents, gunshot wounds, blows to the head or back, etc.), end up in the rupture of membranes that results in severe damage of glial, neuronal, and vascular elements. Nonspecific cellular loss caused by primary insults further activates a series of secondary reactions such as cytotoxicity, inflammation, and neurodegeneration, causing prolonged cellular death, system malfunction, and system-level modifications.

13.4.1.2 PHENOTYPICAL CHANGES OF REACTIVE ASTROCYTES AFTER CNS INSULT

Glial scar development following the CNS insult is managed by complex intracellular as well as extracellular signaling pathways. Under normal circumstances, astrocytes are a common supporting cell inside the CNS which maintain the BBB (blood-brain barrier) as well as neurons. They are stimulated by various kinds of pathological phenomenon such as infection, inflammation, traumatic insult, and ischemia, where they offer a vital part in the pathophysiology of each of them by a phenotypical change, i.e., reactive astrogliosis. Following the SCI, immature astrocytes successively show complementary phenotypes, primarily the reactive astrocytes and afterward the scar-forming astrocytes. Even though the actual mechanism behind the reactive response of immature astrocytes has not been completely elucidated, researchers have found that the microglia are involved in altering astrocytes phenotype via P2Y1 (purinergic receptor) downregulation. The reactive astrocytes are essential in the process of tissue remodeling and severe wound healing; though, they ultimately become the scar-forming astrocytes and produce a compact glial scar. Though these various astrocytic roles are complex and depend on surrounding conditions, yet the term 'astrogliosis' has incompletely defined the actual part of astrocytes in the pathophysiology of the spinal cord injury (SCI) [13].

In the reactive-astrocytes, expression of Axin2, Mmp13, Nes Mmp2, Plaur, Ctnnb1 genes is selectively increased in contrast to immature and scar-forming astrocytes. Therefore, these genes are demarcated as bio-markers for reactive astrocytes. In the same way, Xylt1, Cdh2, Chst11, Sox9, Csgalnact1, Acan, Pcan, Slit2 are found to be bio-markers for scar-forming astrocytes (Figure 13.2). In a previous study, it was seen that during the scar-forming phase in the injured spinal cord, type 1 collagen (Col I) gene was found to be excessively. This advocates that the Col I encourage alteration of reactive astrocytes into the scar forming astrocytes through integrin N-cadherin dependent pathway [14].

Secondary insult to spinal cord triggers necrosis and cascade of inflammatory reactions at the site of injury within minutes to hours. Inflammation is one of the main causes of that continuously increasing degeneration. In the initial aspects of an inflammatory reaction, two types of cells are predominantly involved: macrophages and neutrophils. Neutrophils are the cells that migrate into the tissue and neutrophil influx leads to increase oxidative stress. Activated microglia and macrophages are the next waves of inflammatory cells that enter the damaged tissue in CNS and peripheral tissues.

13.4.1.3 APOPTOSIS

Damage inflicted by any type of injury may lead to anomalies in the apoptotic mechanisms in certain neurons and consequently initiates apoptotic pathways. These pathways remain persistently activated for longer periods after CNS trauma. A plethora of evidence also pointed towards the caspase involvement in spinal cord injury. Previous studies depict that after spinal cord injuries, the level of active caspase-3 and -9 begin to increase for up to three days. Within a short time following primary injury, an increase in intracellular free Ca^{2+} concentration occurs, which can activate the cysteine protease, e.g., calpain (calcium-dependent activation) for neurodegeneration in SCI [15]. It has been seen in the rat model that caspase-2 and-3 mRNA level was upregulated upon unilateral middle cerebral artery occlusion. Mutated SOD1 has shown an age-dependent increase in caspase-1 and caspase-3 activity in the spinal cord of mice having ALS. Similarly, caspase-1 activity has also been found to increase in human patients suffering from ALS [16].

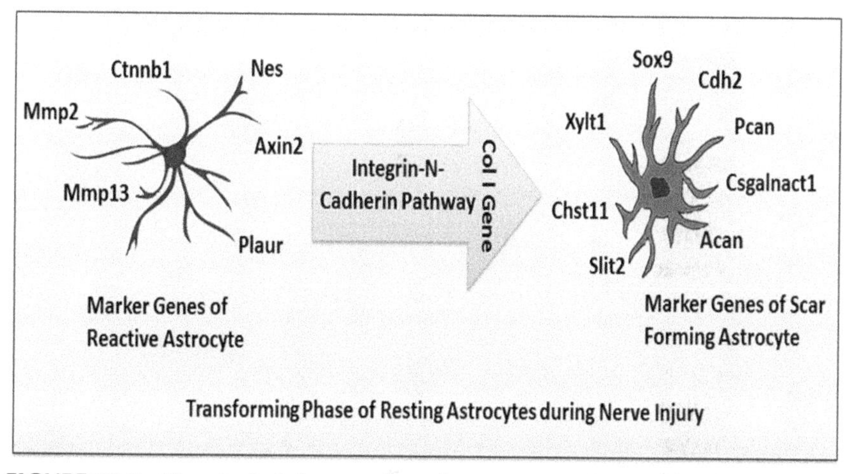

FIGURE 13.2 Phenotypical changes of reactive astrocytes after CNS insult.

Activation of caspases takes place as a result of physical or physiological injury, which in turn induces apoptosis through both external and internal pathways [17]. The extrinsic caspase pathway is commenced by ligation of cytokines, i.e., tumor necrosis factor-alpha (TNF-α) and interferon-gamma (IFN-γ). This ligation further activates caspase-8 or caspase-10 to initiate extrinsic pathway. Both IFN-γ and TNF-α are known for the induction of SCI [18, 19]. According to research, extrinsic caspases signaling has been amplified via caspase-8 (which mediates brake down of Bid to truncated Bid (tBid)) which further is translocated in mitochondria and induces cytochrome c release into the cytosol for the induction of neuronal apoptosis [17, 20].

On the other side, the intrinsic caspase cascade is also commenced to release cytochrome c from mitochondria into the cytosol and serial activation of caspase-9 and -3 for apoptosis following SCI [17, 21, 22]. The final executioner of apoptosis, Caspase-3, is stimulated for both neuronal and glial cells death [23, 24]. The inhibition of caspase signaling acts as an obvious therapeutic goal to prevent neuronal death for functional neuroprotection in SCI [25, 26] (Figure 13.3).

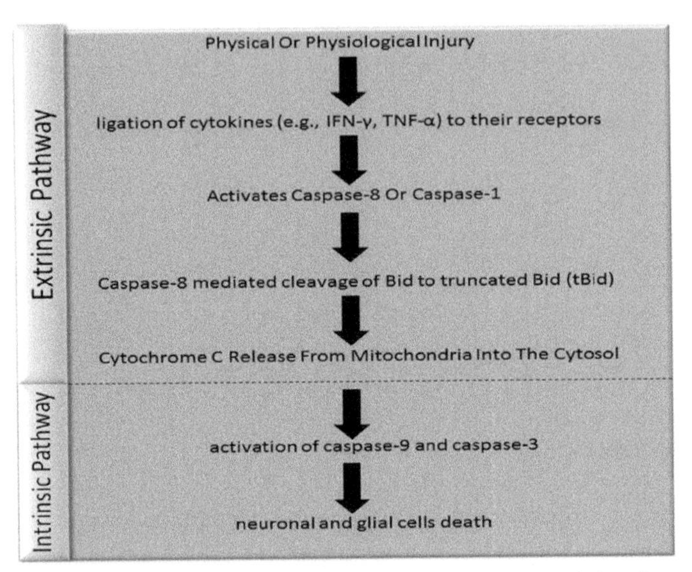

FIGURE 13.3 Apoptosis: Neuronal injury triggers caspase mediated signaling to induce glial and neuronal cell death.

13.4.1.4 INFLAMMATION

After brain injury, several indicators for activation of pro-inflammatory procedures have been described including elevated expression level of pro-inflammatory cytokines, i.e., IL-1β, IL-6, and TNFα [27–29]. After injury, specific receptors that work together with these cytokines are expressed, and this leads to triggering the inflammatory signaling pathways for the expression of the inflammatory genes.

13.4.1.5 EXCITOTOXICITY

A CNS' primary excitatory neurotransmitter is Glutamate. Apart from its basic physiological functions, glutamate possesses a crucial part in the pathology of numerous ailments such as stroke, epilepsy, and NDs [30, 31]. Excitotoxicity can be defined as a pathological process that causes neuronal cell death as a consequence of the toxicity of excitatory amino acids. Studies demonstrate that prolonged exposure of glutamate augments neuronal excitotoxicity and further exceed the influx of ions which lead to the occurrence of injury and ultimately death of neurons. Specifically, the overloaded calcium has also been found neurotoxic, taking towards the activation of the enzymes that further destroy membranes, proteins, and nucleic acids [32].

Glutamate-mediated excitotoxicity is the foremost pathological cause in several acute as well as chronic CNS disorders and injuries. Numerous studies suggested that gray matter cytotoxic edema is produced by spreading depolarization (SD) which is associated with a significant release of glutamate [33–35] Glutamate is excessively released into the extracellular space from presynaptic nerve terminals and astrocytes; causes excessive influx of Ca^{2+} and subsequent overactivation of glutamate receptors, particularly N-Methyl-D-aspartate (NMDA) receptors [36]. The increased concentration of intracellular Ca^{2+} is due to the overactivation of NMDARs and due to activated voltage-dependent Ca^{2+} channels (VDCCs) [37] (Figure 13.4). In most glutamate excitotoxicity, as proposed by some studies, intracellular calcium is restored into the mitochondria [38, 39]. This sequestration of calcium results in mitochondrial dysfunction as well as release of reactive oxygen species (ROS) [40, 41].

Like most or even all insults, excitotoxic glutamate receptor overstimulation produces apoptosis under certain conditions and it has a robust tendency to prompt necrosis [36]. Several investigations have proved that exposure of mild excitotoxic insults to cortical neurons induced apoptosis and provided a link between apoptotic pathways and caspase-3 activation. Apart from the correlation of cellular injury and programed cell death, activation of glutamate receptor can promote apoptosis by following two ways. First, activation of NMDA receptor has been affected directly by the elevated production of ROS by mitochondria. Second, excitotoxicity induced an imbalance in homeostasis of cellular K^+ level via activating NMDA receptor and further initiated cascade of reactions to cause neuronal apoptosis [42].

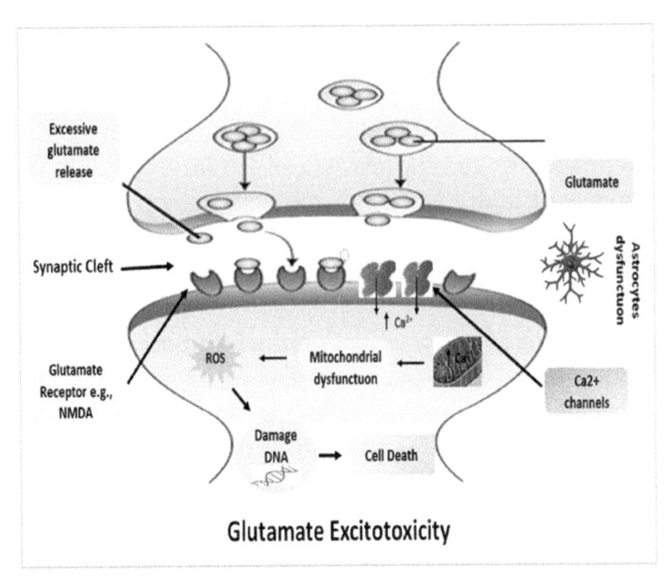

FIGURE 13.4 Mechanism of glutamate excitotoxicity.

13.4.2 PATHOPHYSIOLOGY OF PNS INJURIES

13.4.2.1 WALLERIAN DEGENERATION

Following a neuronal injury in the PNS, the axon's distal portion starts dying within 24 hours and the injured zone is invaded by blood borne macrophages and proliferating Schwann cells (SCs). SCs and macrophages clean the environment around the damaged axons via phagocytosing the axonal debris to facilitate axonal regeneration. SCs exhibit a significant role in Wallerian degeneration. They get activated during 24 hours of injury, having high mitotic rate, and enlarged nucleus and cytoplasm. Differentiated and activated Schwann daughter cells stimulate gene expression of multiple molecules which ultimately assist in the nerve repairing process. First Schwann cell worked to eliminate the degenerated myelin and axonal debris, then clear it towards the activated macrophages. Typically, the macrophages migrate toward the traumatized area via a hemopoietic route. Macrophages and SCs function side by side to clean the injury site. This process lasts for a week or even several months to aid the rapid axonal rejuvenation [43].

Wallerian degeneration is supplemented by the SCs redifferentiation and immune response activation [44]. Amongst the immune cells at the site of peripheral nerve injury, macrophages are the primary remarkable cell type. The macro-phages not only played a chief role for eliminating myelin debris, but also stimulated axonal regeneration by discharging a massive number of factors such as extracellular matrix (ECM) proteins, chemokines, cytokines, and growth factors [45, 46]. It has already been shown that many immune cells, including macrophages, neutrophils, and T cells invaded the damaged area and carry out the pathogenic processes [47, 46].

Besides the major role of the hematogenous macrophages, the resident macrophages are also accompanying Wallerian degeneration. Resident macrophages make up 2–9% of the total cells in the peripheral nerve and are highly blessed with phagocytic skills [48, 49]. In the progression of Wallerian degeneration, inflammation is induced by resident macrophages through the release of IL-13 and IL-1β as well as via increased expression of TLRs. Thus, along with hematogenous macrophages the resident macrophages also play their role in the pathophysiology of PNS injury [49–51].

13.4.2.2 CA2+ OR ELECTROLYTES IMBALANCE

The axonal modifications at the distal end sooner or later take towards the collapse of the nerve stump, thus make a way for neurons to regenerate. The representative marker of this segment is the granular fragmentation of the cytoskeleton. This takes place after an unexpected inflow of extracellular ions (Ca^+ and Na^+). This ionic flow of ions stimulates a sequence of events such as apoptosis, which further serves to employee macrophages and activated SCs. The distal nerve stump is capable of spreading a movement potential hour after transection, a fact used to examine synaptic transmission and muscle characteristics in isolated nerve-muscle preparations. Currently, it has been observed that glial cell line-derived neurotrophic factor (GDNF) and brain-derived neurotrophic factor (BDNF) mRNAs is upregulated in the injured nerve while neurotrophin-3 (NT-3) and ciliary neurotrophic factor (CNTF) are within the distal stump of transected tibial nerves.

13.5 TREATMENT APPROACHES FOR PNS INJURY

The time of recovery of damaged nerve is dependent on various variables comprising nerve repair strategies and early diagnosis. The rate of recovery of the axon is 1–2 mm/day and there is still no such treatment found to speed up the recovery phase of injured axons [137]. Even after 12–18 months, the denervated muscle has to face irreversible degeneration and it completely disrupted after 26 months [6]. Therefore, appropriate surgical treatment should be done on time to prevent permanent muscle loss. The timing of recovery and type of surgical repairing strategy depends on the severity of the damaged area [138].

13.5.1 SURGICAL APPROACHES FOR PERIPHERAL NERVE REGENERATION

13.5.1.1 Fibrin Glue

Fibrin glue may be a sutureless patch-up for PNI with counterfeit or natural cement. The commercially accessible item used for fibrin sealant 'Tisseel' was presented in the 1970s. Patch-up with fibrin sealants guarantees brief regeneration time, less fibrosis, and diminished provocative responses [52]. The foremost advantage of the fibrin stick as a therapeutic approach is that it is a fast and simple to use even in an emergency condition for nerve repair. A perfect fibrin sealant is used extensively because it possesses particular mechanical, auxiliary, and organic properties to speed up recovery. The greatest hindrance of commercially accessible sealants is due to the use of human blood. Taking this into consideration, unused wind venom based heterologous fibrin sealant, abbreviated as HFS, has been found. It can avoid the liquid mishap, lessens the time of surgery, and diminishes hemorrhage. Recent research has provided the latest nerve sealants that can work to bona fide the nerve repair and act as favoring source for those people enduring peripheral nerve injury (Figure 13.5).

13.5.1.2 Nerve Conduits

Nerve conduits are the biological material that act as a bridge like connection between the damaged proximal and distal axonal stumps of the harmed nerve. They give a framework for nerve fiber recovery and utilized as a most selective treatment for the nerve repair [139]. In this procedure, proximal ends are embedded between two endings of a nerve conduit, permitting the axonal recovery from proximal to distal nerve end. The conduits prevent attacking o adjacent tissues into an opening between the damaged ends. Appreciably, these conduits are enriched with neurotrophic variables that promote the recovery of the damaged axon.

13.5.1.2.1 Synthetic Nerve Conduits

They incorporate non-degradable and degradable nerve conduits. The foremost vital benefit of a conduit is its capacity to supply a perfect micro-environment for neuronal rejuvenation. For this reason, a perfect nerve conduit should be permeable, adaptable, lean, biocompatible, porous, biodegradable, and neuroconductive to facilitate better regrowth of damaged nerve.

13.5.1.2.2 Non-Degradable Nerve Conduits

Silicone, Plastic, and Polytetrafluoroethylene Tubes are used as non-degradable nerve conduits. The silica gel canal was known as the most commonly used nerve conduit [53]. In rodent sciatic nerve it has been observed that empty silicon tubes can repair the damage area less than 1 cm. However, silicone tubes that are filled with SCs have the capacity to repair a 1.5 cm imperfection. Even though non-degradable nerve conduits take hold to gather autologous nerves, continuously irritate the encompassing tissues and compressed the nearby nerves that facilitate the recovery of injured nerve axons [140].

13.5.1.2.3 Degradable Nerve Conduits

The frequently utilized degradable materials include collagen [141, 142], chitin [143, 144], Polyglycolic, polylactic, and polylactic corrosive conduits, glycolide trimethylene carbonate conduits [145], polycaprolactone conduit, poly (lactide-co-glycolide) conduit, normal collagen, and hydrogel conduit. Viable nerve recovery with these conduits has been detailed as they exhibit negligible remote body response. The foremost dependable nerve conduit is collagen-based nerve conduit. FDA affirmed that collagen-based conduits (NeuraGen, Neuro Flex, Neuro Lattice, Neuro Wrap, and NeuroMend) are best nerve conduits because they are restorable and adaptable. These conduits are preferred because they minimize the scar accumulation, permit exchange of supplements, and provide an appropriate environment to damage nerve for recovery without using any compression neuropathy. Recently, Neurotube (polyglycolic corrosive) and Neurolac (poly-lactide caprolactone) has been defined to repair the nerve crevices from 8 mm to 3 cm and even more than 3 cm, suggesting their extensive use in the recovery of injured nerve.

13.5.1.2.4 Organic Nerve Conduits

These conduits include autologous supply routes such as veins, umbilical cord vessels, muscle, and human amniotic film. These biomaterials have been used on broad terms for the repair of a nerve fracture of <3 cm. These sorts of conduits are appropriate in case of only when the damaged area is small. However, nerve conduits for bigger fractured area have been manufactured but it is inconvenient to use them.

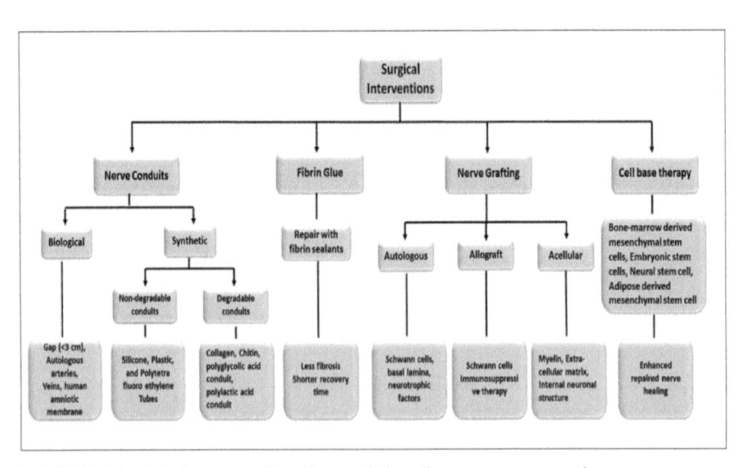

FIGURE 13.5 Surgical approaches for peripheral nerve regeneration.

13.5.1.3 NERVE GRAFTING

Autologous nerve grafting is the gold-standard choice and surgical approach to repair injured nerves. Autologous nerve joints can facilitate the recovery of >3 cm nerve lesions, proximal wounds, and basic nerve wounds. Autologous nerve grafts have been thought to provide the best regeneration outcome, as it includes SCs, basal lamina, neurotrophic components, and grip atoms for the nerve recovery. Nerve allograft is another type of nerve grafting that is collected from a cadaver. These allografts contain reasonable SCs, and immunosuppressive agents. These immunosuppressive agents are required for 18–24 months to inactivate immune cells of the body to support nerve recovery.

Although the immunosuppressive treatment has shown numerous side effects such as causing tumor production. Normally, cadaver-based nerve allografts are recommended in those patients who failed to respond to autologous nerve grafts. This method provides great regeneration but its use is limited due to higher cost and less practical feasibility. Recently, researchers have found out another type of nerve grafting known as acellular human nerve grafts Acellular nerve joints can be obtained from myelin, inner neuronal structure, and ECM molecules, i.e., collagen, and laminin However, these grafts have presented good results in re-growing damaged nerves, but they are still not suggested for long nerve fracture [54].

13.5.1.4 CELL-BASED THERAPY

Cell-based treatments can be considered an auspicious department for regenerative pharmaceuticals. The thought of utilizing refined glial cells (SCs) or mesenchymal cells from bone marrow in combination with nerve exchange strategies would be an alluring and potential source of treatment. Although the reformative probability of the cells may alter marginally in different tissues, the bone marrow still is considered as a favored source of hematopoietic as well as mesenchymal cells. Essential component s used in cell-based therapy for rejuvenating damaged nerves include SCs, bone marrow-based mesenchymal stem cells (BMSCs), adipose-based mesenchymal stem cells (ADSCs) and also pluripotent stem. SCs are the foremost and prior option seed cells. They play a central role in nerve recovery by releasing neurotrophic factors (nerve growth factors (NGF), CNTF, BDNF, platelet-derived growth factors (PDGF), and neuropeptide-Y (NPY) at the injury site.

Interestingly, activated SCs have high mitotic rate and facilitate the fundamental recovery of nerve by multiplying at the damaged area. In this therapy, the best and most commonly used source of SCs are neural crest cells. Embryonic stem cells (ESCs) have ideal preferences for giving a boundless source of cells, great separation potential along with long-lasting expansion capacity. However, moral fears act as main issues upon utilizing these transplanted cells. NSCs include neurons and glial cells. Their utilization is constrained since the gathering of these cells is troublesome. Bone marrow derived stem cells (BMSCs) are capable to distinguish into Schwann cell-like cells (BMSC-SCs). In this notion, it is appeared that the separation potential of BMSCs is not as solid as NSCs.

Adipose derived mesenchymal stem cells (ADSCs) can separate into SC-like cells and discharge different growth factors at injury sites such as BDNF, NGF, and vascular endothelial development figures (VEGF). Use of ADSCs is restricted as they have ability to distinguish instantly into adipocytes. The successful outcome of a cell-based treatment is based on the transplanted cell's capacity to distinguish into SCs like cells, to discharge axonal developing components as well as to initiate myelination of axons. Schwann cell societies have typically appeared as a satisfactory tool for prompt nerve recover, however there is a need to explore a perfect cell-based therapy to treat the nerve injuries. Similarly, Bone marrow-derived mesenchymal cells have also illustrated to provide favorable comes about due to their trophic factors releasing ability at the site of injury.

13.5.2 NON-SURGICAL THERAPIES FOR PERIPHERAL NERVE REGENERATION

Even if surgical approaches are useful to repair the nerve, they are expensive with high risk of treatment failure. So, the non-surgical treatments are also given attention for healing nerve injuries. These approaches include electrical nerve stimulation and medications.

13.5.2.1 Medicinal Therapy

Many types of medicines are presented in the market for healing neuronal pain. Anyhow, selecting the best one is tricky and depends on the cause as well as severity of pain. These medicines include corticosteroids, analgesics, opioids, and gels as shown in Figure 13.6. These can serve as 1st line treatment and as pain relievers as well. Unluckily, the available medicines are not always beneficial to treat every type of PNI as they can only relieve pain but cannot speed up the functional recovery in complicated situations [4].

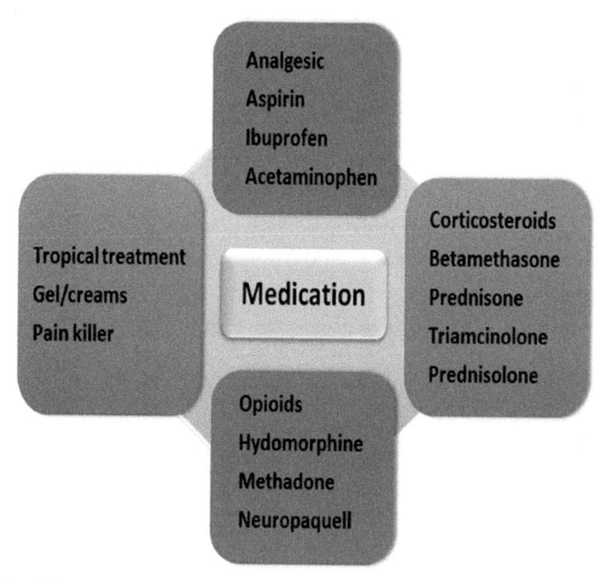

FIGURE 13.6 Non-surgical approaches for peripheral nerve regeneration.

13.5.2.2 Electrical Nerve Stimulation

The denervated nerve disrupts the communication between muscles and nerves until it is completely recovered. The method to diminish the muscular atrophy directly is by stimulating the muscle via electrical stimulus [55]. Such stimulation shows a significant role in the current of disorders related to neuromuscular junction (NMJs) [56]. The electrical stimulation (ES) of NMJs is done by directly electrical current application to the superficial skin and muscles to affected. It provokes a muscular contraction that can prevent muscles from being atrophied [57]. The electric stimuli should apply to the affected muscles several times a day with enough power, pulse duration and frequency that can revive the muscle functions. The schedule for applying electrical stimuli is quite tricky and still under debate.

Different reports suggest that the major enhancement in crushed nerve was observed when electrical stimulus was applied during the middle period, i.e., 12–21 days following the onset of nerve injury. Anyhow, no outcome has been seen on applying electrical stimulus was other than this duration. It points towards the fact that effective electrical stimulus sufficient to revive muscle function, is time-specific with narrow time window [58]. This fact is further supported by findings that daily electrical stimulus can cause inhibitory effects while treating NMJs [59]. ES therapy can be done by a number of methods which are as follows: transcutaneous electrical nerve stimulation (TENS), percutaneous electrical nerve stimulation (PENS), Deep brain stimulation (DBS) and Repetitive transcranial magnetic stimulation (rTMS) (Figure 13.7). Electrical stimuli at a frequency of 100 Hz have a tendency to increase the expression level of neurotrophic factors at denervated muscle [60]. However, this stimulus at a higher frequency of 200 Hz can fasten the myelination process [61].

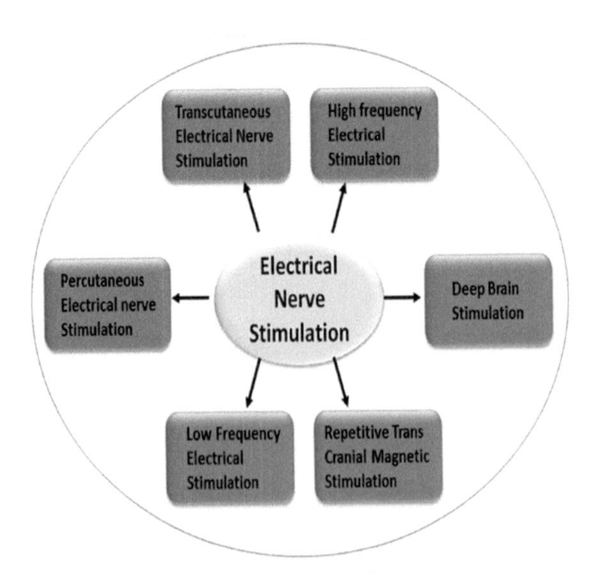

FIGURE 13.7 Electrical nerve stimulation methods.

These electrical stimulus therapies are helpful for nerve regeneration, yet they exhibit a number of harmful effects. The use of TENS can disrupt the axonal morphology which produces dark axoplasm, edema, and messy cytoarchitecture [62]. Furthermore, a decrease in axonal number with weak myelination has also been observed. This therapy can also minimize the muscle excitability and NMJs integrity by reducing the expression of neural cell adhesion molecules and cross-sectional area of muscle fibers [58].

13.5.3 PHYTOCHEMICALS FOR PERIPHERAL NERVE REGENERATION

Phytochemicals are the compounds which are plant-derived chemical and are excessively founds in nature. They are mostly used in the treatment of a large number of diseases since long ago. They are being considered by researchers because of having profound potency against diseases with less or no side effects. Recently, Hussain et al. have briefly described the therapeutic potential of alkaloids, flavonoids, and tannins (highly abundant phytochemicals) against neurological disorders [63, 64]. Moreover, effective involvement of fatty acids (lipids, cholesterol, sphingolipids) has also been well-discussed on a similar note [65, 66]. A number of phytochemicals have been enlisted in Figure 13.8 which are found to play a critical role in accelerating the recovery upon PNIs onset.

13.5.3.1 AMINOPYRIDINE

Aminopyridine is a known for its voltage-gated potassium channel blocking activity. It is reported to be very effective against various neurological disorders. 4-AP (4-Aminopyridine) induces durable re-myelination and recovery following acute traumatic nerve injury. This observation justifies its valuable use in improving nerve repair. It can be potentially used to treat lesions where this drug endorses efficient recovery [67].

13.5.3.2 QUERCETIN

Quercetin is famous for its antioxidant and anti-inflammatory activities that have been reported in numerous reports. Recently, its antioxidant and neuroprotective effects were reported in an animal model of sciatic nerve injury (SCI). This study indicates useful properties of quercetin in quickening the regeneration of nerve and minimizing the retrieval time in mild to moderate type of crush injury injuries [68, 69].

13.5.3.3 URSOLIC ACID

It is a-triterpenoid, found in abundance in herbs' flowers, fruits, and leaves. It offers antimicrobial, antioxidant, anti-inflammatory properties along with immune-modulatory, hepatoprotective, cardioprotective, chemopreventive, hypoglycemic, anti-tumor, and antihyperlipidemic qualities [66]. A study has confirmed its role in enhancing neuro regenerative processes in mice induced with SCI [65].

13.5.3.4 CURCUMIN

Curcumin is obtained from plants of genus Curcuma. It is a polyphenolic compound that is very effective in elevating inflammation, oxidative stress, metabolic disorders and depression. It is very effective in arthritis and hyperlipidemia its neuroregenerative effects following crush injury are quite evident from literature [70, 71]. It has been reported recently Recent reports reveal the potential role of curcumin in stimulating recovery following a complete sciatic nerve amputation [72]. Furthermore, it relieves the neuropathic pain by lessening the activity of p300/CBP HAT which facilitates the expression of BDNF and Cox-2 [73]. Most importantly, it owns the ability to uplift the S100 protein level and decreases the apoptosis of SCs. This upregulated S100 protein fasten the neuroregenerative process and thus facilitate early functional recovery [72, 74].

13.5.3.5 HUPERZINE

Huperzine-A (Hup A) is a sesquiterpene alkaloid and occurs naturally in firmoss Huperzia serrate. It is beneficial for the problem related to the memory (dementia), it is also beneficial for myasthenia gravis because it can cross the BBB and hereby exhibits various neuroprotective activities such as an inhibitor of acetylcholinesterase (AChE) [75, 76] and antagonist of NMDA [77]. Whereas, inhibition of NMDA receptors protects against ischemia-induced SCI by attenuating the neuron cells' demyelination and immune cell infiltration [78]. Furthermore, Hup A is a powerful analgesic with minimum hazardous effects. It improves the spontaneous pain and reduces the mechanical allodynia induced by the peripheral inflammation or PNI.

Meanwhile, peripheral inflammation can also initiate spontaneous pain. The attributed mechanism involved in the attenuated pain condition is the inhibition of AchE; accumulated following the chronic pain. Thus, Hup A serves as a protective agent in the case of PNI [79]. Moreover, it is worth mentioning that being capable of crossing the BBB, it acts as a strong inhibitor of AChE; an enzyme regulating the breakdown of acetylcholine. Therefore, Hup A is accredited as a pivotal neuroprotective agent in AD [80]. Conclusively, it is effective to evaluate the precise function of Hup A in peripheral nerve regeneration to introduce a therapeutic intervention against PNI.

13.5.3.6 BERBERINE (BBR)

Berberine (BBR) is an isoquinoline alkaloid that is exists in various plants such as goldenseal, Phellodendron, tree turmeric, and European barberry. It displays various pharmacological actions such as anti-inflammatory [81], anti-oxidative [82], and anti-tumor [83]. BBR regulates the regeneration of peripheral nerve following an injury by enhancing the proliferation rate of SCs. As the axonal regeneration in the PNS largely depends upon the onset of guiding direction and release of neurotrophic factors, by SCs [84]. Whereas, SCs also proliferate to cover the gap at the injury site as well as to myelinate the new axons. In this way, BBR reveals a significant treatment therapy against SCI. It also improves the axonal regeneration in a SCI rodent model [85]. It inhibits the neuro-inflammation by down-regulating different inflammatory mediators such as IL6, IL1β, and TNF-α. It also reduces the apoptosis via autophagy-associated protein modulation and stimulates autophagy by autophagy-associated protein modulation [86].

Therefore, it can be attributed as a useful therapeutic compound in PNS injuries and needed to be investigated further for its clinical connections. Even though BBR has been reported as a neuroprotective agent against PNIs, but it also exerts toxic effects. Though there is very limited data to evidence the toxic properties of BBR, some significant studies are described here. It exerts neurotoxic effects of the dose of 10–30 μM and inhibits the synthesis of dopamine. It also disturbs cognitive and motor functions by attenuating the dopaminergic neurons in substantia nigra of the dose of 5–15 mg/kg [87]. In the future, to explore the pharmacological profile of BBR, specific studies are required to address its toxicity while looking at its different activities.

13.5.3.7 CENTELLA ASIATICA

Centella asiatica, also recognized as *Hydrocotyle asiatica* L. is a famous Urban herb, is popular for its use as a nerve tonic in Ayurvedic system of medicine [88]. In a study, its ethanolic extract (100 μg mL^{-1}) has been found to stimulate an extraordinary neurite outgrowth in human SH-SY5Y cells in the presence of NGF. This ethanolic extract has been reported to possess Asiatic acid which can increase the neurite growth [89]. Furthermore, studies have revealed speedy functional recovery and uplifts axonal regeneration following treatment with this herb. This provides evidence that the compounds found in its extract would be beneficial for increasing the nerve repair [66].

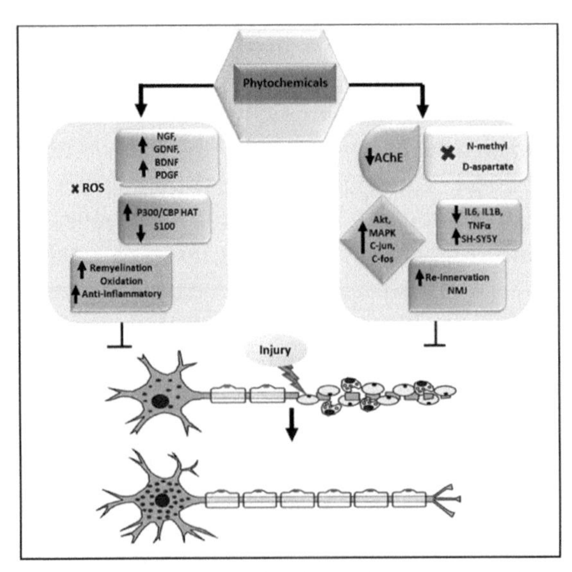

FIGURE 13.8 Phytochemicals mediated peripheral nerve regeneration.

13.5.3.8 VINORINE

A natural monoterpenoid indole alkaloid, vinorine, is attributed as the main compound against peripheral nerve injury. It also exhibits anti-tumor, anti-inflammatory, and antibacterial activities. It enhances the sensory and motor function retrieval after the injured sciatic nerve fiber. It boosts the regeneration process through the regulation of NGF [90] which then stimulates the regeneration of a nerve after the injury [52]. Until now, a very inadequate number of studies have talked about its role in providing neuroprotection in peripheral nerve regeneration, so further studies are required to discover its pharmacological profile.

13.5.3.9 HERICIUM ERINACEUS

It is a well-known edible mushroom with amazing medicinal properties and is helpful against immune-regulatory disorders, cancer, and AD [91]. The aqueous extract of *Hericium erinaceus* fresh fruit bodies accelerates the axonal regeneration as well as re-innervation of NMJs [92]. It has the potential to enhance local protein synthesis at axonal distal segments of crushed nerves. It is observed in a study (conducted in rat model) that a daily oral dose of *Hericium erinaceus* has enhance the restoration of the injured peroneal nerve at an initial step of retrieval [93]. Positive effects of aqueous extract on accelerating nerve regeneration are thought to be due to control of the signaling pathways such as MAPK, Akt, c-Fos, c-Jun, and protein synthesis [93, 94].

13.5.3.10 LUMBRICUS EXTRACTS

It is generally known as the earthworm and it has an ability to redevelop its removed part. This ability of Lumbricus has fascinated the scientists to discover its remedial characteristics [95]. This extract has been used by the Chinese as an integral part of their system of traditional medicines for endorsing the functions of nerve by upgrading the nerve conduction velocity [96]. Furthermore, the oral administration of these extract accelerates functional regain and regeneration following a nerve compression injury [97].

13.6 TREATMENT APPROACHES FOR CNS INJURIES

The prevention of lesions is the foremost measure in the spinal cord injury treatment. Primary steps include the publicizing the knowledge about how to avoid spinal fractures caused by jumping accidents, decrease the frequency of traffic associated accidents, boost vehicle safety; and plans to increase at-home safety, especially to decrease the number of falls in old persons. Secondary preventions are being established for use upon happening of the accident. These preventions include bases of sufficient rescue and availability of transportation to particular health centers. Tertiary health care is the most complicated and includes restoration of affected persons' functions (Figure 13.9).

13.6.1 THERAPEUTIC POSSIBILITIES

In 1700 BC Edwin Smith translated an Egyptian surgical papyrus and, in that papyrus, the frustration triggered due to severe SCI was defined; Most of such studies proceed towards the cure of acute SCI in the following four manners: physical [98], corrective surgery [99, 100], biological treatment [101, 102] and pharmacological treatment [62, 103].

13.6.2 SURGICAL APPROACH

Surgical decompression is the most frequently used approach with or without arthrodesis. On the other hand, it has been noted that 1–1.8% of the patients gone through the thoracic and cervical SCI are able to walk following a surgical process [104]. The purpose of surgery in case of acute SCI is to nerve decompression, stabilization of the spine, and spinal alignment that helps inhibiting any further neurologic injury. A study showed that when decompression is executed within 1 day after trauma, there is an improvement in the chance of functions rehabilitation [105]. These observations highlight the necessity for abruptly decreasing decompressing fractures and dislocations, either by open surgery or through cranial traction [106–108].

13.6.3 BIOLOGICAL THERAPY

Components promoting neuronal recoveries include tissue development variables and totipotent cells (autologous or homologous). These are natural treatments for SCI. Undifferentiated cells are multi powerful cells that can multiply and engender cells of an ancestry or tissue sort. It has already appeared in certain studies that stem cells when transferred into an ordinary or harmed spinal rope become differentiate into neuronal precursor which proliferate there. These cells can be differentiated into both neurons and oligodendrocytes and also astrocytes. Transplanted stem cells do not make bridge between split axonal ends. On the contrary, they work by releasing components responsible to increase axonal reformation and removing harmful immune cells there [102]. Thus, stem cells can be a future plausibility as a remedy for SCI [101].

After central anxious framework harm, transplants may emphatically impact utilitarian recuperation through a wide run of instruments, which incorporate the non-specific results of transplantation, trophic activities, t hormones and transmitters discharge and indeed components including the particular reinnervation. In expansion, these cells can give a populace of cells at the location of damage that act as a substrate/reservoir for re-constructing communication between the levels over and underneath the injury. One thinks about detailed that bone marrow-derived cells managed intravenously after sub deadly illumination brought about in cells communicating neuron-specific qualities. The control of these stem cells is permissible future treatments for serious diseases and disorders.

13.6.4 PHYSICAL APPROACHES

It is on the presumption that low temperature can provide a protection to the tissues of the central anxious framework from hypoxic as well as ischemic impacts. In any case, it is troublesome to use this strategy because of the associated high mortality rate. Cooling treatment does not anticipate potassium misfortune, which happens in steroid treatment [109]. Hyperbaric oxygen treatment may be a therapeutic methodology that is based on accomplishing a tall fractional oxygen conc. Inside tissue by having the quiet breath in clean oxygen interior a hyperbaric chamber at a weight more noteworthy than the barometrical weight). The method of reasoning for this helpful tactic is that a diminished perfusion is recompensed by expanding the halfway weight of oxygen [110]. Improved outcomes above been detailed following the utilization of hyperbaric oxygen in SCI [111].

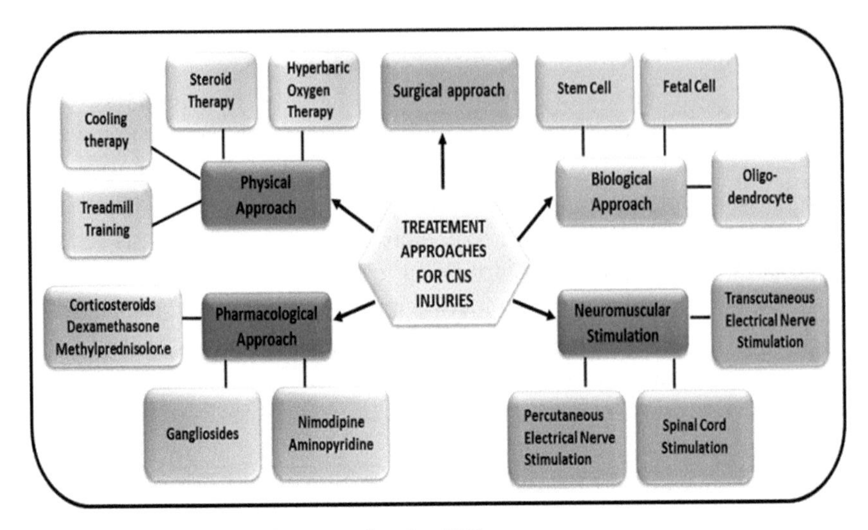

FIGURE 13.9 Therapeutic approaches for CNS.

13.6.5 PHARMACOLOGICAL APPROACHES

Pharmacology plays a substantial role in treating SCI. Both experimental and clinical trials reveal the effective use of medicines in treating secondary SCI [112]. For this purpose, corticosteroids, and gangliosides are already accepted for symptomatic treatment in humans [109]. Dexamethasone and methylprednisolone are the most frequently studied corticosteroids. The effectiveness of methylprednisolone has been evident in patient with SCI. Collectively, these trials are named as NASCIS (National Acute Spinal Cord Injury Study). NASCIS-1 first appeared in public in 1985 [109] while the NASCIS-2 was published in 1992 and the NASCIS-3 in year 2001 [113, 114]. Researchers have experimentally tested the Methylprednisolone to use it as a prophylactic drug before those surgeries having a risk of SCI, but no significant improvement in reducing the risk have been seen [110].

Further studies have been done by adopting the same r methodologies used in NASCIS and reported very harmful outcomes of massive dosage of corticosteroids. These harmful outcomes of corticosteroids, particularly methylprednisolone warn the physicians not to use them in SCI [115]. Gangliosides (glycolipid molecules) are derived from sialic acid and are found to promote neurite growth and protoplasmic expansion of axons which ultimately induce neuronal regeneration [116]. Gangliosides have been well studied to find their potential in healing SCI. Improved sensory and motor functions have been observed in the patients of SCI when treated with glycosides [117]. In another study, glycosides administration to individuals with incomplete SCI along with physical therapy had prominently improved the motor scores as compare to those who were on placebo and those who had been given only physical therapy. These findings support the potential of glycosides in SCI [118]. The recommended dose of gangliosides for patient with SCI is 300 mg (loading dose) followed by 100 mg (maintenance dose) once daily for 30 days [98]. Oxidative stress is one of the major deteriorating factors that happen spontaneously following SCI [119].

Therefore, attenuation is considered as a therapeutic intervention in case of SCI [103, 120]. Antioxidants and free radical scavenging agents are noted to be very active in accelerating the functional repossession in animal model of SCI [101]. Calcium channel blockers are found to increase medullary microcirculation [121]. Nimodipine (calcium channel blocker) has been studied for this purpose, but findings do not commend its medical use in the patients diagnosed with SCI [116]. A selective inhibitor for K channel named as TRAM-34 (triarylmethane-34), has been found to improve the motor functions, lessen tissue loss, and augmented neuronal sparing [122]. Aminopyridine (potassium channel blocker) has the capacity to improve nerve transmission in demyelinated axons [116, 123] with motor and sensory functions regain improvement [111, 124].

13.6.6 NEUROMUSCULAR STIMULATION

The scientific reports for electrotherapy have progressively revealed the opportunity for rational therapy of many aching conditions. These new methods include TENS and PENS.

13.6.6.1 TRANSCUTANEOUS ELECTRICAL NERVE STIMULATION (TENS)

TENS is one of the most commonly employed method of non-pharmacological pain-relieving strategy. It includes a source of electricity along with portable electric unit and electrodes (which are used to attach to the skin). The electrodes will send electrical impulses to the nerve fibers beneath that area. There are 3 kinds of frequency modulations: HF-TENS >50 Hz (high frequency), LF-TENS <10 Hz (low frequency) and VF-TENS (variable frequency) [125]. The power is mostly adjustable. After the application of HF-TENS, power is usually low, accompanying a tingling sensation. Contrarily, the application of LF-TENS is linked with high power following a robust, but usually, a muscle jerking but it is comfortable. After TENS, pain relief may be quick, but in the beginning, the outcomes are not long-lasting. Spinal cord stimulation therapy is an invasive technique. In which electrode is placed in the epidural space (which applies pulsated electrical signals to the spinal cord directly). In the lower abdominal area, the electrical pulse generator is placed. Shealy and Mortinor tried to stimulate the spinal cord's dorsal horn for the very first time in 1967 [126].

13.6.6.1.1 Potential Mechanisms for the Analgesic Action of Electrical Stimulation (ES)

Gate Control Theory of Pain suggesting that "the substantia gelatinosa in the dorsal horn acts as a gate control system, which modulates the synaptic transmission of nerve impulses from peripheral fibers to the central cells" was established by Melzack and Wall in 1967 [127]. This hypothesis explains that the small nociceptive fibers grip the supposed gate in a relatively open position, while large mechano-receptive A-β fibers upon stimulation, shut the gate and prevent the conduction pain message to the brain. Since small fibers responsible for nociceptive stimulus are characterized by a higher action threshold potential than the large mechano receptive fibers. So, the selective activation of the large mechanoreceptors may stop or lessen pain message conduction. The idea of Melzack and Wall has taken towards applying gentle ES to the skin (later on, this method was named as TENS [128]. The action of this pathway has been highlighted by two interpretations. At first, the analgesic effect of HF-TENS has partly been decreased following spinal cord cut (spinalization). It removes the descending inhibitory components [129].

Secondly, the analgesic effect of TENS persisted for several hours following the stimulation time survives, implicates the involvement of extra-segmental factors [125]. The downward pathway (originating inside midbrain and passing through periaqueductal gray (PAG)) directs roots towards the rostral ventral medulla (RVM), trailed by roots propagating towards spinal dorsal horn [130, 131]. It is believed that TENS induces activation of PAG and RVM which has been considered to be intricate in the lessening of hyperalgesia in rat model of arthritis [131]. The most important thing is the boosted discharge of endogenous opioids and serotonin. This happens via PAG-RVM pathway [57]. Enhanced levels of μ-opioids in arthritic rat's spinal fluid following treatment with LF-TENS as well as increased concentration of in the δ-opioids following HF-TENS justify the analgesic capability of these therapies [132]. While, GABA is an essential contributor in TENS-induced analgesia. HF-TENS are found to enhance the discharge of GABA in spinal cords. Both HF- and LF-TENS diminish the hyperalgesia by activating spinal GABA-A receptors [133]. Only HF-TENS are found to suppress some excitatory amino acids levels, i.e., glutamate, and aspartate, in the dorsal horn [135]. Spinal cord stimulation is found to enhance the synthesis of acetylcholine in the dorsal horn. This indicates that initiation of the cholinergic system may play a role in giving analysis effect following spinal cord stimulation [134].

13.6.6.2 PERCUTANEOUS ELECTRICAL NERVE STIMULATION (PENS)

It has been documented in a study that PENS decreased pain and physical activity of patients with diabetic peripheral neuropathy. Also, the mental components of the SF-36 exhibited betterment as compared to the sham group. In another sham-controlled crossover, study PENS was found to be more effective than TENS in the patients having lower back pain [136].

13.7 SUMMARY

Nerves comprise a complex network of cables of our nervous system and coordinate all the fundamental functional activities of the body. Nervous system comprised of Central and PNS. Any external stimuli even a blow to the head or accidents can cause severe damage to nerves. Injured nerves in CNS are not able to exhibit regeneration whereas, peripheral nerves can recover themselves better after an appropriate treatment. Phytochemicals are the natural sources and used to ameliorate nerve injuries in PNS. Different research studies have been reported that phytochemicals could potentiate the rejuvenation rate following injury. Moreover, Surgical, physical, and pharmacological approaches have also provided extensive output in regrowing the damaged nerves. This chapter provides all fundamental strategies that are used now a days for recovery of nerve injuries. However, there is further need to find out the therapeutic markers which will obviously facilitate the physicians to treat the injury related disorders.

KEYWORDS

- **excitotoxicity**
- **injuries**
- **nerve stimulation**
- **nervous system**
- **pathophysiological changes**
- **pharmacological approaches**
- **phytochemicals**

REFERENCES

1. Rea, P., (2014). *Introduction to the Nervous System.* In Clinical Anatomy of the Cranial Nerves.
2. Tortora, G. J., & Derrickson, B., (2014). *Principles of Anatomy & Physiology* (14th edn.). In Wiley.
3. Campbell, W. W., (2008). Evaluation and management of peripheral nerve injury. *Clinical Neurophysiology, 119*(9), 1951–1965.

4. Hussain, G., Wang, J., Rasul, A., Anwar, H., Qasim, M., Zafar, S., & Sun, T., (2020). Current status of therapeutic approaches against peripheral nerve injuries: A detailed story from injury to recovery. *International Journal of Biological Sciences*. Effect of neurotrophic factors on peripheral nerve repair.
5. Tubbs, R. S., Rizk, E., Shoja, M. M., Loukas, M., Barbaro, N., & Spinner, R. J., (2015). *Nerves and Nerve Injuries.*
6. Lee, S. K., & Wolfe, S. W., (2000). Peripheral nerve injury and repair. *The Journal of the American Academy of Orthopaedic Surgeons, 8*(4), 243–252.
7. Gerald, J. F., & Fawcett, J., (2007). Repair in the central nervous system. *Journal of Bone and Joint Surgery - Series B.*
8. Barha, C. K., Nagamatsu, L. S., & Liu-Ambrose, T., (2016). Basics of neuroanatomy and neurophysiology. In: *Handbook of Clinical Neurology, 138*,53–68.
9. Mietto, B. S., Mostacada, K., & Martinez, A. M. B., (2015). Neurotrauma and inflammation: CNS and PNS responses. *Mediators of Inflammation.*
10. Burn, D. J., & Jaros, E., (2001). Multiple system atrophy: Cellular and molecular pathology. *Journal of Clinical Pathology - Molecular Pathology.*
11. Conforti, L., Gilley, J., & Coleman, M. P., (2014). Wallerian degeneration: An emerging axon death pathway linking injury and disease. *Nature Reviews Neuroscience.*
12. Menorca, R. M. G., Fussell, T. S., & Elfar, J. C., (2013). Nerve physiology. Mechanisms of injury and recovery. *Hand Clinics, 29*, 317–330.
13. Okada, S., Hara, M., Kobayakawa, K., Matsumoto, Y., & Nakashima, Y., (2017). Astrocyte reactivity and astrogliosis after spinal cord injury. *Neuroscience Research.*
14. Hara, M., Kobayakawa, K., Ohkawa, Y., Kumamaru, H., Yokota, K., Saito, T., & Okada, S., (2017). *Interaction of Reactive Astrocytes with Type I Collagen Induces Astrocytic Scar Formation through the Integrin - N-Cadherin Pathway After Spinal Cord Injury.*
15. Ray, S. K., Samantaray, S., Smith, J. A., Matzelle, D. D., Das, A., & Banik, N. L., (2011). Inhibition of cysteine proteases in acute and chronic spinal cord injury. *Neurotherapeutics: The Journal Of the American Society for Experimental NeuroTherapeutics, 8*(2), 180–186.
16. Li, M., Ona, O. V., Guegan, C., Chen, M., Lewis, V. J., & Friedlander, R. M., (2000). Functional role of caspase-1 and caspase-3 in an ALS transgenic mouse model. *Sciencemag,* 335–339.
17. Ray, S. K., Hogan, E. L., & Banik, N. L., (2003). Calpain in the pathophysiology of spinal cord injury: Neuroprotection with calpain inhibitors. *Brain Research. Brain Research Reviews, 42*(2), 169–185.
18. Koopmans, G. C., Deumens, R., Buss, A., Geoghegan, L., Mu, A., Honig, W. H. H., & Brook, G. A., (2009). Acute rolipram / thalidomide treatment improves tissue sparing and locomotion after experimental spinal cord injury. *Experimental Neurology, 216*(2), 490–498.
19. Souza, J. M., Pithon, K., Ozahata, T., Oliveira, R. T., Teo, F. H., Blotta, M. H., & Jr, W. N., (2010). Subclinical atherosclerosis is related to injury level but not to inflammatory parameters in spinal cord injury subjects. *Spinal Cord, 48*(10), 740–744.
20. Yu, W. R., Liu, T., Fehlings, T. K., & Fehlings, M. G., (2009). Involvement of mitochondrial signaling pathways in the mechanism of Fas-mediated apoptosis after spinal cord injury. *The European Journal of Neuroscience, 29*(2008), 114–131.

21. Keane, R. W., Kraydieh, S., Lotocki, G., Bethea, J. R., Krajewski, S., Reed, J. C., & Dietrich, W. D., (2001). Apoptotic and anti-apoptotic mechanisms following spinal cord injury. *Journal of Neuropathology and Experimental Neurology, 60*(5), 422–429.
22. Wingrave, J. M., Schaecher, K. E., Sribnick, E. A., Wilford, G. G., Ray, S. K., Martin, D. J. H., & Banik, N. L., (2003). Early induction of secondary injury factors causing activation of calpain and mitochondria-mediated neuronal apoptosis following spinal cord injury in rats. *Journal of Neuroscience Research, 104*(1), 95–104.
23. Mcewen, M. L., & Springer, J. E., (2005). A mapping study of caspase-3 activation following acute spinal cord contusion in rats. *Journal of Histochemistry Cytochemistry, 53*(7), 809–819.
24. Nottingham, S., Knapp, P., & Springer, J., (2002). FK506 treatment inhibits caspase-3 activation and promotes oligodendroglial survival following traumatic spinal cord injury. *Experimental Neurology, 251*(1), 242–251.
25. Festoff, B. W., Ameenuddin, S., Arnold, P. M., Wong, A., Santacruz, K. S., & Citron, B. A., (2006). Minocycline neuroprotects, reduces microgliosis, and inhibits caspase protease expression early after spinal cord injury. *Journal Of Neurochemistry, 97*(5), 1314–1326.
26. Okutan, O., Solaroglu, I., A, E. B., & Taskin, Y., (2007). Recombinant human erythropoietin decreases myeloperoxidase and caspase-3 activity and improves early functional results after spinal cord injury in rats. *Journal of Clinical Neuroscience, 14*(4), 364–368.
27. Holmin, S., Schalling, M., Höjeberg, B., Nordqvist, A. C. S., Skeftruna, A. K., & Mathiesen, T., (1997). Delayed cytokine expression in rat brain following experimental contusion. *J. Neurosurg., 86*(3), 493–504.
28. Kinoshita, K., Chatzipanteli, K., Vitarbo, E., Truettner, J. S., Alonso, O. F., & Dietrich, W. D., (2002). Interleukin -1 messenger ribonucleic acid and protein levels after fluid -percussion brain injury brain temperature. *Neurosurgery, 51*(1), 195–203.
29. kossman, M. C. M., Lenzlinger, P. M., Hans, P., Stahel, P., Csuka, E., Ammann, E., Stocker, R., et al., (1997). Production of cytokines following brain injury : beneficial and deleterious for the damaged tissue. *Mol Psychiatry, 2*(2), 133–136.
30. Derouiche, A., (2004). The perisynaptic astrocyte process as a glial compartment-immunolabeling for glutamine synthetase and other glial markers. *Advances in Molecular and Cell Biology, 31*, 147–163.
31. Kostandy, B. B., (2012). The role of glutamate in neuronal ischemic injury : The role of spark in fire. *Neurological Sciences, 33*(2), 223–237.
32. Berliocchi, L., Bano, D., Nicotera, P., Berliocchi, L., Bano, D., & Nicotera, P., (2005). Ca^{2+} signals and death programs in neurons Ca2C signals and death programs in neurons. *Philosophical Transactions of the Royal Society of London. Series B, Biological Sciences, 360*(1464), 2255–2258.
33. Harreveld, A. V. A. N., (1959). Compounds in brain extracts causing spreading depression of cerebral cortical activity a n d contraction of crustacean muscle. *Journal of Neurochemistry, 3*, 300–315.
34. Harreveld, A. V. A. N., & Fifkova, E. V. A., (1970). Glutamate release from the retina during. *Journal of Neurobiology, 2*(1), 13–29.
35. Harreveld, V. A. N., Harreveld, V. A. N., Harreveld, V. A. N., Harreveld, V. A. N., Harreveld, V. A. N., & Harreveld, V. A. N., (1965). Amino Acid release from

the cerebral cortex during spreading depression and asphyxiation. *Journal of Neurochemistry, 12*, 431–439.

36. Zipfel, G. J., Debra, J., Lee, J., & Choi, D. W., (2000). *Neuronal Apoptosis After CNS Injury : The Roles of Glutamate and Calcium, 17*(10), 857–869.
37. Yagami, T., Ueda, K., Sakaeda, T., & Itoh, N., (2004). Protective effects of a selective L-type voltage-sensitive calcium channel blocker , S -312- d , on neuronal cell death. *Biochemical Pharmacology, 67*(6), 1153–1165.
38. Duchen, M. R., (2000). Topical review mitochondria and calcium: From cell signaling to cell death. *Journal of Physiology,* 57–68.
39. Thayer, S. A., & Wang, G. J., (1995). Glutamate-induced calcium loads : Effects on energy metabolism and neuronal viability. *Clinical and Experimental Pharmacology and Physiology, 22*, 303–304.
40. Li, C., Chin, T., & Chueh, S., (2018). Rat cerebellar granule cells are protected from glutamate-induced excitotoxicity by S -nitrosoglutathione but not glutathione. *American Journal of Physiology. Cell Physiology, 286*(4), 893–904.
41. Schild, L., Huppelsberg, J., Kahlert, S., Keilhoff, G., & Reiser, G., (2003). Brain mitochondria are primed by moderate Ca rise upon hypoxia / reoxygenation for functional breakdown and morphological disintegration. *The Journal of Biological Chemistry, 278*(28), 25454–25460.
42. Bloom, R. D. S., & Sah, R., (2001). Gamma-aminobutyric acid A neurotransmission and cerebral ischemia. *Journal of Neurochemistry, 77*(2), 353–371.
43. Burnett, M. A. R. K. G., & Zager, E. R. I. C. L., (2004). Pathophysiology of peripheral nerve injury: A brief review. *Neurosurg. Focus, 16*(5), 1–7.
44. Namikawa, K., Okamoto, T., Suzuki, A., Konishi, H., & Kiyama, H., (2006). Pancreatitis-associated protein-III Is a novel macrophage chemoattractant implicated in nerve regeneration. *The Journal of Neuroscience, 26*(28), 7460–7467.
45. Gaudet, A. D., Popovich, P. G., & Ramer, M. S., (2011). Wallerian degeneration: Gaining perspective on inflammatory events after peripheral nerve injury. *Journal of Neuroinflammation, 8*(1), 110.
46. Martini, R., Fischer, S., & David, S., (2008). Interactions between Schwann cells and macrophages in injury and inherited demyelinating disease interactions between Schwann cells and macrophages in injury and inherited demyelinating disease. *Research Gate,* 1566–1577.
47. Liou, J., Liu, F., & Mao, C., (2011). Inflammation confers dual effects on nociceptive processing in chronic neuropathic pain model. *Anesthesiology, 114*(3), 660–672.
48. Griffin, J. W., George, R., & Ho, T., (1993). Macrophage systems in peripheral nerves: A review. *Journal of Neuropathology and Experimental Neurology, 52*(6), 553–560.
49. Lindborg, J. A., Niemi, J. P., & Zigmond, R. E., (2014). The neuroimmunology of degeneration and regeneration in the peripheral nervous system. *Neuroscience*, (302), 174–203.
50. Dieter, C., Pierre, S., Weigert, A., Weichand, B., Altenrath, K., Schreiber, Y., & Scholich, K., (2014). Prostacyclin mediates neuropathic pain through interleukin 1 b -expressing resident macrophages. *Pain, 155*(3), 545–555.
51. Ydens, E., Cauwels, A., Asselbergh, B., Goethals, S., Peeraer, L., Lornet, G., & Janssens, S., (2012). Acute injury in the peripheral nervous system triggers an alternative macrophage response. *Journal of Neuroinflammation, 176*(9), 1–17.

52. Wood, R. L., Karlinsey, K. S., Thompson, A. D., Rigby, M. N., Boatright, G. D., Pitt, W. G., & Cook, A. D., (2018). Baseline effects of lysophosphatidylcholine and nerve growth factor in a rat model of sciatic nerve regeneration after crush injury. *Neural Regeneration Research, 13*(5), 846–853.
53. Guidi, L., Antico, L., Bartoloni, C., Costanzo, M., Errani, A., Tricerri, A., & Frasca, D., (1998). Changes in the amount and level of phosphorylation of p56(lck) in PBL from aging humans. *Mechanisms of Aging and Development, 102*(2, 3), 177–186.
54. Nadim, W., Anderson, P. N., & Turmaine, M., (1990). The role of Schwann cells and basal lamina tubes in the regeneration of axons through long lengths of freeze-killed nerve grafts. *Neuropathology and Applied Neurobiology, 16*(5), 411–421.
55. Willand, M. P., (2015). *Electrical Stimulation Enhances Reinnervation After Nerve Injury, 25*(4), 243–248.
56. Wong, J. N., Olson, J. L., Morhart, M. J., & Chan, K. M., (2015). Electrical stimulation enhances sensory recovery: A randomized controlled trial. *Annals of Neurology, 77*(6), 996–1006.
57. Heidland, A., Fazeli, G., Sebekova, K., Hennemann, H., Bahner, U., & Iorio, B. D., (2012). *Neuromuscular Electrostimulation Techniques: Historical Aspects and Current Possibilities in Treatment of Pain and Muscle Waisting, 78.*
58. Su, H. L., Chiang, C. Y., Lu, Z. H., Cheng, F. C., Chen, C. J., & Sheu, M. L., (2018). Late administration of high - frequency electrical stimulation increases nerve regeneration without aggravating neuropathic pain in a nerve crush injury. *BMC Neuroscience,* 1–12.
59. Benato, D. G., Russo, T. L., Geuna, S., Domingues, N. R. S. R., Salvini, T. F., & Parizotto, N. A., (2010). Electrical stimulation impairs early functional recovery and accentuates skeletal muscle atrophy after sciatic nerve crush injury in rats. *Muscle & Nerve, 41*(5), 685–693.
60. Willand, M. P., Rosa, E., Michalski, B., Zhang, J. J., Gordon, T., Fahnestock, M., & Borschel, G. H., (2016). Electrical muscle stimulation elevates intramuscular BDNF and GDNF mRNA following peripheral nerve injury and repair in rats. *Neuroscience, 334*, 93–104.
61. Kao, C., Chen, J. J., Hsu, Y., Bau, D., Yao, C., & Chen, S., (2013). *High-Frequency Electrical Stimulation Can Be a Complementary Therapy to Promote Nerve Regeneration in Diabetic Rats, 8*(11).
62. Murina, F., & Francesco, S., (2015). *Transcutaneous Electrical Nerve Stimulation,* 105–117.
63. HUSSAIN, G., Huang, J., Rasul, A., Anwar, H., Imran, A., Maqbool, J., & Sun, T., (2019). Putative roles of plant-derived tannins in neurodegenerative and neuropsychiatry disorders: An updated review. *Molecules, 24*, 2213.
64. Hussain, G., Zhang, L., Rasul, A., Anwar, H., Sohail, M. U., Razzaq, A., & Sun, T., (2018). Role of plant-derived flavonoids and their mechanism in attenuation of Alzheimer's and Parkinson's diseases: An update of recent data. *Molecules, 23*, 1–26.
65. Hussain, G., Anwar, H., Rasul, A., Imran, A., Qasim, M., Zafar, S., & Umair, M. A., (2019). Lipids as biomarkers of brain disorders. *Critical Reviews in Food Science and Nutrition,* 1–24.

66. Hussain, G., Wang, J., Rasul, A., Anwar, H., Imran, A., Qasim, M., & Iqbal, J., (2019). Role of cholesterol and sphingolipids in brain development and neurological diseases. *Lipids in Health and Disease,* 1–12.
67. Tseng, K. C., Li, H., Clark, A., Sundem, L., Zuscik, M., Noble, M., & Elfar, J., (2016). 4-Aminopyridine promotes functional recovery and remyelination in acute peripheral nerve injury. *EMBO Molecular Medicine, 8*(12), 1409–1420.
68. Lesjak, M., Beara, I., Simin, N., Pintać, D., Majkić, T., Bekvalac, K., & Mimica-Dukic, N., (2017). Antioxidant and anti-inflammatory activities of quercetin and its derivatives. *Journal of Functional Foods, 40*, 68–75.
69. Türedi, S., Yuluğ, E., Alver, A., Bodur, A., & İnce, İ., (2018). A morphological and biochemical evaluation of the effects of quercetin on experimental sciatic nerve damage in rats. *Experimental and Therapeutic Medicine, 15*(4), 3215–3224.
70. Hewlings, S. J., (2017). *Curcumin: A Review of Its' Effects on Human Health,* pp. 1–11.
71. Ma, J., Liu, J., Yu, H., Wang, Q., Chen, Y., & Xiang, L., (2013). Curcumin promotes nerve regeneration and functional recovery in rat model of nerve crush injury. *Neuroscience Letters, 547*, 26–31.
72. Liu, G., Xu, K., Li, J., & Luo, Y., (2016). Curcumin upregulates S100 expression and improves regeneration of the sciatic nerve following its complete amputation in mice. *Neural Regeneration Research, 11*(8), 1304.
73. Zhu, X., Li, Q., Chang, R., Yang, D., Song, Z., & Guo, Q., (2014). *Curcumin Alleviates Neuropathic Pain by Inhibiting p300 / CBP Histone Acetyltransferase Activity-Regulated Expression of BDNF and Cox-2 in a Rat Model, 9*(3).
74. Zhao, Z., Li, X., & Li, Q., (2017). Curcumin accelerates the repair of sciatic nerve injury in rats through reducing Schwann cells apoptosis and promoting myelinization. *Biomedicine and Pharmacotherapy, 92*, 1103–1110.
75. Wang, R., Yan, H., & Tang, X. C., (2006). Progress in studies of Huperzine A, a natural cholinesterase inhibitor from Chinese herbal medicine. *Acta Pharmacologica Sinica, 27*(1), 1–26.
76. Wang, B., Wang, H., Wei, Z., Song, Y., na, L., & Chen, H., (2009). Efficacy and safety of natural acetylcholinesterase inhibitor huperzine A in the treatment of Alzheimer's disease: An updated meta-analysis. *Journal of Neural Transmission, 116*(4), 457–465.
77. Coleman, B. R., Ratcliffe, R. H., Oguntayo, S. A., Shi, X., Doctor, R. P., Gordon, R. K., & Nambiar, M. P., (2008). [+]-Huperzine A treatment protects against N-methyl-d-aspartate-induced seizure/status epilepticus in rats. *Chemico-Biological Interactions, 175*(1), 387–395.
78. Ke, T., Li, R., & Chen, W., (2016). Inhibition of the NMDA receptor protects the rat sciatic nerve against ischemia/reperfusion injury. *Experimental and Therapeutic Medicine, 11*(5), 1563–1572.
79. Zuo, Z. X., Wang, Y. J., Liu, L., Wang, Y., Mei, S. H., Feng, Z. H., & Li, X. Y., (2015). Huperzine A alleviates mechanical allodynia but not spontaneous pain via muscarinic acetylcholine receptors in mice. *Neural Plasticity,* 1–11.
80. Venkatesan, R., Ji, E., & Kim, S. Y., (2015). Phytochemicals that regulate neurodegenerative disease by targeting neurotrophins: A comprehensive review. *BioMed Research International, 2015*.

81. Mo, C., Wang, L., Zhang, J., Numazawa, S., Tang, H., Tang, X., & Xiao, H., (2014). The crosstalk between Nrf2 and AMPK Signal Pathways Is Important for the anti-inflammatory effect of berberine in LPS-stimulated macrophages and endotoxin-shocked mice. *Antioxidants & Redox Signaling, 20*(4), 574–588.
82. Hasanein, P., Vahed, M. G., & Khodadadi, I., (2017). Effects of isoquinoline alkaloid berberine on lipid peroxidation, antioxidant defense system, and liver damage induced by lead acetate in rats. *Redox Report, 22*(1), 42–50.
83. Jin, P., Zhang, C., & Li, N., (2015). Berberine exhibits antitumor effects in human ovarian cancer cells. *Anti-Cancer Agents in Medicinal Chemistry, 15*(4), 511–516.
84. Scheib, J., & Höke, A., (2013). Advances in peripheral nerve regeneration. *Nature Reviews. Neurology, 9*(12), 668–676.
85. Han, A. M., Heo, H., & Kwon, Y. K., (2012). Berberine promotes axonal regeneration in injured nerves of the peripheral nervous system. *Journal of Medicinal Food, 15*(4), 413–417.
86. Zhang, Q., Bian, H., Guo, L., & Zhu, H., (2016). Pharmacologic preconditioning with berberine attenuating ischemia-induced apoptosis and promoting autophagy in neurons. *American Journal of Translational Research, 8*(2), 1197–1207.
87. Ahmed, T., Gilani, A. H., Abdollahi, M., Daglia, M., Nabavi, S. F., & Nabavi, S. M., (2015). Pharmacological reports berberine and neurodegeneration : A review of literature. *Pharmacological Reports, 294*, 1–10.
88. Soumyanath, A., Zhong, Y. P., Yu, X., Bourdette, D., Koop, D. R., Gold, S. A., & Gold, B. G., (2005). Centella asiatica accelerates nerve regeneration upon oral administration and contains multiple active fractions increasing neurite elongation in-vitro. *Journal of Pharmacy and Pharmacology, 57*(9), 1221–1229.
89. Rio, D., & Rio, D., (2015). *Role of GSK3 in Peripheral Nerve Regeneration, 10*(10), 2014–2015.
90. Guo, D., Lu, X., Xu, X., Gou, H., Wang, Z., & Cao, Y., (2018). Therapeutic effect of vinorine on sciatic nerve-injured rat. *Neurochemical Research, 43*(2), 375–386.
91. Jiang, S., Wang, S., Sun, Y., & Zhang, Q., (2014). Medicinal properties of *Hericium erinaceus* and its potential to formulate novel mushroom-based pharmaceuticals. *Applied Microbiology and Biotechnology, 98*(18), 7661–7670.
92. Sabaratnam, V., Wong, K. H., Naidu, M., David, P., Abdulla, M. A., Abdullah, N., & Kuppusamy, U. R., (2011). Peripheral nerve regeneration following crush injury to rat peroneal nerve by aqueous extract of medicinal mushroom *Hericium erinaceus* (Bull.: Fr) Pers. (Aphyllophoromycetideae). *Evidence-Based Complementary and Alternative Medicine.*
93. Wong, K. H., Kanagasabapathy, G., Naidu, M., David, P., & Sabaratnam, V., (2016). *Hericium erinaceus* (Bull.: Fr.) pers., a medicinal mushroom, activates peripheral nerve regeneration. *Chinese Journal of Integrative Medicine, 22*(10), 759–767.
94. Bastami, F., Vares, P., & Khojasteh, A., (2017). Healing effects of platelet-rich plasma on peripheral nerve injuries. *Journal of Craniofacial Surgery, 28*(1), 49–57.
95. Trisina, J., Sunardi, F., Suhartono, M. T., & Tjandrawinata, R. R., (2011). DLBS1033, a protein extract from *Lumbricus rubellus*, possesses antithrombotic and thrombolytic activities. *Journal of Biomedicine & Biotechnology,* 519652.

96. Zhang, P., Wang, Z., Kou, Y., Han, N., Xu, C., Yin, X., & Feng, X., (2014). Role of *Lumbricus* extract in the nerve amplification effect during peripheral nerve regeneration. *American Journal of Translational Research, 6*(6), 876–885.
97. Wei, S., Yin, X., Kou, Y., & Jiang, B., (2009). Lumbricus extract promotes the regeneration of injured peripheral nerve in rats. *Journal of Ethnopharmacology, 123*(1), 51–54.
98. Cristante, A. F., Damasceno, M. L., Filho, T. E. P. B., Oliveira, R. P. D., Marcon, R. M., & Rocha, I. D., (2012). Evaluation of the effects of hyperbaric oxygen therapy for spinal cord lesion in correlation with the moment of intervention. *Spinal Cord,* 502–506.
99. Marcon, R. M., Fogac, I. A., Teixeira, I. W. J., & Kenji, I. I. D., (1988). *Fractures of the Cervical Spine,* (5), 1455–1461.
100. Middendorp, J. J. V., Barbagallo, G., Schuetz, M., & Hosman, A. J. F., (2012). *Design and Rationale of a Prospective, Observational European Multicenter Study on the Efficacy of Acute Surgical Decompression After Traumatic Spinal Cord Injury : The SCI-POEM Study,* 686–694.
101. Cristante, A. F., Barros-filho, T. E. P., Tatsui, N., Mendrone, A., Caldas, J. G., Camargo, A., & Alexandre, A., (2009). Stem cells in the treatment of chronic spinal cord injury : Evaluation of somatosensitive evoked potentials in 39 patients. *Spinal Cord,* 733–738.
102. Orientador, T., & Oliveira, R. P., (2008). *Análise funcional e histológica da ação da neurotropina-4 sobre a narazakilesão medular em ratos Livros Grátis*. Retrospective study of the results of the use of halo.
103. Narazaki, D. K., Eloy, T., Barros, P. D., Regina, C., Oliveira, G. C. M. D., Cristante, A. F., & Rp, O., (2006). *Basic Research Spinal Cord Regeneration : The Action of Neurotrophin-3 in Spinal Cord Injury in Rats, 61*(5), 453–460.
104. Janssen, L., & Hansebout, R. R., (1989). Pathogenesis of spinal cord injury and newer treatments. A review. *Spine, 14*(1), 23–32.
105. Fehlings, M. G., Vaccaro, A., Wilson, J. R., Singh, A., Cadotte, D. W., Harrop, J. S., & Rampersaud, R., (2012). *Early versus Delayed Decompression for Traumatic Cervical Spinal Cord Injury : Results of the Surgical Timing in Acute Spinal Cord Injury Study (STASCIS), 7*(2).
106. Cer, U. N. A., (2006). *Risco De Tr Ação Ex Cessiv A Nas Lesões Tipo Ação -Flexão D A Col Distração Ação-Flexão Cervical Risk of Excessive Traction on Distraction-Flexion-Type Injuries of the Low Cervical Spine, 14*(2), 75–77.
107. Damasceno, M. L., Letaif, O. B., Cristante, A. F., Marcon, R. M., Iutaka, A. S., Oliveira, R. P., & Barros, F. T. E. P. D., (2010). Estudo retrospectivo dos resultados da utilizaÃ\Sa_o do halo craniano nas fraturas-luxaÃ\So_es subaxiais. Choice of surgical route for treatment of cervical fractures. *Coluna/Columna, 9*, 376–380.
108. Letaif, O. B., Damasceno, M. L., Cristante, A. F., Marcon, R. M., Iutaka, A. S., Oliveira, R. P., Barros, F. T. E. P. D., (2010). Escolha da via cirÃ\textordmasculinergica para tratamento das fraturas cervicais. *Coluna/Columna, 9*, 358–362.
109. Schwab, M. E., & Bartholdi, D., (1996). Degeneration and regeneration of axons in the lesioned spinal cord. *Physiological Reviews, 76*(2), 319–370.
110. Cristante, A. F., Filho, T. E. P. D. B., Marcon, R. M., Letaif, O. B., & Rocha, I. D. D., (2012). Therapeutic approaches for spinal cord injury. *Clinics, 67*, 1219–1224.

111. Fogac, A., Barros, P. D., Marcon, R. M., & Biraghi, O., (2012). *Therapeutic Approaches for Spinal Cord Injury*, *67*(10), 1219–1224.
112. Blight, A. R., & Zimber, M. P., (2001). Acute spinal cord injury: Pharmacotherapy and drug development perspectives. *Current Opinion in Investigational Drugs* , *2*(6), 801—808. London, England.
113. Bracken, M. B., Shepard, M. J., Collins, W. F., Holford, T. R., Baskin, D. S., Eisenberg, H. M., & Young, W., (1992). Methylprednisolone or naloxone treatment after acute spinal cord injury: 1-year follow-up data. *Journal of Neurosurgery, 76*(1).
114. Bracken, M. B., Shepard, M. J., Holford, T. R., Leo-Summers, L., Aldrich, E. F., Fazl, M., & Young, W., (1997). Administration of methylprednisolone for 24 or 48 hours or tirilazad mesylate for 48 hours in the treatment of acute spinal cord injury: Results of the third national acute spinal cord injury randomized controlled trial. *JAMA, 277*(20), 1597–1604.
115. Hall, E. D., & Springer, J. E., (2004). Neuroprotection and acute spinal cord injury: A reappraisal. *NeuroRX, 1*(1), 80–100.
116. Gebrin, A. S., Cristante, A. F., Marcon, R. M., Da Silva, C. F., & Filho, T. E. P. D. B., (1997). Intervenções farmacológicas no trauma raquimedular: Uma nova visão terapêutica. *Acta Ortopedica Brasileira*.
117. Marcon, R. M., Cristante, A. F., Filho, T. E. P. D. B., Oliveira, R. P. D., & Santos, G. B., (2010). Potentializing the effects of GM1 by hyperbaric oxygen therapy in acute experimental spinal cord lesion in rats. *Spinal Cord,* 808–813.
118. Domingo, A., Yahya, A. A. A., Asiri, Y., Eng, J. J., & Lam, T., (2012). A systematic review of the effects of pharmacological agents on walking function in people with spinal cord injury. *Journal of Neurotrauma, 29*(5), 865–879.
119. Jia, Z., Zhu, H., Li, J., Wang, X., Misra, H., & Li, Y., (2012). Oxidative stress in spinal cord injury and antioxidant-based intervention. *Spinal Cord, 50*(4), 264–274.
120. Hall, E. D., (2011). *Antioxidant Therapies for Acute Spinal Cord Injury,* 152–167.
121. Shi, R., & Sun, W., (2011). Potassium channel blockers as an effective treatment to restore impulse conduction in injured axons. *Neuroscience Bulletin, 27*(1), 36–44.
122. Bouhy, D., Ghasemlou, N., Lively, S., Redensek, A., Rathore, K. I., Schlichter, L. C., & David, S., (2011). Inhibition of the Ca^{2+}-dependent K^+ channel, $KCNN_4$/KCa3.1, improves tissue protection and locomotor recovery after spinal cord injury. *Journal of Neuroscience, 31*(45), 16298–16308.
123. Nehrt, A., Rodgers, R., Shapiro, S., Borgens, R., & Shi, R., (2007). The critical role of voltage-dependent calcium channel in axonal repair following mechanical trauma. *Neuroscience, 146*(4), 1504–1512.
124. Domingo, A., Yahya, A. A. A., Asiri, Y., Eng, J. J., & Lam, T., (2011). A systematic review of the effects of pharmacological agents on walking function in people with spinal cord injury. *Journal of Neurotrauma, 29*(5), 865–879.
125. Sluka, K. A., & Walsh, D., (2003). Transcutaneous electrical nerve stimulation: Basic science mechanisms and clinical effectiveness. *The Journal of Pain, 4*(3), 109–121.
126. Mortimer, J. T., S, M., Reswick, J. B., & Sc, D., (1967). *Electrical Inhibition of Pain by Stimulation of the Dorsal Columns: Preliminary Clinical Report,* 489–491.
127. Melzack, R., & Wall, P. D., (1965). *Pain Mechanisms: A New Theory, 150*(3699), 971–979.

128. Gildenberg, P. L., (2006). *History of Electrical Neuromodulation for Chronic Pain, 7,* pp. S7-S13.
129. Woolf, C. J., Mitchell, D., & Barrett, G. D., (1980). Antinociceptive effect of peripheral segmental electrical stimulation in the rat. *Pain, 8*(2), 237–252.
130. Santana, J. M. D., Silva, L. F. S. D., De Resende, M. A., & Sluka, K. A., (2009). Transcutaneous electrical nerve stimulation at both high and low frequencies activates ventrolateral periaqueductal grey to decrease mechanical hyperalgesia in arthritic rats. *Neuroscience, 163*(4), 1233–1241.
131. Rakel, A., & Sluka, K. A., (2009). *NIH Public Access, 10*(6), 492–499.
132. Kalra, A., Urban, M. O., & Sluka, K. A., (2001). Blockade of opioid receptors in rostral ventral medulla prevents anti-hyperalgesia produced by transcutaneous electrical nerve stimulation (TENS). *Journal of Pharmacology and Experimental Therapeutics, 298*(1), 257–263.
133. Maeda, Y., Lisi, T. L., Vance, C. G. T., & Sluka, K. A., (2007). Release of GABA and activation of GABAA in the spinal cord mediates the effects of TENS in rats. *Brain Research, 1136*, 43–50.
134. Schechtmann, G., Song, Z., Ultenius, C., Meyerson, B. A., & Linderoth, B., (2008). Cholinergic mechanisms involved in the pain-relieving effect of spinal cord stimulation in a model of neuropathy. *PAIN, 139*(1).
135. Sluka, K. A., Vance, C. G. T., & Lisi, T. L., (2005). High-frequency, but not low-frequency, transcutaneous electrical nerve stimulation reduces aspartate and glutamate release in the spinal cord dorsal horn. *Journal of Neurochemistry, 95*(6), 1794–1801.
136. Ghoname, E. A., Craig, W. F., White, P. F., Ahmed, H. E., Hamza, M. A., Henderson, B. N., & Gatchel, R. J., (1999). Percutaneous electrical nerve stimulation for low back pain a randomized crossover study. *JAMA, 281*(9), 818–823.
137. Pfister, B. J., Gordon, T., Loverde, J. R., Kochar, A. S., Mackinnon, S. E., & Cullen, D. K. (2011). Biomedical engineering strategies for peripheral nerve repair: surgical applications, state of the art, and future challenges. *Crit Rev Biomed Eng. 39*, 81–124.
138. Li, W., Han, L., Yu, P., Ma, C., Wu, X., & Xu, J. (2014). Nested PCR-denaturing gradient gel electrophoresis analysis of human skin microbial diversity with age. *Microbiol. Res. 169*, 686–692. doi: 10.1016/j.micres.2014.02.008.
139. Wolford, L. M., & Rodrigues, D. B. (2013). Nerve grafts and conduits. In: *Trigeminal Nerve Injuries*. 271–290.
140. Battiston, B., Geuna, S., Ferrero, M., & Tos, P. (2005). Nerve repair by means of tubulization: Literature review and personal clinical experience comparing biological and synthetic conduits for sensory nerve repair. *Microsurgery 25*, 258– 267.
141. Daly W., Yao L., Zeugolis D., Windebank A., & Pandit A. (2012). A biomaterials approach to peripheral nerve regeneration: bridging the peripheral nerve gap and enhancing functional recovery. *Journal of the Royal Society Interface. 9*(67), 202–221. doi: 10.1098/rsif.2011.0438.
142. Wangensteen K. J., & Kalliainen L. K. (2010). Collagen tube conduits in peripheral nerve repair: a retrospective analysis. *Hand. 5*(3), 273–277. doi: 10.1007/s11552-009-9245-0.

143. Liu, B. S., Huang, T. B., & Chan, S. C. (2014). Roles of reinforced nerve conduits and low-level laser phototherapy for long gap peripheral nerve repair. *Neural Regen Res. 2014 Jun 15; 9*(12), 1180–1182. doi: 10.4103/1673-5374.135323
144. Ramburrun, P., Kumar, P., Choonara, Y. E., Bijukumar, D., Toit, L. C. D., & Pillay, V. (2014). A Review of Bioactive Release from Nerve Conduits as a Neurotherapeutic Strategy for Neuronal Growth in Peripheral Nerve Injury. *BioMed Research International, vol. 2014*, Article ID 132350, 19 pages, 2014. https://doi.org/10.1155/2014/132350.
145. Xie, F., Li, Q. F., Gu, B., Liu, K., & Shen, G. X. (2008). *In vitro* and in vivo evaluation of a biodegradable chitosan-PLA composite peripheral nerve guide conduit material. *Microsurgery. 28*(6), 471–479. doi: 10.1002/micr.20514. PMID: 18623157.

INDEX

F

Q

R